Eco-Design of Buildings and Infrastructure

T0136143

Editors

Bruno Peuportier
MINES ParisTech

Fabien Leurent
Ecole des Ponts ParisTech

Jean Roger-Estrade
AgroParisTech

CRC Press
Taylor & Francis Group
Boca Raton London New York

CRC Press is an imprint of the
Taylor & Francis Group, an **informa** business

A BALKEMA BOOK

Photo credits (cover): Luc Benevello, Augusto da Silva/Graphix Images, Pascal Le Doaré, Frédéric Stucin, Photothèque VINCI and subsidiaries, DR.

Originally published in French
'Livre blanc sur les recherches en énergétique des bâtiments' by Bruno Peuportier
© 2013 Presses de Mines, Paris

CRC Press
Taylor & Francis Group
6000 Broken Sound Parkway NW, Suite 300
Boca Raton, FL 33487-2742

First issued in paperback 2019

© 2016 by Taylor & Francis Group, LLC
CRC Press is an imprint of the Taylor & Francis Group, an Informa business

No claim to original U.S. Government works

Typeset by MPS Limited, Chennai, India

ISBN-13: 978-0-138-02967-5 (hbk)
ISBN-13: 978-0-367-88949-4 (pbk)

Library of Congress Cataloging-in-Publication Data

CIP data applied for

i it t e Ta or   ra ci   e ite at
tt :   ta ora d ra ci com

a d t e   re   e ite at
tt :   crc re   com

Eco-Design of Buildings and Infrastructure

Sustainable Cities Research Series

ISSN: 2472-2502 (Print)
ISSN: 2472-2510 (Online)

Book Series Editor:

Bruno Peuportier

Centre for Energy Efficiency of Systems, MINES Paristech, Paris, France

Volume 1

Table of contents

7   Eco-operation of road traffic   99
Vincent Aguiléra

Part 3: Application to a case study, Cité Descartes

Foreword

Xavier Huillard
Chairman and Chief Executive Officer of VINCI

Ecological transition, energy performance, sustainable towns, urban densification, eco-mobility: the concepts abound and at times even contradict themselves. The challenges are real enough, technical progress and knowhow are increasingly convincing, yet we still lack the tools and simple, easy-to-use reference systems we need to move forward together!

More than five years ago, it struck us that eco-design, which had worked well in the industrial sphere, could bring the right solutions to energy and environmental issues applied to the construction and concessions sector, because we needed to get an overall vision. In our profession, eco-design means involving the environment right from the design of towns and infrastructures with a vision of their entire life cycle. Housing, schools and subways have a function, and the idea is not to overlook or diminish that function, as if the environment must be protected in an absolute way, but to ask the right questions on reducing impacts, preserving resources and maintaining quality of life and usage.

We therefore took the gamble of investing in R&D through our eco-design Chair in order to define these strategic tools. This took the shape of a totally new patronage of three ParisTech universities, MINES ParisTech, l'Ecole des Ponts ParisTech and AgroParisTech. The aim was to develop scientific knowledge on buildings and infrastructures to share with all urban stakeholders.

The Chair's target was to bring engineering schools and business closer together to foster new concepts and tools aimed at eco-urbanism. This cross-fertilization will also train new generations of engineers working on eco-design applied to construction and concessions.

The gamble paid off: global eco-design applied to buildings and transport, in addition to a simple kWh/m^2 criterion and various certifications, is confirmed. Tools and reference systems are now a reality. Researchers have produced concepts, new methods and application tools tested by VINCI companies that have accepted to share their experiences, projects and inquiries.

This joint work, competently edited by Bruno Peuportier, Fabien Leurent and Jean Roger-Estrade, reports on this progress in line with the Chair's three research axes: evaluation of the environmental quality of buildings and neighbourhoods; life cycle assessment of transport infrastructures and their impacts; organization of buildings and transport, and regulation of their usage to optimize protection of the environment.

When knowledge feeds into collective wisdom, the result is a satisfaction that is well worth sharing.

Introduction: background, new targets resulting from corporate social responsibility

Bruno Peuportier, Fabien Leurent & Jean Roger-Estrade

One of the major development trends in the transport infrastructure and buildings domain is a shift in the value chain from the construction itself to the operation, although economic and governance models are still far from stable. The complexity of transport systems and land use systems requires a detailed understanding of how they work. This kind of understanding is based on analytical and digital models, which can simulate the way they work and the effects of diverse management strategies on sustainability, including not just economic aspects, but also social and environmental impacts. Concern for environmental issues is increasing around the world, especially in developing countries, so that deciders and companies are under pressure to act. The concern is global and mainly involves the production of greenhouse gases; but it is also local, with a demand for better quality of water, air, atmosphere, landscape and protection of local ecosystems. Lastly, sustainable development is becoming an economic issue for business, in particular since economic models are shifting towards the long term to include the rising cost of energy and soon CO_2 emission permits. These facts bring new risks, but also opportunities to change technological and operating compromises, especially in sectors connected to the built environment or transport, for which different types of public utilities propose frameworks for design and concrete action.

In the building sector, which represents almost half of France's total energy consumption, energy and environmental constraints are imposing far-reaching changes. The long lifespan of installations results in very high inertia: new buildings constructed in one year only correspond to around 1% of the existing stock, and so decisions have a greatly delayed impact. This is bringing about change in public policies in terms of regulations and tax incentives, and mobilizing increasing numbers of contracting clients. Banks now offer appropriate financial packages like mortgages linked to environmental quality targets. Debate is underway on the notion of responsible social investment, and an energy-saving certificate market is starting to emerge. It is thus important to produce supply that corresponds to these new demands.

Environmental constraints are also very high in the transport sector. On the one hand, impact factors resulting from buildings (especially energy and CO_2) and recovering materials at the end of their lifespan pose increasing problems since building materials weigh significantly in the overall environmental balance. On the other hand, transport infrastructures have local impacts on the environment, either due to their actual existence (i.e. dividing effect, runoff or ground waters, impact on biodiversity or farming), or through their operation (i.e. impact on air or water quality, noise). Along with these environmental impacts integral to transport infrastructure, other sustainability issues are linked to using transport for moving people and goods, in synergy with the location of residential and production buildings on a territory: the distribution in space of buildings devoted to various residential, production and service functions results in flows of movement and thus traffic of various means of transport. This raises the question of the environmental impacts of these traffic flows, and of land use at the scale of a territory rather than just a zone: these questions in turn have impacts on the ridership of transport infrastructure and facilities, and thus on their economic viability.

Half of the greenhouse gas emissions generated by human activity are due to buildings and transport. The Group, which is active in construction (VINCI Construction, Eurovia, VINCI Énergies, etc.) and transport (VINCI autoroutes, VINCI park, etc.), can play an important role in reducing emissions, but also endures the consequences of climate change. In its conviction that it is responsible for anticipating stakeholders' expectations as well as the economic and social consequences of this change, the Group takes a proactive position in this domain.

ParisTech universities have built up expertise on these questions and are keen to continue developing research and education activities whose results could help companies make progress with their strategy. The aim of these studies is to provide long-term support for implementing life cycle assessment and eco-design measures in companies, with implications for public policies, integration at the level of environmental performance by the contracting client, and awareness amongst customers/users of a more virtuous operation of buildings and infrastructures. These universities have strong ties with specialized French research institutions like CSTB and IFSTTAR (merger of INRETS and LCPC) and international ones, and the competitiveness cluster Advancity (sustainable cities cluster).

The ambition shared by VINCI and the ParisTech universities is to create decision-aid tools available to diverse stakeholders and with a scientific character that guarantees ideological neutrality. These tools should be able to measure and quantify the economic impacts (i.e. production costs, operating costs, merchant revenue and more broadly inter-stakeholder transfers) and environmental impacts (i.e. greenhouse gas emissions, impacts on health and biodiversity, water and energy consumption, waste, etc.) of French and international projects that interest VINCI. The Chair's preferential geographic perimeter for the research study is Europe, with a focus on France.

This book presents the research carried out by the Chair and the results obtained over five years. The first part gives an overview of basic knowledge, which is used in the more operational developments presented in the second part, in turn applied to a joint case study in the third part.

Part I

Methodological bases

Chapter 1

Using eco-design tools as a cognitive support for creating the urban fabric

Rebecca Pinheiro-Croisel & Franck Aggeri
MINES ParisTech

The climate emergency facing humanity has had a transforming impact on those working in engineering and urban creation. In building projects, sustainability issues are now integral to infrastructure, urban spaces and landscape. Indeed, some actors make them a priority in the design, construction and operation processes of their projects. Clients and contractors have had to rework their knowledge, concepts and action protocols to meet with the demands of sustainable development. The building sector, town planners and urban system operators, have always taken a systemic approach to projects. The presence of sustainability issues, which are often measured by performance levels, reinforces this practice. Design and evaluation tools are valuable aids for bringing stakeholders and issues together around an object to be designed. More particularly, in the sector of sustainable urban planning and sustainable buildings and structures, these tools act as a guideline that can help to create a common language as well as disseminate and create new knowledge. Elements of sustainable urban planning, such as eco-neighbourhoods and highly sustainable groups of buildings, are not yet totally stabilized. Some uncertainties remain regarding the performance of materials, the possibilities and the means of reaching certain objectives, as well as the financial and legal resources needed to do so.

This underlines the importance, in the creation process, of using devices that encourage dialogue between stakeholders, specify performance targets in contracts, and disseminate the knowledge needed for designing totally or partially unknown objects.

Eco-design is not simply a tool. It involves a set of engineering tools organized to be used in an approach that aims to increase the sustainability performance of a service, object, concept or project. Eco-design can thus be defined as a method or approach to implementing devices that facilitate the design of sustainable projects. It is the result of practical and academic research on devising measurement and evaluation tools in different sectors of activity.

This methodological approach is systemic, integrating different fields of knowledge and aiming to take into account mainly environmental problems – and occasionally sustainability – during the design processes of products and projects. In innovative projects, or when some aspects of the project are little known, this approach can be a way of guiding functional analysis and prioritizing objectives.

It is important to remember that the origins of this approach are not the design of urban components, but rather industrial ones. Nor did it emerge from a quest for sustainability; the initial aim was to devise an instrument to increase the environmental performance of products and later services. Research responding to the global environmental issues set out in the Meadows Report (Stockholm Conference, 1972), and written after the 1973 oil crisis, led researchers and industrialists to develop tools for taking these environmental issues into account. For example, the first studies using an eco-design tool, i.e. life cycle assessment, which were carried out in 1969 in the United States[1] at the packaging sector level, focused only on waste and energy aspects (Blouet and Rivoire, 1995, Hunt and Franklin, 1996, Abrassart and Aggeri, 2002).

In the course of experiments in different sectors, eco-design opened out to the global problem of sustainability, no longer considering only the environmental side, but also economic and social aspects.

The design efforts of city stakeholders are no more focused on the demand for sustainability; however, these issues underline the systemic approach to urban creation. This factor, associated with the need to create a space for dialogue and understanding between those involved in the project's solutions and requirements, along with new legal and financial demands for including performance and results in the contract, can allow eco-design to play an integrating role.

Since 2008, the issue of the commercial position occupied by the manager of a complete project for a sustainable town has been frequently raised; especially by forecasting, strategic marketing, sales forces and development teams at major French building companies, urban system operators and real estate promoters. In 2009, a group of American entrepreneurs wanted to build a large ecological village with a broad range of functions in the Paris region. They drew up a project including design, construction and operation and approached a French industrial group that offered all three trades. The group's property director went round the company questioning his colleagues on whether they could fulfil the order. After a few weeks, he realized that the group's companies could not organize themselves; although they did possess complementary skills, they had neither the engineering tools nor the management tools to bring them together. His colleagues were also reluctant to accept this "new way of doing things", preferring the comfort of a trade-based approach, given that they knew little about the overall object (i.e. eco-neighbourhood/eco-village). The company therefore ended up turning down the client, even though the contract would have represented a significant share of its turnover for several years.

Some companies – keen to secure a strategic position as integral parts of the town, i.e. designers, builders and operators – have tried to work with other businesses possessing complementary skills. They soon realized that their alliances remained locked within the perimeter of each of the project's stages rather than spanning its totality.

Lastly, one actor that clearly takes a very active role in the three stages of an urban development project – from its creation through to its working life – is the local authority. Most of the time, it is both the project's pilot and client, except for state-managed activities. However, we know that all skills required to see through the entire life cycle of an urban project cannot be found within a single local authority, even the best endowed.

---

[1] *Resource and Environmental Profile Analysis* (REPA).

This guiding and managing party, which represents the general interest, therefore needs devices to help it build and share its vision of the town, to make this vision manifest and understandable by all, and to have it applied, modified, put into contract form, and assessed over time.

Whether via engineering consultants or project ownership assistance, local authorities are increasingly making use of collective action in urban project design processes.

At the higher level of steering and disseminating a vision, the state uses, and experiments with, tools that spread the doctrine of a sustainable town, make it possible and evaluate it. These are obviously instruments for establishing public policy at local level, but not only that, since the approach the ministry of housing takes to eco-neighbourhoods (the "ÉcoQuartier" label) has spurred a genuine eco-neighbourhood movement in France, in both small and large towns.

Design-aid tools are thus indispensible for designing urban projects that require feedback on performance levels, especially environmental ones, and providing for them in contracts. This emerging need is not just due to rigorous environmental quality requirements for objects and measures, but also feeds into learning, producing knowledge and building a common language in sustainable urban planning. Thus, in many urban design and management sectors, tools for decision-making and evaluation can help stakeholders optimize costs, time and impacts while reducing errors throughout the creation chain of built and vacant urban environments.

For around thirty years, French research teams have been looking closely at engineering tools to design towns of high environmental quality. This research involves actors in the field confronted with ecological and sustainable issues. It started off at building scale and then expanded to cover all building complexes and transport networks, along with other systems like biotopes that are present or need restoring in an urban space. Several researchers – including those involved in the ParisTech-Vinci EEBI chair – develop and apply tools with an eco-design approach in the domains of biodiversity, transport, infrastructure and buildings. Studies are also carried out to understand how urban design activities work, so as to find a way of integrating an eco-design approach into these processes.

Presenting some of this progress to the largest possible number of urban planners will hopefully result in its continual evolution, reflecting the actual object of the research, i.e. the town.

Since 2009, through observing the design processes of eco-neighbourhood projects in France, and particularly national EcoQuartier competitions (2009 and 2011), we have noted that these projects are put together in a joint creative process. Engineering tools play a role throughout this process. Those working in urban design and construction, as well as in the functioning of towns, do not always share the same objectives or have the same vision of sustainability and urbanity for built and vacant areas. Eco-design tools thus play a significant part in helping create cohesion between stakeholders and in getting these processes organized (Segrestin, 2003). Stakeholders can "step out" of their mono-expert role and take part in the systemic design of the urban area. Design-aid tools can thus support a joint design (Pinheiro-Croisel & Hernes, 2012, Pinheiro-Croisel, 2013) shared between actors, whereby all participants in the design process play a part in the creation. A common language understood by all, along with visible results and the means to achieve sustainability performances, combine to reinforce stakeholders' design capacity, which is a major asset in an eco-design approach.

A META-REFERENCE FOR CITY STAKEHOLDERS

The practices that have changed the most over the last decade, apart from technological requirements, are the management of urban design activities, and the reliable organization of sustainability targets expressly provided for in contracts.

Mostly in Europe, we have observed that local authorities organize themselves "differently" to implement reference operations in terms of urban sustainability. Sustainable development projects in central and northern Europe include public/private partnerships, and can involve researchers, unconventional professionals or residents. Sustainable development experiments carried out in the projects *InnovationCity Ruhr* (Germany), *DigiEcoCity* (Finland) and *SymbioCity* (Sweden) provide examples of changes in the way that urban project design is run and contracted. In these three examples, the eco-design approach focuses more on developing project governance to encourage financing that allows for innovative ecological solutions to achieve fixed targets. Stakeholders design "more sustainable projects", not so much by using integrated engineering tools, but rather through partnership-based collaboration and the simultaneous integration of one or several stakeholders. This eco-design approach, usually taken by local authorities, starts by ranking and prioritizing sustainability objectives that are ambitious and compatible with national state policies; it is only then that a call for partners is launched. "Recruits" are invited to take part in the design process, and the project then evolves, as do sustainability targets. If measurement devices are used with this eco-design management approach, results can be compared more easily and written into contracts.

Eco-design can thus be seen as an approach based on two main axes:

–   Design process management method;
–   Design-aid and measurement tools.

A recent analysis by the Siemens group compared 130 towns carrying out sustainable design and development projects throughout the world. It found that the common denominator of these towns, on different continents with high-quality sustainable urban projects, was strategic governance[2]. Using tools in the design process of an innovative project managed in a conventional way (i.e. sequential design) can simply weigh down the process and compartmentalize stakeholders. If the manager of an urban creation process only uses eco-design tools to make the process more rigorous, then the risk is to end up with a project that lacks overall priorities and perhaps even meaning. These tools should be used in a project whose governance is compatible with an innovative design process (Le Masson, Weil and Hatchuel, 2006). In other words, the object should not be set in stone with over-precise specifications; its design devices should be built or adapted, and it should not respond to a single axis or sustainability target that may be disconnected from the urban project local context.

Although eco-design at urban project scale is an approach that uses a certain number of engineering tools, it does not simply involve coordinating their application. It involves a combination of coordinating and consolidating (Segrestin, 2003) both projects and stakeholders during a design process.

---

[2]Study report "GreenCity Index" (Siemens, 2012).

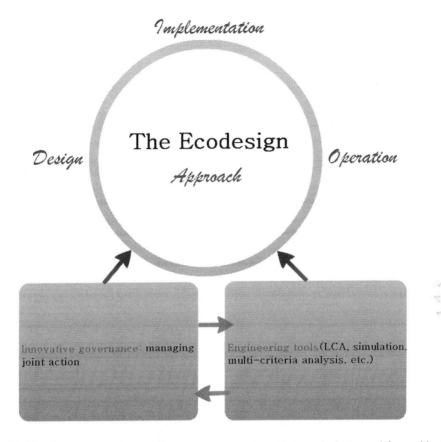

Figure 1.1 Eco-design approaches require governance systems that are inclusive and favourable to innovation as well as a set of tools.

Eco-design thus acts as a cognitive support for the design process. The tools and action recommended by an eco-design approach can involve a common language shared by stakeholders, and the emergence of innovative solutions thanks to direct, reliable, comprehensible interaction resulting from using these devices.

Action carried out during an eco-designed project is visible throughout its life cycle, starting with forecasting and diagnosis, and including the definition, contracting, design, construction and working life of the object.

ECO-DESIGN, OR ORGANIZING A "PROJECT LANGUAGE" UNDER CONSTRUCTION

A number of towns, in France and around the world, have opted to manage central-ized projects. Enterprising political leaders (Arab and Lefebvre, 2012), who are well informed and sometimes visionaries, show a strong capacity for setting out a project's values and priorities. These decision-makers, whether they are involved in Brasilia, Haussmann Paris, or in projects with collective design processes (like the redevelopment of Bordeaux or Dunkirk) leave a specific, often long-lasting, mark on their city. This capacity and potential for centralizing and maintaining a framework of priorities and challenges broken down into an urban form cannot be applied to all of the politico-technical contexts of urban design.

Building a long-term urban strategy, defining which type of project to construct, choosing a location, and setting up an urban operation or building programme, all require skills in managing design activities along with expert knowledge of the challenges that need to be tackled and met.

Leading sustainable urban designers could do more to include stakeholders that are indispensible yet often excluded due to a lack of legal flexibility, and sometimes for want of a long-term, urban, systemic vision. These stakeholders, who are responsible for making urban quality last over time, could use various engineering tools to structure their reflection and actions.

More than ten years ago, when the first French eco-neighbourhoods (such as the Bonne "ZAC" comprehensive planning area and the Grand Large neighbourhood) were in the early phase of designing building areas and envelopes, the field of measuring environmental performance was considerably vague. It proved difficult to define targets and the means to succeed, and to evaluate protocols.

For the *Bonne Zac*, which was linked to the European Concerto programme, evaluation protocols and targets came under European funding programme specifications. Stakeholders were assisted in constructing the environmental performance by a legal-financial tool.

Equipping the urban design process with tools is a way of giving stakeholders the means to exceed expectations, or simply to make progress in an organized way in face of the unknown. This involves not just instilling greater rigour to move away from conceptual vagueness, and the poor results of so-called sustainable projects. Measuring devices, such as life cycle assessments of neighbourhoods; or methods, like defining neighbourhoods' functional units, directly contribute to constructing a sustainable "object", with a view to systematically integrating sustainability into design.

On the other hand, in the urban planning field, it is important to adapt the way that design-aid tools are used (Peuportier, 2008). From a design point of view (Le Masson *et al.*, 2010), a neighbourhood as a design object cannot be seen as a fixed product when it is delivered to the consumer. A neighbourhood can be viewed over the long term as an evolving "object", since its design process is not finished at the point of delivery; on the contrary, it continues to evolve throughout the project's lifespan, and users play an even more important role in this process of ongoing design (Pinheiro-Croisel, 2013). Even if a neighbourhood does not change its main functional usage over the years, its identity can be transformed as a result of the social phenomena it contains. In Paris, the Marais quarter has been highly residential for several centuries, but it has undergone significant transformations in its uses, due to the way it is utilized and its significance. It has changed from a very run-down neighbourhood into an exclusive, wealthy area.

A neighbourhood's programme and physical perimeter do not always change over the years, but the way functions and uses evolve plays an important role in the architectural design process. Architects interviewed on design practices talk about making spaces flexible to reflect how people lay claim to them and make changes in usage and significance.

Although the evolution of neighbourhoods should be put into perspective, projects' environmental quality studies do not always anticipate the different ways in which people lay a claim to a place, or the unexpected transformations of a space or building. However, the way it is used will condition the quality and efficiency of the recommended technical solutions. Usages are not only an expression of the claim made on a place, they also participate in the transformation of the object itself. After all, a neighbourhood is not a set of buildings and urban areas, it is the result of interactions between people, and between people and space, as inhabitants continually and spontaneously construct their quality of life. Thus, residents and users of a neighbourhood can transform it, simply by turning the space in which they live into a genuine living environment. This occurs once the neighbourhood responds to the characteristics of urban quality that wholly connect it to the local context (town), its memory and local resources (territory).

Thus, since a neighbourhood is an evolving object, it is important that design-aid tools do not designate it with an over-specific functional unit whose attributes are fixed from the start of the design process. The creation of an urban project itself evolves. Creating a piece of a town is a joint action organized in a concerted process (Garel, 2003) that is more likely to generate innovation and societal evolution.

In LCA terms, even though "functional analysis involves shifting the point of view of the analysis" (Gobin, 2003) instead of just thinking in terms of responding to a user's given requirements, it should integrate the evolving character of the object and its uses. Simulations and hypotheses justify taking a rigorous approach using a tool like the neighbourhood LCA (Gomes et al., 2012). Usages take into account the realities of human interactions that affect the "neighbourhood system".

The implicit limits of the LCA's functional unit should not however rule out resorting to this kind of tool. The benefits of an eco-design approach remain positive for an urban project's ecological performance. The best precaution is to apply (a) a more general functional unit (e.g. develop a twenty-hectare neighbourhood) and (b) extend the boundaries of the system to consider movements of services (e.g. developing thirty residences in the town avoids building thirty extra residences on the outskirts). Thus, these limits could be reduced by leaving a great deal of freedom to the design community. This would also make it possible to evaluate the benefits of a development scenario and/or undesirable impacts on other territories.

Thus, a neighbourhood LCA study can be combined with a *consequential life cycle assessment* (CLCA). This approach has the potential to, for example, avoid urban sprawl, by showing that for each hectare of housing created in a town, one hectare will remain vacant on the outskirts. By considering the negative (i.e. environmental) impacts to obviate, this approach aims to define the functional equivalence that can be avoided for a given project.

Similarly, the eco-design tools described in the following chapters should be coupled with project governance that prioritizes durability issues as a whole, including the collective aspect of the design process and considering the capacity for change of the object to be created.

When eco-design is tackled in a multi-disciplinary way, it can contribute to achieving the sustainability of the final object, however complex. Reducing eco-design to a methodological and instrumental tool so that a project can achieve high ecological performance means limiting the project potential in terms of overall impacts to be avoided, along with the quality of life that it can bring about.

An eco-design approach gives stakeholders an opportunity to create a project ecosystem that produces knowledge. Measurement tools, using a universal language of performance, can set up a direct dialogue between players from all horizons that do not share a language. These tools also make it possible to strengthen systemic actions and think of the design in terms of overall cost. Many urban projects suffer from insufficient coordination and tools for managing the creation, construction and operation process. Money and time are wasted for want of vision, openness and rigorous systematization. When eco-design is integrated into informed, specific project governance, it can help build up these three components of design activities organization.

In addition, urban projects can be compared with each other, and organizations like the World Bank, ISO, industries and states, as well as local authorities, can understand, extend and disseminate design processes and performances more easily. These funding and partnership arrangements can mobilize the tools and their results to build up the legal framework.

Urban design today is no more complex than before; the current challenges are different but just as urgent. However, time may have brought us greater knowledge, experience and hindsight. These factors have helped build the eco-design approach, which will constantly draw from them in its inevitable evolution.

ORCHESTRATING THE COMMUNITY TO MANAGE A SUSTAINABLE, INNOVATIVE DESIGN PROCESS

Our observations of the design and implementation process of sustainable urban projects have shown us the importance of participative, open governance, ready to take on new practices and types of stakeholder. Although an urban project does not rely on a single person, principals are responsible for design processes and can either open up the fields of technological, managerial and social innovation, or confine these processes to rules. Several political trends exist on this subject; however, there is a clear design movement that recognizes joint action as part of urban creation and management.

We can analyze the design process of an urban project from different angles, depending on our priorities and study topics. We can also analyze them from the point of view of a stakeholder's practices. Here, we have chosen to look at the possibilities and impossibilities that a strategic and political stakeholder can create during these processes.

Local authorities (towns, urban communities, etc.) are now recognized as crucial to the design, construction and operation process. However, since mayors are responsible for initiating, motivating and building the vision of the town and making it a reality, they must have the capacity to manage the design process. Just as professional trades possess specialist skills, municipal technical services need the physical resources and knowledge to respond to new sustainability requirements.

We know that mayors, as political personalities central to upholding the general interest at local level, cannot be expected to personally manage the design process of urban projects in their towns. They must, however, be capable of encouraging,

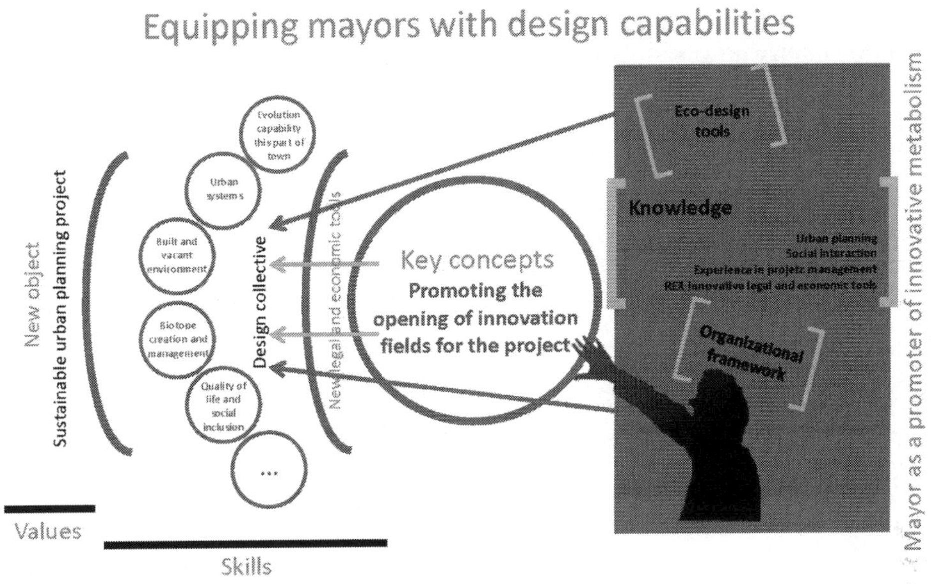

Equipping mayors with design capabilities

Figure 1.2 Give local authorities the capacity to design and manage a collective, innovative design process (Pinheiro-Croisel, 2013).

delegating and establishing the operating conditions of the process so that the town's urban and sustainability objectives can be met. They must open up innovation fields by launching concept and design challenges. To do so, they must foster a participative governance system with an open project and allocate the required legal, political and financial resources. On the other hand, if mayors, their councillors and the town's technical services tie the process into an over-precise set of specifications and restrict the diversity of designers, the sustainability aspect of the performance could be greatly threatened, along with the project's outcome in terms of users' quality of life.

Eco-design can facilitate an "ecosystem of innovative, sustainable design" that mayors can help give shape to. This system is based on the values and objects expressed in a town's sustainable urban planning policy. An eco-design approach can support the integration, cohesion and coordination of this system and act as a device for controlling results.

Sometimes, technical and territorial engineering teams become aware – before politicians – of changes and requirements in projects and governance. In such cases, mayors should welcome initiatives and changes in routine project management practices. Politicians need knowledge to make the decision of letting innovative design processes emerge. They need to be aware that a number of innovations are possible if they wish to promote investigating and developing them.

By opting for an eco-design approach, local authorities can explore solutions suited to their territorial context and their vision of the town, and can help stakeholders provide for contractual performance specifications, control them and capitalize on practices and results. This is possible if, as we have pointed out, the approach is part of a collective form of governance that is keen to concretely integrate sustainability into the creative process of launching, implementing and managing the town.

REFERENCES

Abrassart C. & Aggeri F., *La Naissance de l'éco-conception*, Responsabilité environnement, Annales des Mines, 2002.

Blouet A. & Rivoire E., *L'Écobilan. Les produits et leurs impacts sur l'environnement*, Dunod, Paris, 1995.

Garel G., *Le Management de projet*, Repères, La Découverte, 2003.

Gobin C., "Analyse fonctionnelle et construction", *Techniques de l'ingénieur. Traité de construction*, CD1 (n°C3052, C3052.1-C3052.15), 2003.

Gomes F., Herfray G., Boujnah H., Henry A., Rivallain M. & Trocmé M., "Bâtiments, transports, biodiversité : les apports de la chaire d'éco-conception ParisTech-VINCI à la ville durable", *Travaux*, 886, 52–53, 2012.

Le Masson P., Weil B. & Hatchuel A., Strategic management of inno*vation and design*, *Cambridge University Press*, 2010.

Peuportier B., *Éco-conception des bâtiments et des quartiers*, Paris, Mines ParisTech Les Presses, 2008.

Pinheiro-Croisel R., "Innovation et éco-conception á l'échelle urbaine : émergence et modèles de pilotage pour un aménagement durable", PhD Mines ParisTech, Paris, 2013.

Pinheiro-Croisel, R. & Hernes, T., *Innovation without design? The dynamics of role making and the gradual emergence of the collective designer*. EGOS Conference. Helsinki, 2012.

Segrestin B., *La Gestion des partenariats d'exploration : spécificités, crises et formes de rationalisation*, Paris, Mines ParisTech, 2003.

Chapter 2

Life cycle assessment applied to neighbourhoods

Bruno Peuportier
MINES ParisTech

Decision-makers' awareness of environmental issues has prompted various initiatives, such as a "high environmental quality" approach to buildings and the creation of numerous "eco-neighbourhoods", although the corresponding concepts are not always clearly substantiated. However, the considerable risks, both local and global, call for more rigorous management of these issues. Decisions on urban planning strongly influence the building and transport sectors, which make a significant contribution to most environment impacts.

The building sector, for example, is the highest energy-consuming sector in France, as well as Europe, where it is responsible for almost half of total consumption, or twice as much as industry. Drinking water consumption represents around 25% of net water withdrawals, but almost as much is used to produce electricity, of which building consumes 60%. From one to two tonnes of materials are used per m^2 built, which makes the building industry one of the most important outlets for industrial products. Demolition, retrofit and construction sites generate 48 million tonnes of waste annually in France, compared to 28 tonnes of household waste. Buildings emit high levels of pollutants, both into the air (e.g. 22% of greenhouse gases) and water (quarter of emissions in phosphate equivalent). It is therefore essential to mobilize all professionals in the sector to preserve future generations and biodiversity.

The concept of sustainable development initially meant satisfying present-day requirements while preserving future generations. It is not therefore the vague compromise between economic, social and environmental aspects that it is all too often seen as today. Satisfying present requirements would, for example, involve responding to a housing demand that is largely unfulfilled, and also reaching certain comfort standards. Preserving future generations implies protecting certain common goods, like energy, water, raw materials and soil, health and biodiversity, and so particularly climate and milieus – air, water, and soil. These examples indicate the extent of the challenge of satisfying all sustainable development criteria, which can moreover be contradictory.

To tackle this challenge involves transforming professional practices and technologies. The decisions that influence the performance of urban projects the most are taken during the early phases. This means that programming is very important and should include environmental performance and eco-design requirements, which involves taking environmental aspects into account in the design process.

This way of thinking is not new and is now widely accepted, yet most stakeholders seem to avoid putting it into practice. One of the main reasons is probably that the economy takes little consideration of these aspects, even though some tools are available, such as internalization of external costs, emissions markets and eco-tax. Regulations are another lever, and Europe is fairly active in this area. Proactive approaches also play a role through labelling and decision-making tools. In any case, whether the motivation is to evaluate external costs, justify a regulatory measure, or accompany a proactive approach, it is crucial to know more about the environmental impacts of the activities concerned.

Yet gaps persist in this knowledge due to the sheer number of activities and pollutants emitted – over 100,000 chemical substances are officially commercialized – and the complexity of chain phenomena, i.e. pollutants are transported in different ecological compartments (air, surface waters, water tables, soil, etc.), these pollutants degrade over time, transfer into food, are ingested by living organisms, with consequences on health and biodiversity.

Individual stakeholders only know about a small portion of a whole that is necessarily inter-disciplinary and inter-sectorial. Evaluating environmental impacts calls for knowledge of ecology, medicine, process engineering and energy, etc. In addition, human activities interact, e.g. energy is required to produce cement, steel and concrete are needed to produce energy, etc. The study of urban complexes illustrates these inter-sectorial issues, since it combines the sectors of building, transport, energy, processes (water, waste treatment, etc.), industry (building materials), and even agriculture (local production of food, revegetation).

To include environmental aspects when making decisions therefore requires tools that capitalize on a vast set of knowledge and make it attainable to stakeholders who possess limited time to make this kind of study. One of the tools developed to satisfy these specifications is the life cycle assessment (LCA). This engineering tool is designed to evaluate a system's environmental impacts on its life cycle, in other words from its manufacture to end of life, including any recycling, from whence the notion of cycle, which corresponds to the idea of reasonable management of resources in the long term. This method was initially developed in industry during the 1970s (Darnay, 1971), followed by the building sector (Kohler, 1986), before being extending to cover all human activities. Its application to urban complexes has been studied more recently (Kohler, 2003; Popovici, 2006; Herfray, 2010, 2011).

The environmental performance evaluation is another environment engineering tool that complements the LCA. These two tools are subject to standards, the ISO 14040 series for LCAs, and 14030 for environmental performance evaluation (ISO, 1999 and 2006). They meet different objectives: environmental performance evaluation, for example, concern factories, while LCAs are used with the aim of reducing the impact of products. Factories have precise locations, which means their impacts can be assessed taking their location into account: e.g. for equal emissions, the health consequences are increased on a densely populated site. However, we do not necessarily know where a product will be transported and used. LCAs thus consider average impacts, independent from the location of emissions.

A building, neighbourhood or urban complex is located precisely, so that environmental performance evaluation might seem a better option than LCA. This was the opinion of some stakeholders during early work on the environmental quality of buildings. However, the majority of impacts resulting from buildings and urban systems are produced outside of these systems: in factories for manufacturing building materials, during electricity and drinking water production, in equipment for treating waste and wastewater, etc. In the design phase, manufacturers, and thus the manufacture sites of building materials, are generally still not known, in particular during the early phases of a project. Yet it is in these phases that the decisions are made that have the biggest influence on the environmental balance. Average data, also known as generic data, are thus used, and the LCA method is the most appropriate in the current state of knowledge.

This does not of course rule out undertaking a complementary impact study, especially if the building or neighbourhood includes a significant source of pollution like a large heating system or a very busy road. Specific studies can also be pertinent, for example on the preservation of local biodiversity. In parallel, some research studies aim to refine LCAs by distinguishing emissions according to population density or the altitude of the emission site: pollution emitted from vehicles' exhaust pipes has a greater impact on health than the same pollution emitted from a very high chimney.

The present chapter describes the method and its limitations, followed by several examples that illustrate its application.

PRESENTATION OF THE METHOD

The performance of buildings, and a fortiori neighbourhoods, depends on the programme drawn up by the client; the choice of a site; the skills of the architect, urban planner and engineering firms; the use of suitable technologies; the care taken by companies in their work; and the civic behaviour of residents. This therefore means that the occupants need to be made aware, and that the professionals involved need to share common performance criteria to compare the various urbanistic, architectural and technical solutions.

Buildings' multiple functions make it more difficult to characterize this performance, which includes ecological, economic and socio-cultural aspects. Single criterion approaches bring the risk of replacing one environmental problem with another: e.g. preserving the ozone layer resulted in selecting new refrigerant fluids that contributed to the greenhouse effect; the current focus on the carbon balance brings the risk of shifting problems towards health risks from nuclear energy or biomass. The idea is therefore to draw up a performance criteria grid. This kind of grid has been studied in several European and national projects, such as Eco-housing (Peuportier, 2005), Adequa (Cherqui, 2005) and Lense (Lense, 2006). Stakeholders pay more attention to performance criteria if the underlying objectives are explained to them. A list of targets resulting from a participative process involving engineers, architects-urban planners, companies and residents as part of a European project (Stoa, 2005) is shown in the table 2.1 below as an example.

The next step is to define indicators for evaluating to what extent a project meets with these different targets. Life cycle assessments concern environmental aspects, but a similar approach can be used to evaluate economic performance, and research is being done on evaluating the social dimension.

To meet these criteria, evaluations require information from materials manufacturers, energy and water producers, waste treatment companies, etc. Yet manufacturing products calls for raw materials and energy, and producing energy requires materials

Table 2.1 Example of target grid resulting from a participative process.

Dimensions	Aims	Targets
1 Ecological	1 Preserve resources	1 Preserve raw materials
		2 Save energy
		3 Save water
		4 Manage land use
	2 Protect ecosystems	1 Limit toxic emissions
		2 Protect the climate
		3 Protect fauna and flora
		4 Protect rivers and lakes
		5 Improve outside air quality
		6 Reduce waste
		7 Reduce radioactive waste
		8 Preserve the ozone layer
		9 Limit risks linked to flooding
2 Economic	1 Reduce overall cost	1 Reduce building cost
		2 Reduce operating cost
		3 Reduce maintenance cost
		4 Reduce renovation cost
		5 Reduce demolition cost
	2 Increase value	1 Facilitate adaptation of space
		2 Facilitate adaption of uses
3 Social	1 Preserve residents' health	1 Improve indoor air quality
		2 Improve water quality
		3 Reduce electromagnetic fields
		4 Reduce risks (fire, explosion, etc.)
	2 Improve comfort	1 Improve visual comfort
		2 Improve thermal comfort
		3 Reduce noise
		4 Reduce odours
		5 Improve well-being (e.g. useful area)
	3 Improve social value	1 Improve usage quality
		2 Increase social equality
		3 Increase gender equality
		4 Facilitate social relations
		5 Improve participation
4 Cultural	1 Increase aesthetic value	1 Improve architecture and image
		2 Improve site integration
	2 Preserve knowledge and history	1 Respect historical sites
		2 Integrate memory
		3 Increase cultural value

to build production facilities. The environmental impacts can be evaluated by solving a matrix system that takes account of this set of interactions. This is the approach implemented in the *Ecoinvent* database (Frischknecht, 2007), which was devised in Switzerland but contains European data on some materials, and even data from different countries including France for some processes (e.g. electricity production). Environmental impacts linked to services (insurance banks, etc.) can be attributed to the different products along the lines of the matrix systems used in economics.

In France, manufacturers of building materials and components have established a standard for characterizing the environmental performance of these products.

Individual manufacturers know about the pollution emitted from their own manufacturing processes, but they need access to other data (in general from the Swiss database) on the raw materials and energy they consume. The advantage of this approach is that evaluations are not managed centrally, with industrials free to choose their service provider, but it could be useful to take a more general look at pooling and processing certain information.

The French database *Inies* (INIES, 2011) only concerns building products. However, the choice of a product often influences the energy consumption of the building in which it is used, and so data are needed on the impacts linked to energy production. Perspectives in terms of outlets for recycled products could facilitate this kind of inter-sectorial management of environmental issues, which could go beyond local optima.

For each product or process, the databases provide a life cycle inventory indicating the quantities of different substances drawn from or released into the environment (several thousands of flows in the Ecoinvent base, around 160 in the Inies base). A simplification of the analyses makes the approach more accessible to the numerous SMEs in the building product market, but requires validation. Environmental indicators are then evaluated. For example, the global warming potential is defined by the intergovernmental panel on climate change. This indicator expresses a CO_2 equivalent for the different greenhouse gases according to their optical properties and lifespan in the atmosphere.

Some environmental aspects require more sophisticated models that still have a significant margin of uncertainty. This is particularly the case for human toxicity. A European-level model involves gridding the territory and studying the outcome of pollutants released into each grid, including the transportation of the substances between ecological compartments (i.e. air, surface waters, water tables, soil, oceans and sediment), their transfer into drinking water and food, the doses received by inhabitants through inhalation and ingestion, and the health risks according to these doses, expressed in disability adjusted life-years lost (DALYs) (Goedkoop, 2001). A similar indicator is defined for biodiversity through the percentage of extinct species on an area of land over a certain time period (PDF.m^2 · year, PDF: Potentially Disappeared Fraction) (Goedkoop, 2001).

Life cycle inventories and impact indicators are related to a "functional unit", defined by a quantity (e.g. 1 m^2), a function (e.g. thermal insulation of a façade), the function's quality (e.g. the insulation's thermal resistance, its acoustic properties, etc.) and a time period (e.g. 30 years).

Thanks to these different data, LCAs can then be applied to more complex systems, in particular buildings. However, a building's impacts are not ascertained by simply adding the impacts of its components. This is because components interact, e.g. solar radiation transmitted by glazing is partially stored in the envelope and, depending on the heating system and its setting, will result in a drop in heating consumption and the corresponding impacts. The significance of energy aspects in the environmental balance justifies coupling the LCA with a dynamic thermal simulation, which also makes it possible to assess the degree of comfort in the building under study.

The objective during the design phase is generally to respect the programme defined by the client – the building must house a number of planned activities, with a certain degree of comfort, quality of life, etc., while reducing external environmental impacts and costs. The object of the study, i.e. the functional unit in the sense of the life cycle assessment, is thus a building or a neighbourhood that satisfies these requirements considered over a certain time period.

For example, the programme can require performance levels based on the economic, social and cultural criteria listed in table 2.1. The functional unit will then integrate these requirements, and the LCA can be used to compare the environmental impacts of the alternatives, with the aim of looking for an urbanistic, architectural and technical solution that minimizes these impacts. However, the programme can also fix the environmental, social and cultural performance, in which case the objective could be, for example, to determine the least expensive solution for reaching these performance levels. In this case, the functional unit integrates the social and cultural aspects, and the LCA is used to check whether the environmental targets have been met. The cost of the various alternatives that respect the programme can then be compared to choose the less costly option. Fixing all of a programme's criteria is more risky since we probably do not yet have sufficient hindsight to guarantee that a solution exists.

However, carrying out an analysis requires modelling a much broader system, which will vary depending on the study's objectives. If the objective is to compare different sites for a building, the study should include the resulting transportation (e.g. home-work), household waste management, energy networks (electricity, gas, and perhaps heat) and water. If the study is restricted to the building's envelope and equipment, human transportation could be left aside if all of the alternatives compared are equivalent from this point of view.

Process inventories take any upstream phases into account, e.g. an inventory of useful kWh supplied by a gas boiler considers gas extraction and distribution. The impacts linked to infrastructures are counted if they are significant. For example, transport inventories include the network's construction and maintenance, whereas the impacts of building factories to manufacture the materials and components might be considered negligible. Household waste can be recycled or incinerated, with or without heat recovery. The corresponding inventory is calculated according to the local context (yield from energy recovery, substituted energy, etc.).

The grid of sustainable development objectives (cf. table 2.1) is matched with a set of indicators used to evaluate a project and look for ways to improve it. Table 2.2, for example, gives the environmental indicators considered in EQUER developed in the 1990s (Polster, 1995) and updated since (Peuportier, 2008).

The acidification indicator relates to the objective of protecting forests and involves diminishing the "acid rain" phenomenon. The Eutrophication indicator is linked to "blue-green algae" production and so protecting rivers and lakes. The "photochemical ozone" theme concerns tropospheric ozone and its impacts on health (respiratory diseases) – ozone concentration is one of the parameters of urban air quality. The ozone layer issue was piloted by the Montreal Protocol and including an indicator on this theme appears less useful today. The "resource depletion" indicator takes into account a matter's global reserves and how fast they are becoming scarce. Odour generation is expressed by dividing emissions (in mg) by the respective thresholds (in mg/m^3) at which the substances can be detected by 50% of a representative sample of the population. The indicator is thus expressed in m^3 of contaminated air.

The EQUER model is based on a life cycle simulation. Environmental impacts are evaluated over a certain time period, e.g. 80 years, in one-year time steps. These include

Table 2.2 List of environmental indicators considered in EQUER.

Environmental indicator	Unit	Reference
Cumulative energy demand	GJ	(Frischknecht, 2007)
Water used	m^3	(Frischknecht, 2007)
Abiotic resource depletion	kg antimony eq.	(Guinee, 2001)
Waste production	t	(Frischknecht, 2007)
Radio active waste	dm^3	(Frischknecht, 2007)
Greenhouse effect (100 years)	t CO_2 eq.	(Forster, 2007)
Acidification	kg SO_2 eq.	(Guinee, 2001)
Eutrophication	kg PO_4^{3-} eq.	(Guinee, 2001)
Damage to biodiversity	PDF $*m^2$.an	(Goedkoop, 2001)
Damage to health	DALY	(Goedkoop, 2001)
Photochemical ozone production	kg C_2H_4 eq.	(Guinee, 2001)
Odour	m^3 air	(Guinee, 2001)

the manufacture, transportation, and replacement of components depending on their life span, end of life and any recycling, along with processes linked to the building's operation stage (i.e. heating, water and electricity consumption, waste management, transport, etc.). Due to the importance of energy aspects, as mentioned above, this LCA model has been linked to a dynamic thermal simulation tool, COMFIE, which puts it one step ahead of other similar tools.

Based on this experience with buildings, the LCA has been extended to apply to neighbourhoods. This model takes into account several types of building, as well as public spaces (roads, parking areas, green spaces, etc.) and networks (water, energy). The advantages are twofold: on the one hand, this scale can be used to study a number of urban planning choices that have a very strong influence on environmental performance, in particular the direction of roads (and so buildings) and compactness, linked to density; on the other hand, some technical options are decided at this level, e.g. domestic heating systems, transport and waste treatment. Here too, a neighbourhood's performance is not ascertained by adding together the performance of its components, because of shared facilities and interactions between buildings (e.g. shading masks).

LIMITATIONS OF THE METHODOLOGY

LCAs only relate to the quantifiable aspects of environmental quality. More subjective appreciations such as aesthetics and quality of life are therefore not covered. For health, as seen above, the human toxicity indicator is evaluated taking the average impacts on the European territory, and so does not depend on the site of the emission (population density, dominant winds, etc.). Yet the emissions site is generally not known because the designer does not know, for example, in what factory the building materials will be manufactured. If the study's aim is to evaluate the health impact on the building's occupants or neighbouring residents, other types of model will be needed that include mass transfers into flooring and walls, and air movements, etc.

Buildings and neighbourhoods are defined by functional units. When usage evolves over time, an evolution scenario needs to be defined and several LCAs carried out corresponding to the successive functional units. Alternatively, sensitivity studies can be done that vary certain usage characteristics.

One limitation that all indicators share is the imprecision of evaluations. It is often difficult to ascertain the margin of uncertainty on data and results, but it can be high. In domains like thermal simulation, the results of a calculation can be compared to an energy invoice or a temperature measurement. However, for CO_2, for example, measurements can only be made on an isolated process and not on the building's entire life cycle. The first level of imprecision concerns evaluating flows of matter and energy (inventory data). The second level corresponds to the aggregation into effects (potential impacts), for example greenhouse gases' global warming potential has been estimated to be 35% uncertain. The third and last level concerns the move from effects to damage (years of life lost, for example). Another cause of error is linked to the duration of the analysis period. For instance, it is difficult to anticipate changes in waste treatment techniques, especially for the demolition, which might take place in the distant future. It might be preferable to make a statistic analysis based on scenarios to which a probability has been attributed. The evaluation's multi-criteria character also requires tackling the issue of priorities, which depend on local context (e.g. scarcity of water in some regions) and decision-makers' choices.

To understand these limitations, eight European tools were compared as part of the European thematic network PRESCO (Peuportier, 2004). Taking the example of a Swiss, wooden-framed house, its contribution to the greenhouse effect calculated over 80 years differed by +/−10% depending on the tool. These European studies are continuing within new projects, with a focus on finding ways to simplify the analyses. Unlike industrial products, buildings are generally one-off items. It is therefore impossible to spend several months working on an LCA as might be the case for a mass-produced product. Yet over-simplifying the method brings a risk of inaccurate results. It is therefore worth validating simplified hypotheses in relation to a more detailed reference.

Other current focuses of research concern more specific aspects like modelling recycling, integrating biogenic CO_2 into the balance, in particular when the material used is wood, or taking into account emissions inside buildings. The current practice of so-called attributional LCAs could also evolve towards "consequential" LCAs. The use of an LCA can in some cases lead to developing a technology that results in a change in the environmental impacts considered as the initial hypothesis. For example, the carbon balance encourages the development of electric heating, but this development leads to a high peak demand, which modifies the production mix and thus the impacts of using electricity. It is no doubt therefore a good idea to develop models that take these dynamic effects into account. Lastly, at neighbourhood level, it would be useful to understand the influence on movements and the corresponding environmental impacts of choices regarding urban morphology and parking.

EXAMPLES OF APPLICATION

LCAs date from the 1990s and have been used in projects for new buildings and retrofits in both the residential and tertiary sectors (i.e. offices, school buildings, etc.).

These applications have demonstrated the significance of the operation stage: in a contemporary standard house, for example, the construction stage only represents around 20% of the energy consumed, with the remaining 80% consumed for its operation (i.e. heating, hot water, lighting and other electricity uses). For a house corresponding to current best practices in terms of energy efficiency (i.e. German "passive house" label), the construction represents 50%. Compared to a standard house heated by gas, a passive house heated by a heat pump emits 3.5 times less CO_2 but generates 30% more radioactive waste. If electricity is produced using renewable energy (e.g. solar roof panels), all of the impacts are reduced in comparison to the standard reference.

These studies have also illustrated how occupants' behaviour impacts the environmental performance of buildings. Eco-design is not sufficient, and residents need instructions for use and greater awareness to optimize the management of buildings.

The approach has been tested on several operations extended to the scale of a small neighbourhood, e.g. the Lyon Confluence neighbourhood. A comparison of several urban morphologies is presented below as an illustration (Vorger, 2011). In a given context, taking the climate in the Parisian area (Ile de France) and the corresponding industrial processes (e.g. French electricity production), several building complexes were compared based on the same functional unit: a usable surface of 33 m^2 corresponding to accommodation for one resident, and with a certain comfort level (heating at 19°C, no air conditioning) for eighty years. Because interactions (shading masks) were taken into account, a group of buildings had to be modelled, but the impacts were then reduced to the functional unit.

In real-life case studies, the morphologies compared correspond to design alternatives adapted to fit the context, however, this intentionally simplified example considers highly typical configurations:

- 16-storey tower blocks with sides of 20 m, spaced 20 m apart on an east-west axis and 40 m on the north-south axis, each with 2000 m^2 of street 200 m^2 of access road,
- 9-storey housing blocks, 10 m wide and 50 m long, spaced respectively 10 m apart (or 20 m) along an east-west axis (or north-south), each with 1300 m^2 of street and 100 m^2 of road,
- Single-storey individual houses of 130 m^2 surrounded by 200 m^2 of garden, each with 120 m^2 of street and 20 m^2 of pedestrian pathway,
- A step-terrace morphology as represented in figure 2.1 below, with significant glazing on the south-east and south-west façades to maximize solar gains,
- Part of a town corresponding to the Solar City in Freiburg (Germany), also shown in figure 2.1, which attains a positive energy balance thanks to its solar-panel roofs.

These five morphologies were compared with equal technologies, based on the "passive house" label, i.e. very high level of insulation, triple glazing, ventilation with heat recovery on exhaust air, high efficiency gas boiler. To make it easier to interpret the results, shown in figure 2.2, transport aspects were not initially considered.

Each axis corresponds to an indicator. The alternatives are represented in relative value compared to the Individual House reference. For example, greenhouse gas emissions are reduced by 30% in Freiburg's positive energy neighbourhood (solar city), but

Figure 2.1 Left, view of the "step-terrace" morphology (ALCYONE software), south and north sides; right, Freiburg's Solar City (architect: Rolf Disch).

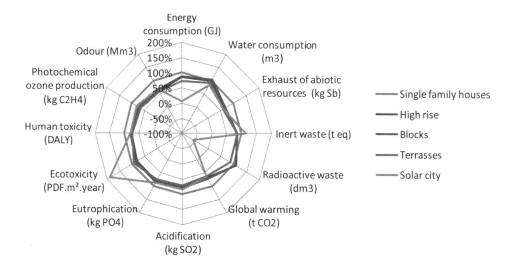

Figure 2.2 Results of a life cycle assessment over 80 years, comparison of alternatives, EQUER software.

impacts are higher in terms of eco-toxicity, due to the production of the solar panels. However, this impact remains very limited in absolute terms. The indicator is negative for radioactive waste production: the electricity produced by the photovoltaic system avoids the standard 80% nuclear production and its corresponding impacts. The more compact layouts reduce most impacts compared to individual houses. This example shows that urban and architectural design has a large influence on performance.

According to the results of the thermal simulation, heating requirements can vary by a factor of 3 depending on a building's compactness and its exposure to the sun, which shows the benefits of taking a so-called "bioclimatic" approach and of integrated design that brings the architect and engineer together at an early design phase. The environmental benefits of technological choices like renewable energy, water savers, and combining materials according to their thermal mass, can be quantified by an LCA and thus contribute to promoting innovation. The main advantage of this approach is to move all professions forward together, and not to impose a single building or technology model.

Urban density can strongly influence transport modes. In the simplified case study above, this aspect was tackled in a second stage. An identical home to work distance of 14 km was considered in all of the compared alternatives, with the journey being made by car for the individual houses and by train for the other cases. According to these hypotheses and in order of size, including transport increases the primary energy consumed by around 80% for the houses and 40% in the other cases. For greenhouse gas contribution, the increase is respectively 140% and 20%.

MOVING TOWARDS A DYNAMIC SIMULATION OF THE LIFE CYCLE

Current LCA tools are based on static methods. However, some processes evolve over time. This is particularly the case for electricity production, since additional means of production need to be used during peak demand periods. We can thus observe seasonal variations due to heating and air conditioning, weekly variations due to professional usage, and hourly variations with a peak at 7 pm, for example. Improving buildings' thermal performance incurs a rise in the relative contribution of electricity consumption in the overall environmental balance. On the other hand, positive energy buildings and neighbourhoods require studying the influence of local electricity production, which is also variable over time depending on solar or wind resources. In order to evaluate more precisely the environmental impacts of usages and electricity production, these variations should therefore be taken into account.

A model was thus developed to evaluate the electricity production mix over time according to the temperature, based on data supplied by the electricity transport network, RTE (and so only in a French context, but including electricity imports). In addition, the impacts of electricity production were allocated among the different uses (heating, air conditioning, hot water, other professional uses and other domestic uses). The results obtained from a first case study reveal a gap of up to 30% compared to a static calculation based on an average annual mix.

This model should be refined once several years of measurement are available from RTE. Hypotheses concerning the production of imported electricity and the procedure for allocating impacts according to usage should also be validated. It would also be useful, taking a consequential LCA approach, to consider the building's influence on electricity production as well as other background system processes.

PERSPECTIVES: MOVING TOWARDS INTER-SECTORIAL MANAGEMENT OF THE ENVIRONMENT

Significant work remains to improve the precision and performance of these tools, but initial applications show potential. The stakes in terms of preserving the environment and industrial developments justify the efforts, which could be partly shared via international partnerships.

A building's performance depends on: a programme drawn up by the client that at the very least respects regulations; the choice of the site; the capabilities of the architect and engineering firm; the use of suitable technologies; the care taken by companies in construction; and the civic behaviour of occupants. As we have seen, the building sector interacts with the sectors of construction materials and component industries, energy and transport. Some consider that the existing building stock cannot be improved and that efforts should focus on supplying "clean" energy. At the other extreme, constraints concern buildings, which are becoming energy producers. Technical innovation could thus centre more on envelopes or equipment.

Urban densification is sometimes proposed to reduce environmental problems linked to transport. In the example above, it would be useful to assess how urban morphology influences residents' choice of transport, but this would require additional developments.

New buildings each year only represent around 1% of the existing buildings, which are therefore a key issue in long-term policies. Demolition only represents about 0.03% of existing buildings each year, so that renewal is much slower than that of e.g. vehicles. It is clearly not sufficient to act on new construction to reach targets like quartering greenhouse gas emissions by 2050. LCAs therefore have very useful applications in retrofit projects.

These few examples illustrate the wide range of questions to which LCAs could provide part of the answer. To this end, inter-sectorial cooperation is essential, in particular when it comes to standardizing databases.

REFERENCES

Cherqui, F., Méthodologie d'évaluation d'un projet d'aménagement durable d'un quartier, method ADEQUA, Doctoral thesis at La Rochelle University, December 2005.

Darnay, A. and Nuss, G., Environmental impacts of Coca Cola beverage containers, Midwest Research Institute report, 1971.

Forster, P.M., Changes in Atmospheric Constituents and in Radiative Forcing, In: Solomon, S., D. Qin, M. Manning, Z. Chen, M. Marquis, K.B. Averyt, M.Tignor and H.L. Miller (ed), Climate Change 2007: The Physical Science Basis. Contribution of Working Group I to the Fourth Assessment Report of the Intergovernmental Panel on Climate Change, Cambridge University Press, 2007.

Frischknecht, R., Jungbluth, N., Althaus, H.-J., Doka, G., Dones, R., Heck, T., Hellweg, S., Hischier, R., Nemecek, T., Rebitzer, G., Spielmann, M., and Wernet, G., Overview and Methodology: Ecoinvent report No. 1, www.ecoinvent.ch, Dübendorf: Swiss Centre for Life Cycle Inventories, 2007.

Goedkoop, M.J. and Spriemsma, R., The Eco-Indicator 99, A damage oriented method for life cycle impact assessment, methodology report, methodology annex, manual for designers, Amersfoort, June 2001.

Guinée, J.B., (final editor), Gorrée M., Heijungs R., Huppes G., Kleijn R., de Koning A., van Oers L., Wegener Sleeswijk A., Suh S., Udo de Haes H. A., de Bruijn H., van Duin R., Huijbregts M. A. J., Lindeijer E., Roorda A. A. H., Weidema B. P.: Life cycle assessment; An operational guide to the ISO standards; Ministry of Housing, Spatial Planning and Environment (VROM) and Centre of Environmental Science (CML), Den Haag and Leiden, The Netherlands, 2001.

Herfray, G., Contribution à l'évaluation des impacts environnementaux des quartiers, doctoral thesis, MINES ParisTech, 329p, October 2011.

Herfray, G. and Peuportier, B., Life Cycle Assessment applied to urban settlements, Sustainable Building Conference 2010, Madrid, April 2010.

INIES, base de données française concernant les caractéristiques environnementales et sanitaires des produits de construction, www.inies.fr, accessed on 22/08/2011.

EN ISO 14031, Environmental management – Environmental performance evaluation – Guidelines, International Organisation for Standardization, Geneva, 1999.

EN ISO 14040, Environmental management – Life cycle assessment – Principles and framework, International Organisation for Standardization, Geneva, 2006.

EN ISO 14044, Environmental management – Life cycle assessment – Requirements and guidelines, International Organisation for Standardization, Geneva, 2006.

Kohler, N., Analyse énergétique de la construction de l'utilisation et de la démolition de bâtiments, doctoral thesis, Ecole Polytechnique Fédérale de Lausanne, 1986.

Kohler, N., Rapport as part of the European project SUIT Sustainable development of Urban historical areas through an active Integration within Towns, Task 3.3 c – Integrated Life Cycle Analysis in the Sustainability Assessment of Historical Areas, European Commission, March 2003.

LEnSE, Stepping stone 1, Sustainability assessment of buildings, issues, scope and structure, LEnSE partners, St Stevens Woluwe (Belgique), November 2006.

Peuportier, B., Kellenberger, D., Anink, D., Mötzl, H., Anderson, J., Vares, S., Chevalier, J., and König, H., Inter-comparison and benchmarking of LCA-based environmental assessment and design tools, Sustainable Building 2004 Conference, Warsaw, October 2004.

Peuportier, B., Towards sustainable neighbourhoods, the eco-housing project, IV International Conference "Climate change – energy awareness – energy efficiency", Visegrad (Hungary), June 2005.

Peuportier, B., Popovici, E. and Trocmé, M., Analyse de cycle de vie à l'échelle du quartier, Seminar ADEQUA Quartiers Durables, Chambéry, October 2006.

Peuportier, B., Eco-conception des bâtiments et des quartiers, Presses de l'Ecole des Mines, Paris, November 2008.

Polster, B., Contribution à l'étude de l'impact environnemental des bâtiments par analyse du cycle de vie, doctoral thesis, MINES ParisTech, December 1995.

Popovici, E., Contribution to the Life Cycle Assessment of settlements, doctoral thesis, MINES ParisTech, February 2006.

Støa, E. and Nesje, A., Supporting participation in the development process of sustainable neighbourhoods, IV International Conference "Climate change – energy awareness – energy efficiency", Visegrad, June 2005.

Vorger, E., Application de l'analyse de cycle de vie à la comparaison de morphologies urbaines, UPMC Science and Technology Master's report, June 2011.

Chapter 3

Methods for evaluating environmental impacts. Analysis and proposals

Jean Roger-Estrade, Samuel Maurice, Géraldine Thomas-Vallejo, Marie Aurenche & Joël Michelin
AgroParisTech

INTRODUCTION

Impact studies were established by the French Act of 10 July 1976, and have changed the way that infrastructure projects are designed and constructed by including an analysis of their effects on the environment. These systematic, regulated studies can be used to anticipate impacts and propose measures to avoid, reduce and if possible, offset them. The studies are available for general consultation in the form of public enquiries.

Several texts have followed on from this Act, providing details and refining the content of environmental analyses. Decree No. 2011-2019 of 29 December 2011 in particular reforms the legal regime for impact studies of building and development projects. It describes their content and defines the operation mode according to the nature of the project. Since April 2009, the *Autorité Environnementale* (environment authority – EA) has been responsible for officially evaluating these studies, which is a genuine step forward.

In parallel, public authorities, citizens and associations have used impact studies to make progress on a number of projects. The scope of impact studies therefore extends beyond the simple client/public authority tandem.

These changes, although beneficial for making sure projects fit into their environment, do have some limitations. In particular, some impacts remain difficult to evaluate because assessment methods are unstable and subject to controversy. The potential consequences include harm to the environment if methods are not suitable; obstacles to accomplishing major projects; and difficult debates with local citizens.

This observation led the eco-design Chair on buildings and infrastructure to produce an inventory of the methods used to evaluate all of the impacts on the environment. The objective of this study is twofold:

- To produce an overview of the methods used to evaluate the environmental impacts of transport infrastructures (and related facilities);
- To propose research avenues and action for implementation by all stakeholders. These propositions could in the future lead to more efficient environmental assessments and more pertinent recommendations as a result of the studies.

Although this analysis centres on linear transport infrastructures, most of the information and proposals presented concern a broad range of projects, such as exploiting quarries, constructing car parks or airports, and installing systems to produce renewable energy.

After a reminder of current regulations, we summarize an overview produced by the chair resulting from the analysis of around thirty impact studies corresponding to different transport infrastructure projects in France, and then conclude with some suggestions for improvement.

REGULATIONS RELATING TO IMPACT STUDIES

Impact study contents are subject to Article R122-4 of the French environment code, modified by Decree No. 2011-2019 of 29 December 2011. The client is responsible for the contents of the study and the quality of its execution. The contents must be in proportion to the size of the project, adapted to its nature and appropriate in terms of the fragility of the affected zone and local environment issues (in the broad sense, e.g. including cultural aspects).

Starting with the study's design, the client clearly possesses a degree of discretion. The law however regulates the contents of studies, which must include:

- a description of the project (design, dimensions, land use, etc.);
- an analysis of the initial state of environments likely to be affected by the project (e.g. state of biodiversity, water courses and ponds, biological balances, cultural heritage, air quality);
- an analysis of the project's impacts: negative and positive, direct and indirect, temporary and permanent, in the short and long term, taking into account how they interact with each other;
- the measures anticipated by the client to avoid, reduce, and if possible offset the project's significant negative effects on the environment, biodiversity and human health, coupled with an estimate of the corresponding expenditure;
- the main substitution solutions proposed by the client, the methods, and any technical or scientific difficulties.

The competent administrative authorities can give indications as to how the impact study should be carried out via circulars, instructions or guidelines, or at the time of the preliminary scoping procedure. Since 30 April 2009, impact studies and environmental studies have been controlled by the **Autorité Environnementale du Conseil Général de l'Environnement et du Développement Durable (CGED[1])**. The creation of this environmental authority (EA) brings France in line with legislation: since the creation of MEEDDAT (French ministry for ecology, energy, sustainable development and territorial planning), the same minister has been in charge of environment and development policies, which has led to conflicts of interest with European principles. The EA thus makes publicly available its recommendations on evaluations of the impacts of major projects and programmes on the environment along with management measures aimed at avoiding, reducing or compensating these impacts. "These provisions aim to make

---

[1]Decree No. 2009-496 of 30 April 2009 relating to the state's competent administrative authority regarding the environment provided for in articles L. 122-1 and L. 122-7 of the French environment code.

it easier for the public to participate in drawing up decisions that concern it, and to improve the quality of projects before decisions are made". The EA gives its opinion on the impact study carried out by the petitioner fairly early on in the process (in practice before the public inquiry) so that the project can be improved.

The client must thus integrate the time period for carrying out impact studies into the project's overall schedule in order to take into account changes in the content of the impact studies and the recommendations of the planning department, including fauna/flora inventories and analyses (air quality, etc.). **Impact studies must be carried out and validated at a very early stage.** The client must undertake the following:

– Establish flora and fauna inventories using specialized consulting firms right from the project's opportunity research phase;
– Watch for new texts and themes that are not yet legally enforceable;
– Be aware of existing administrative documents (guides, circulars, instructions, etc.);
– Keep track of administrative jurisprudence on impact studies in order to: understand, e.g. the notions of deficiency or absence; respond in a suitable way to observations from the planning department, to the Environmental Authority, and to the public; and increase legal security surrounding the project;
– Consider the financial consequences of delays resulting from an inadequate impact study.

However, no general, up-to-date document currently exists to help clients carry out impact studies in line with legal and regulatory requirements. The present chapter makes a contribution in this area.

The following table sets out the administrative and technical process of making an environmental assessment of a linear infrastructure project.

• Implementation stages.	• Comments
• Project planned by the project initiator/client.	• The project initiator looks closely at the opportunity studies. He calls on internal or external expertise to identify the main environmental issues. These are not yet impact studies.
• Scoping of the impact study by the authorities (optional procedure).	• The project initiator consults the state planning department to obtain general guidelines as to the content of the impact study: methodologies, themes to be studied, definition of study scope, cumulated effects between project initiators, etc.
• Impact study by engineering firm in liaison with project initiator.	• The project initiator draws from the guidelines given by the planning department to define the specifications for carrying out the impact study to be followed by service providers (i.e. main contractor, specialized consulting firms).
• Consultation procedures.	• State departments take note of the impact study and make their recommendations on the document's form and content. This is a pre-assessment.

- Modifications/ additions made to the impact study.
- Application sent to the government representative ("prefect") (public inquiry, etc.).
- Government departments consult with each other.
- Modifications/ additions made to the impact study if necessary.
- Referral to the Environmental Authority by the prefect.

- Recommendation by the Environmental Authority considering the project.

- Public inquiry: observations, counter-propositions.

- Project initiator responds to public inquiry observations.

- Act authorizing/ approving the project.

- The project initiator has the impact study changed, explaining how the recommendations are taken into consideration.
- The project initiator explains the changes made to the impact study following the planning department's pre-assessment.

- The state services check that the pre-assessment proves effective.

- The state services take note of the impact study and make their recommendations on the document's form and content. This is a second pre-assessment.

- The Environmental Authority may be central (General Council for Environment and Sustainable Development at the ministry of ecology) or local (regional authority delegates to regional environment directorate). It assesses the impact study.
- The Environmental Authority makes a recommendation that is the official assessment of the impact study. This assessment relates to the impact study's formal content (respect of the content defined by the environment code) and its substance (the environmental assessment strictly speaking). It highlights the manner in which the project takes the environment into consideration. The recommendation may indicate shortcomings or require that the client improve the application and the impact study in certain areas. If the recommendation is negative, the prefect will not launch a public inquiry.
- The Environmental Authority's recommendation is included in the public inquiry paperwork and therefore accessible to the public. The inquiry gives the public an opportunity to make observations, add any additional information or request complementary studies.
- The project initiator provides all of the responses required and may have to undertake complementary studies. A public inquiry may be suspended or a new public inquiry organized to produce the required complementary/additional studies.
- The prefect justifies the authorization to carry out the project in terms of the environmental assessments and questions asked during the public inquiry. It will only authorize the project if the responses made are judged to be satisfactory.

The project initiator is invited to use all available preliminary consultation possibilities to perfect the impact study, especially for themes that will not be standardized or officialised in the analysis of the effects.

OVERVIEW OF THE ANALYSIS OF THIRTY IMPACT STUDIES

An impact study project covers the various domains set out in the scoping note in more or less detail depending on the context of the project: the initiator must produce an impact study that is in proportion to the challenges and effects anticipated (i.e. application of the proportionality principle provided for in the French environment code), while analyzing all domains. Each of these will require a variety of methods and methodologies to carry out the impact study (e.g. cartography, field surveys, analyses of these surveys or previous data, prospective modelling).

The projects analyzed included the following study areas:

- Physical environment (surface waters, ground waters, geology, geomorphology, air, climate, waste, acoustic environment);
- Fauna and flora integrating habitats' characteristics;
- Landscape and heritage;
- Human environment (activities, land use, health, socio-economics, traffic).

The main conclusions of the analysis reveal that, although no particular remarks pertain to the assessment of impacts on heritage and human activities, and although methods for assessing impacts on air and acoustic quality meet with standards, the same cannot generally be said for assessments of impacts on surface water quality, which often draw from a wide range of documents (guides or administrative circulars). In addition, assessing impacts on biodiversity involves recognized scientific methods that are not obligatorily subject to standards. They clearly vary according to species: some are still being developed and are subject to scientific debate. Three questions arise: (i) the choice of indicator species: over the last few years, more species have been studied and the associated evaluation methods are not all established (e.g. in the PACA region in south-east France, a methodology for the Hermann tortoise has only recently been adopted); (ii) the lack of consideration for impacts' temporal dimension; assessments essentially look at immediate impacts, yet these can change over time, which is only very rarely considered; (iii) the uncertainty regarding the effect of proposed compensation measures is never raised. Yet these effects are not certain and it would be worth matching the proposals with a level of risk of failure (or success), along with taking measures for monitoring efficiency.

Along with these general remarks on content, the analysis revealed shortcomings in the projects' form: maps were often difficult to interpret because scales, directions and keys were imprecise; data sources were frequently not precise enough, whether they came from literature or measures undertaken for the study. This latter point makes it difficult to check results. However, the law makes a number of requirements: e.g. an impact study must be accessible and understandable to the public, thus the document should be easy to interpret. Measurements require description and justification of the methods, sampling, references and tools used.

PROPOSALS

In the previous section, we highlighted a number of flaws that occurred in most, if not all, of the studies. These flaws concerned both the content and form of the projects.

Regarding the form, maps or plans presented in the studies should be checked and include the correct information, along with a concise presentation of the principle behind producing the maps so that readers can understand the overview and analysis of the data presented. The source of information and data used should systematically be provided to make it easier to check. It is crucial that the scale be clearly indicated, whether or not it is adapted to the study being undertaken. Adequate details on the study scales should be provided. In addition, scales depend on the species targeted and the inhabitants described, and should therefore be in proportion to the objects studied, which will save time and money and make it easier to obtain results that are more representative of projects' impacts.

Regarding the content, some minimum information should accompany methods. Although in some very standardized domains with established methods, bibliographic references are sufficient for some administrative departments, it is useful to provide a systematic minimum description (e.g. principles, tools, main parameters measured, sampling method), to inform non-specialists for instance. That said, two key components condition the precision of the information. Firstly, the expertise of engineering offices needs to be maintained, and some develop in-house methods to stand apart from their competitors that need to be protected. In addition, some themes use standardized methods that are sometimes subject to a fee with detailed protocols that cannot be made available to the public.

Post-project monitoring measures (or guidelines for organizing monitoring) should be systematically provided in assessment projects. They are a way of indicating how the client considers long-term evaluation. When monitoring is planned as early as possible, it creates an opportunity to anticipate with experts the main points for vigilance and set up suitable methods in good time. Lastly, *ex post* environmental balances can be produced by engineering firms different from those that produced the pre-project studies. If the monitoring to be carried out after the project is pre-planned, it can be systematically based on the results obtained in the initial state produced *ex ante*. Any changes resulting from the project's presence can thus be analyzed more efficiently, and so management methods can be adapted and/or a successful construction can be developed on solid scientific bases. Lastly, monitoring can also create an opportunity to interact with research organizations to validate and/or perfect any methods for evaluating the impacts.

Lastly, we suggest that studies should be more cross-cutting, rather than restricted to a particular domain. Successfully integrating all issues on a territory requires dialogue between specialists from the different departments. One person should also work on a systematic overview of the different areas: this could be the project manager for road construction projects, or assemblers for quarry projects.

No method is perfect. Most studies do not mention the limitations of the methods used, and yet indicating the limitations is key to a rigorous scientific approach. Physical and natural environments are particularly complex and a critical review of the methods and results can often show the reader that a study's imperfections result from lack of knowledge or technical means (which is very different). Knowledge of these limitations should be included in the analysis and proposals. What is more, indicating limitations is a way of identifying what research needs to be done to improve the efficiency of assessments. VINCI has everything to gain from such an approach in showing itself to be a driver of progress in assessment methods.

A final point related to content is particularly important: the studies we consulted did not sufficiently consider ordinary biodiversity; they put too much focus on preserving remarkable, rare and native species. This seems detrimental to us for several reasons. Firstly, ordinary biodiversity appears to be a major determinant of the services rendered by ecosystems (ecosystem services as defined by the Millennium Ecosystem Assessment). Consequently, we might expect the legal status of ordinary biodiversity to evolve over the coming years. Impact studies should take this into account. We might also expect a precautionary principle to be applied to some extent: there is much that we do not know about the value of ordinary biodiversity, but that does not mean that what we do not know has no value. The loss of ordinary biodiversity could involve the loss of ecosystem services that are very important although still little understood[2]. A new branch of ecology called reconciliation ecology views ordinary nature as fundamental to ecosystem services. Lastly, ordinary biodiversity supports "remarkable" diversity, providing it with a refuge, movement and feeding passages, and reproduction areas[3]. Field studies should therefore put particular focus on groups that have never or rarely been evaluated. For some groups (e.g. insects), thorough inventories are difficult to establish because of the very high numbers of species. Research studies should therefore look at developing methods based on ascertaining bio-indicator species that are routinely useable and centred on biological characteristics (features of species).

Biodiversity is defined according to three aspects: function, organization and composition[4]. Noss (1990) insists that ecological processes should be considered when evaluating biodiversity: it is not enough to evaluate the populations present. We should also try to understand what determines species' capacity to feed, reproduce and protect themselves from predators or climate hazards. All of these processes should be evaluated in partnership with research bodies, with a view to developing experimental methods or so that certain methods can be applied to groups that are currently rarely evaluated.

Assessing a project's impact on biodiversity is a delicate issue. Making efficient recommendations often means going beyond simply responding to regulatory constraints. Several initiatives make a move in the right direction. For example, in his book "Road Ecology"[5] Forman puts forward a method for working out the distance from the road of major types of impacts (e.g. chemical pollution, noise), distinguishing

[2]European Community, 2008. The Economy of Ecosystems and Biodiversity. Interim report. 68p.

[3]Biotope, 2007. Etat des lieux de la connaissance et des attentes des acteurs sur l'impact des infrastructures de transport terrestres sur les paysages et les écosystèmes. Rapport final. Compréhension des impacts des infrastructures de transport terrestre sur le milieu naturel : vers une prise en compte de la dynamique des territoires. 183p.

[4]Noss, R. F. 1990. Indicators for Monitoring Biodiversity: A Hierarchical Approach. Conservation Biology, volume 4, December 1990.

[5]Forman, R. T. T. et al., 2003. Road Ecology. Science and Solutions. 481p.

between short-, mid- and long-distance effects. He suggests locating the inventory of species and environments within these various zones to produce a stratified sampling. This kind of method involves developing knowledge on the sensitivity of each level of biodiversity to the various impacts.

The theme of biodiversity is constantly evolving. These changes will necessarily influence the way we apprehend biodiversity in building projects and how we modify them.

Chapter 4

The planning of territorial facilities taking an eco-design approach: principles and methods for land use and transportation

Fabien Leurent
Paris-East University, City Mobility Transport Laboratory, Ecole des Ponts ParisTech, Ifsttar, UPEM

HOW CAN ECO-DESIGN BE APPLIED TO TRANSPORTATION AND LAND USE PLANNING?

For a territory, made up of a geographical area whose inhabitants form a society and which is endowed with political power of its own, transport modes and networks are facilities, i.e. resources invested in the long term for lasting usage. Similarly, buildings and other support items (e.g. roads, public squares, car parks) are facilities that equip the territory so that it can be used for a range of human activities of a social and economic nature. In practice, these facilities are technically designed to last in order to amortize the financial cost over time.

However, this intrinsic durability needs a rethink in the face of the ecological challenges of sustainable development, i.e. withdrawals of resources (matter, energy) and emissions of all kinds (pollutants, greenhouse gases, noise). As a result, a great number of territorial planners are enthusiastically involved in the "empirical eco-design" of territorial facilities, such as modes of transport and buildings. Concerning transport, the emphasis is on modes that are "friendly" to the environment and "active" for users, like walking and cycling, along with mass transit, whose ecological impacts per functional unit are much lower than those of private cars, and all kinds of car sharing in real time (car pooling) or delayed time (car sharing, vehicle sharing), and even private vehicles that are less damaging to the environment than standard cars (i.e. motorcycles, electric cars). Concerning land use planning, the emphasis is on "sustainable neighbourhoods", "sustainable towns", the sustainable management of infrastructures, and green areas, etc.

In these pragmatic applications, design that respects environmental impacts, i.e. eco-design, tends to be based on intuition rather than reason, even when a formal scoring system is involved (Saheb, 2008). The risk is to design less efficient systems, i.e. within the limits of an overall budget, to prioritize certain courses of action rather than other potentially more fruitful ones; and even, to come up with counter-productive solutions in the presence of complex effects in the system under study. For instance, in some towns, constructing a major roadway corridor through the city to support efficient mass transit services brings the risk that dense car traffic will move towards smaller roads, where it can produce more negative impacts than before.

In this context, the aim of the *Eco-design Chair of Buildings and Infrastructure* is to establish rational methods for engineers, developers and planners so that they can correctly integrate functional, material, energetic and other systemic issues: carry-over effects, overall effects like congestion phenomena that impair operations and increase the financial and environmental costs.

This chapter's *objectives are to recognize transport and development as eco-design fields, and to situate and qualify the current state of simulation and evaluation methods for designing territorial facilities.*

Apart from the present introduction, this chapter is *organized* into 5 parts. We start by defining the notion of territorial facilities and listing their major features. We then successively deal with transport and land use, treating each as an eco-design domain, and making a distinction between objects and themes as well as potential action. Next, we cover design methods, with a brief look at simulation methods and a more detailed focus on assessment methods, comparing the respective principles of Life Cycle Assessments (LCA) and standard socio-economic assessments. On this subject, in order to revive the assessment methodology for the territorial planning assessment, we recommend integrating into the socio-economic assessments both the actual situation of the territorial system concerned and the propagation of value flows between stakeholders and in particular the community: this propagation of socioeconomic impacts will match that of material and energetic impacts in an LCA.

TERRITORIAL FACILITIES: DEFINITION AND CHARACTERISTICS

We define a "territorial facility" as an artificial object set up in geographical space in order to fulfil a social or economic function for a long period of time, ranging from several years to several centuries. In this definition, the *social or economic function may be of a material nature* to contain, house or accommodate an activity; and/or *of a cultural, symbolic nature*, if the facility has a particular meaning, i.e. power, religion, community or geographic identity; and/or *of an aesthetic nature*, if the item is remarkably beautiful to residents and visitors.

Its position in space determines the extent of the function fulfilled, service rendered, and impacts generated. Its technical lifespan ranges from several years (e.g. urban furniture, transport equipment) to several centuries (e.g. major roads, remarkable edifices), or perhaps several decades (e.g. housing, public transport vehicles).

a) The eco-design approach

When taking an eco-design approach, territorial facilities offer remarkable characteristics:

1 Their very long lifespan puts the emphasis on the usage phase in the environmental impact assessment over the life cycle.
2 Since objects of this kind are necessarily fixed and stationary, their substance (dimension, layout, capacity) and functional destination largely determine their usage. A client who orders an object of this kind from a builder is thus ultimately responsible for its environmental performance. In addition, construction works are localized production operations, and so their environmental impacts are relatively easy to trace.

3   The usage of the object depends on its position in space, and its situation within a configuration, i.e. the system of settlements for a building, the multimodal transport network for a transport facility. The latter offers a more fluid, variable usage in the short and mid term than a building. In other words, the situation strongly influences the usage of an object, which is always an installation in context.

4   Reciprocally, an infrastructural object strongly influences the territory that it fits into. Allocating a building to certain kinds of usage (i.e. types of activity) generates flows and movement; in the longer term it may lead to relocations (of households or companies). More obviously, when a transport facility becomes available, its direct impact is to drain flows by capturing them and diverting them from other means of transport: the facilities contribute to organizing traffic flows.

5   A residential building, both through its intrinsic quality and its location in a neighbourhood, strongly influences the everyday lives of its residents. Environmental quality can interact with the social make-up of the resident population, on building, neighbourhood and town scales. Thus the social dimension is added to the ecological dimension in a sustainable development design approach.

b) Essential notions: usage and system, capacity and congestion

Two key words emerge from the above analysis regarding the design of buildings and infrastructures: *Usage* and *System*. From the specific features observed, we can define a broad design field that covers the land use system, the transport system and the interaction between transport and land use.

On a territory, the land use system includes land use by stakeholders and activities: building and other superstructure facilities as well as infrastructure facilities; the qualitative and quantitative distribution of types of activity depending on the place and in terms of configuration; the locations of microeconomic agents living on the territory; their movements between places in order to carry out their activities in more suitable places; structural set-ups by the authorities or major stakeholders in terms of allotted space, lay-out in space, and facility location.

A transport system includes traffic flow and parking for individual modes (e.g. private car, motorbike, by foot), public modes (e.g. bus, train), and mixed modes (e.g. taxi). Each mode uses movement and parking facilities and involves vehicles (except for walking), depending on the services made available by operators if the transport is public, or self-produced for private modes; and with certain usage protocols that determine access and usage. The concrete characteristics of modes, for their users, depend on the way that infrastructures and services are organized, their flowing capacity that is faced to the movement volumes required by all of the users, and the ergonomics and quality of services.

Interaction between transport and land use, or better still, between the transport system and the land use system, is reciprocal. On the one hand, transport as a means to cross space influences the configuration of buildings and the location of agents. Transport's performance has an influence on the productivity of economic activities and determines producers' potential for obtaining effects of scale. The accessibility from one place to a set of attractive places will depend on transport's performance. Real estate and land prices capitalize accessibility and reflect its variations in space. These prices influence the settlement and development of the building supply.

On the other hand and reciprocally, the overall location of a population of agents will depend on the land and real estate capacities distributed over the territory: the ratio of agent population to building stock constitutes the occupation rate of the building system and measures how congested it is. Where agents are located will determine the demand for activity according to the type of agent, the motives for activities and the configuration of their locations. The demand for movement results from the demand for activities: it is faced with flow and parking capacities in line with quality and price modalities. The relation between flow intensity and capacity determines the capacity occupation rate, which measures the congestion of the transport facility.

Two new key words can thus be added to our analysis: capacity and congestion, which determine the quality of life for residents in the land use system, and the quality of services for users in the transport system.

c) Decisions about space: a field of decision-making

Land use and transport involve complex systems. Their development and daily management imply different decisions of a general and specific nature, that are global and local, long and short term, bearing in mind that through the play of interactions, a change in some "decision variables" can lead to a chain or even cascade of direct and indirect effects. Making an effort to preserve the environment requires watching out for not just economic and social effects, but ecological ones.

For a territorial stakeholder that holds several responsibilities, the diversity of objects and potential changes result in another coupling, this time of a financial kind: the stakeholder chooses what action to take within the limits of a budget, which means it must give priority to certain projects.

To explore, recognize and mark out the decision-making domain, we can identify the following four typical problems:

The overall sizing of spatial capacity: settlement capacity (for housing, economic activity ...) for a land use system, or flow and parking capacity for a transport system. These fundamental decisions involve the organization and operation of the system as a whole, just like a decision to mark out the territory or to regulate its relation with the external world in terms of access and exchanges.

What technologies should be used to offer a certain capacity? Let us define a technology as a combination of material set-ups and procedural set-ups. As an example we could mention technical generations in buildings, or modes of transport like Bus Rapid Transit or shared car systems. For an equivalent usage performance, several technologies might perform in diverse ways from a financial and ecological viewpoint, yet each will only achieve economic efficiency and noticeable ecological effects if it is used in a sufficient pool of objects, constituting a sub-system. For this reason, selecting a technology involves the producer-operator of the sub-system, as well as the authority responsible for the territory subject to the impacts.

The development of a local capacity, in a determined place and within an established configuration. In this kind of problem, the decision-making domain is limited. It is up to the stakeholder, as a client, to choose a technology or establish the performance requirements: the client's specifications mark out the decision-making sub-domain, that is the margin for manoeuvre left to the initiative of project management candidates. The client remains responsible for choosing the performance requirements and integrating the system effects – or at least certain major effects.

The conversion of local facilities with a fixed capacity. This involves a management decision regarding the operation of an object or system that has already been established and configured. This category includes things like rehabilitating territorialized objects, managing the flows of infrastructures, and managing the quality and tariffs of a car park. Once again, the question arises of choosing a sub-group whose performance could be transformed. This choice constitutes a problem for the client. To facilitate the decision, using a methodical approach, an information base should be established that distinguishes and qualifies the different sub-groups on the territory.

To sum up: we have set out four types of decision-making problems resulting from an eco-design approach at territory scale or within a territory. The first two types mainly depend on the territorial stakeholder, which is in theory a local authority or an organization appointed by a local authority. The fourth type also mainly involves the stakeholder, which varies depending on the spatial scale and the size of the object: it may be an individual stakeholder for an object like a building, or a collective stakeholder depending on the size of the urban block or neighbourhood. Project management parties are in a subordinate position: they can only take the initiative for problems of the third and fourth kind, in response to performance requirements specified by a client.

To extend the scope of initiative and increase the general influence of the different stakeholders, a collective client can make a call for proposals to help it with its decision problems. This kind of approach is valid for all types of problem, and in particular for choosing where to develop a local capacity or convert and transform facilities to improve their ecological performance. Once again, taking a rational approach to this kind of decision problem requires an information base, a support for territorial diagnosis.

TRANSPORT AS AN ECO-DESIGN FIELD

Transporting a person or a good is the function of *transferring the entity concerned from its origin location to its destination in specified time and packaging conditions.* The transfer function is essential and goes hand in hand with the container, protection, movement and conduct functions. Underlying this function is the need to maintain the physical integrity of the person or product, ensure its safety, and reach the destination effectively and, as far as possible, time and cost efficiently, with sufficient reliability to permit insurance and planning. On a technical level, a transport service is based on technical means and demand for use. Transportation's technico-economic system can be analyzed in four sub-systems relating respectively to infrastructures, protocols, vehicles and usages (cf. Figure 4.1).

This technical breakdown allows us to link environmental, economic and social impacts to components, and to identify design problems according to sub-system: each of these problems relates to a specific eco-design (cf. Figure 4.2).

In this chapter, we focus on identifying a set of potential solutions and indicating several avenues. Other chapters explore different avenues: public transport, eco-circulation of road traffic, and parking.

Infrastructure sub-system	Vehicle sub-system
Depends on mode of transport: road, rail... Includes track and diverse equipment Stationary and very long lasting	Depends on mode, by foot or mechanized: two-wheeler, car, bus, coach, train, lorry... Mobile and long lasting
Protocol sub-system	**Usage sub-system**
Organization of service: respective situation of server and user, especially in holding and using a vehicle Planning, maintenance, current operation Traffic management Real-time driving of vehicle	Travel and dispatch (freight) practices Trip-makers: individuals, freight senders, transporters, logisticians Travel needs Activity motives The settlement of activities in the territory

Figure 4.1   Systemic analysis of transport.

Infrastructure	Vehicle
Layout design, profiles, interchange, insertion Facility design: components, protection Building worksite organization: materials, procedures, removal-deposit balance Ecological offset between two states of territory	Specification: dimensions, mass, braking, motorization, aerodynamic profile, wheels and road contact Material composition: parts, materials Industrial chain, manufacturing process, distribution End-of-life recycling
Protocols	**Usages**
Maintenance Supplies (including energy, biofuel) Fleet management Freight management, logistics Traffic management: capacity, speed, orientation, fare setting Soft driving	Eco-gesture: eco-behaviour Eco-decision: choice of vehicle, mode, path, time of departure, etc. Conditioning of usages and needs - Multimodality - Transport-telecom interaction - Land-use

Figure 4.2   Eco-design problems per "object".

In general, eco-design should involve different operational stakeholders working in their own particular field (cf. Figure 4.3) and in relation with their partners. Thus, a car manufacturer relies on its equipment suppliers, while an infrastructure constructor relies on its materials and energy suppliers. Each stakeholder has to examine the options available, putting the priority on a "circular economy", local recycling or as near

Infrastructure builders	Vehicle constructors
Eco-materials	Eco-vehicles
Eco-worksite	
	Users
Infrastructure Operators	Eco-gesture: eco-behaviour
Eco-equipment	Eco-decision: choice of vehicle, mode, path,
Eco-circulation	time, etc.
Parking	Holding of equipment: vehicles, parking
Local protection	spaces, travel cards, location
Suppliers	**Planner**
Vehicle services	Organization of mobility
Energy	+ multimodal conditions
	Action on mobility needs
Service operators	+ spatial planning
Passenger transport	+ temporal planning
Freight transport	**Promoter**
Logistics	

Figure 4.3 Stakeholders concerned and their action area.

Infrastructure builders	Vehicle constructors
Given the layout:	Fixed vehicle model
Impacts per product unit, e.g. m2 of road	LCA per vehicle
Infrastructure/vehicle fleet operators	**Users**
In local traffic management: impacts per veh.km	Impacts per trip, per unit of distance travelled
If alternative routes: impacts per path between	Impacts of a set of practices, per period
choice points	**Mobility organizers**
Suppliers	Multimodality
Impacts per energy unit	Quality of service
Service operators	**Spatial planners**
Passengers: impacts per pass.km	Accessibility
Freight: impacts per tonne.km	Opportunities, attractiveness

Figure 4.4 Individual performance.

as possible, like balancing removals (demolition debris) and deposits (embankments) during earthworks on a civil engineering worksite.

By analyzing transport as a technical system, following the actual composition of the activity and the service rendered, a whole series of operational problems emerge that are as many opportunities for eco-design. It is up to the stakeholders to exploit these opportunities (Figure 4.4). Many opportunities are already exploited: by subdomain, the extent to which the life cycle assessment (LCA) is implemented gives an excellent indication of whether opportunities are exploited.

However, problems with strong territorial features, e.g. mobility management and the planning of the related transport system in a territory, are intrinsically complex and require developing frameworks and methods for a specific eco-design.

LAND-USE PLANNING AS AN ECO-DESIGN FIELD

Land-use planning is the activity, or the result, of arranging a given area to provide favourable living or activity conditions for human beings. Specific land-use planning settings include:

- Artificialization, building, equipping with infrastructures and services.
- Shaping and sharing out densities: promoting land use by the settlement of activities.
- Laying out, integrating, and matching objects into the space.
- Attracting activities and channelling them into buildings.
- Connecting: making activity opportunities accessible and improving them.

a) Space, usage and context

Land-use planning thus involves *bringing together and articulating a space with its operating functions*, by working with shapes in space (e.g. the building's ground space, height, underground space, respective positions) and determining the operating functions, e.g. a residence could target certain household profiles, or certain social activities, economic or administrative activities. Clearly, the stakeholder-occupants of the space interact with the act of urban planning, at least at a later stage, by adapting the concrete result. They may also be involved at an earlier stage through involvement in the design.

The planning must also *combine its actual object and a particular context* whose conditions range from micro-local scale to a potentially much larger one. The micro-local conditions of a building concern its position, its terrain (i.e. type of soil, ruggedness), its exposure to local conditions including weather, climate and sun (which may involve local shading or masking) and hydrology, among other material flows.

Position is particularly important for a transport infrastructure because it determines its situation in the network configuration – hence the size of its traffic flows, or in other words its intensity of use.

The particular context also determines the exposure to the usage demand: in transportation the demander chooses his or her destination and means (mode, equipment, path and time of travel), while for a building, the location demander chooses a premises and a position that will ensure certain interactions. Through this choice he or she puts the object in relation with others and so personally participates in establishing a context for the object. In addition, the usage demand finances the object, either directly or indirectly. And the demanders interact, either competing to obtain a usage (e.g. transport congestion, property auctions), or constructively to constitute a consolidated whole with a degree of social power (e.g. club, coalition effect).

b) Fundamental problems of spatial planning

To plan land use by activities and facilities, the ultimate problems are *sizing capacities and locating them*:

- What size should be assigned to the capacity of the territorial facilities (building or infrastructure), and on what spatial scale? This scale must be selected in a range extending from the actual object to the territorial system and including a sub-system. Recommending one particular capacity rather than another obviously constitutes a decision about opportunity. The capacity must respond to a certain volume of demand, which will depend on the territory as well as on the quality and price of the service rendered.
- What places on a territory should be determined as appropriate places to take action, invest in facilities or in facility conversions? This problem of selecting in space is subject to the extent of interactions between territorialized objects, due to agents and activities: and among these interactions, to effects of spatial transfer and substitution in space, not just for flows but for buildings.
- What modalities should be chosen, or how should they be put together, to supply a spatial capacity in optimized conditions? This generic expression is applied in terms of functional allocation for the land use system (nature of activity, type of occupancy, e.g. split between private and social housing), and in terms of transport modes for the transport system.
- An obvious additional problem is: What is the ideal capacity occupancy rate and what capacity reserve should be permanently maintained for use in emergencies? For the time being we will consider this problem as secondary.

The ecological aspect is present in each problem, adding to and backing up the usage destination that constitutes the direct concrete challenge.

The ultimate challenge of an eco-design approach is to accord the same or nearly the same level of importance to the environmental impact, which is an indirect concrete issue, as to the direct issue, rather than relegating it to the background.

c) Axioms for eco-design in spatial planning

Eco-design in spatial planning is a fertile subject, both due its varied aspects and its diverse applications, i.e. geo-diversity. Beyond this diversity, some principles are so crucial that they can constitute general axioms:

1 The complexity of the object depends on the "territory volume": in terms of its scope in space and "content mass", which involves the size of the population and activities.

2 The usage model must integrate the object's situation in an "organic configuration": that of the particular territory that contains the object, a territory comprising diverse components that ensure operations, with a morphology (configuration) and interacting metabolism (circulation).

3 The notion of territory "consubstantially" involves social, economic and environmental impacts. This is because a territory is constituted by a certain population established in a geographic area, and this population forms a society that is

endowed with political power. The function of territorial facilities is fundamentally of a social or economic nature, and a facility project must necessarily be evaluated on these registers. The environmental impacts are important too: the spatial scale specific to each one opens up specific management possibilities, which a territory as an entity can trade with other territories.

The last axiom puts into perspective the local management of an environmental impact: the idea of a territory being self sufficient in a particular resource (e.g. energy) is unrealistic, since in a closely connected world, the vital challenge of durability is that trading impacts should be equitable between particular territories.

ECO-DESIGN METHODS FOR SPATIAL PLANNING

In the introduction to this chapter, we pointed out the intuitive, rather than rational, character of pragmatic eco-design in spatial planning. In fact, the diverse aspects in a defined territory, both within it and in relation with the outside, requires that every planning operation should be assigned with selected specific goals and performance targets, rather than examined in all of their facets and impacts.

The Cité Descartes planning project provides a good illustration: the planners started by establishing their comprehensive perception of the whole territory, its potential and needs, including the necessity of improving north-south links to overcome fracture lines caused by the Marne River and large infrastructures on the east-west route (lines A and E of the commuter railway and A4 motorway). An overall project was then put forward, broken down into land use and urban development, networks and local modes of transport, green and blue corridors, and energy management, along with identity to be symbolized by an emblematic monument. This overall, multi-facetted design was supported by specialized impact studies for each facet, one on land use, another on transport, and others on different environmental aspects. However, the planning was not comprehensive and the project could be improved upon. For example, one important component is the local reorganization of the A4 motorway to moderate car traffic and offer a high-level service bus liaison on a specific bus lane for each direction. However, an earlier simulation of regional transport supply and demand showed that this kind of line would not be used by enough people to justify the cost.

Ideally, an array of technical tools should be used combining sensitive simulation models with assessment methods to compare alternatives and progressively point to a preferential planning option.

Currently, the *simulation models for planning* are mainly transport models. The models are based on hypotheses regarding land use in terms of localized inhabitants and jobs, and statistic regularity on the generation of trips, from which they produce trip flows between places. These flows are then assigned to modal networks according to behaviour principles, in order to establish local traffic loads and travel conditions, i.e. local speeds, costs for the user and local impacts. For an urban development plan, the simulation is operated for all or part of the built-up area. This scale is broad, and the conditions are micro-local, so that the simulation does not pick up on transport requirements within the neighbourhood or local set-ups to foster walking and cycling. "Microscopic" traffic models exist but their current applications still mainly concern road traffic. At the other extreme, a simple simulation of traffic based on land use does not process feedback on diverse local conditions (the puzzle) on local use (the puzzle piece), as determined by residential locations, whether residences or firms. This deficiency needs to be compensated using models integrating land use and transport, but their application in France is still experimental and fairly restricted.

To sum up, planning simulations of the usage of buildings and transportation facilities still require considerable development.

Regarding *assessment methods*, it is clear that the current state of the art could be improved on, by better integration of LCA methods, impact studies and cost-benefit analyses of projects. Although LCA methods for buildings and built areas propagate a wide range of environmental impacts in long chains of consequence, they reduce local conditions to the object itself, with local support and specific interactions such as shadows falling between buildings: in general, occupants' exposure to air quality or noise is not dealt with, nor interaction with local means of transport (e.g. parking for two- or four-wheeled vehicles, access to public transport stations).

Impact studies, however, focus on a specific type of impact, or a particular target, often for a defined area. These specificities mean that the scope is restricted, and so partial.

Last but not least, to study development schemes and transport policies, possibly including a land use component, we can use a technico-economic method that benefits from long-standing, international experience, i.e. the *cost-benefit analysis*. This method is based on simulating supply and demand for transport in a given territory, which means it is subject to the sensitivity limitations mentioned above in terms of spatial detail and microeconomic usage behaviour. The method originally included the benefits and costs for the supplier or demander, e.g. production costs, consumer surplus. Over time, other impacts have been identified and integrated into the analysis in the form of specific indicators that can be incorporated into a multi-criteria table, or if applicable in a monetary form that can be part of an overall assessment.

Notable environmental impacts concern noise, greenhouse gases, air quality and local pollution of water and soil. These themes can be the object of local impact studies: some of the items are also addressed by LCAs, which can also look at additional impacts that are still absent from territorial planning studies (e.g. eutrophication of water courses). Lastly, each cost and advantage directly concerns a particular stakeholder – although it is standard practice to aggregate the cost-benefit analyses of the various stakeholders to obtain a community analysis, it is also worth evaluating impacts per recipient and emitter to specifically study the acceptability by the local agents. This brings us to the subject of socio-economic impacts, which also constitutes an important research field.

RENEWING EVALUATION FOR TERRITORIAL PLANNING

Comparing LCAs with cost-benefit analysis methods puts into perspective certain simplifications that have become standard but that are too simplistic in some applications. This is because the development of traffic and congestion, and the need to protect the environment, have led to diverse transport policies that have planned several modal networks in an integrated way in order to diminish road flows and the presence of cars in the town, or to manage demand. Up until now, the approach used to evaluate these policies has been to extend the method for an infrastructure project to a set of modal networks, considering pluri-modal demand and adding environmental impact criteria. However, mobility policies that constrain cars have more far-reaching effects than the simple choice of a mode of transport: it is considered positive to reduce the holding and usage of private cars by households in dense urban centres, or to restrict parking places in buildings or on roads. However, these profound effects for users have direct repercussions on the economic activity of producing equipment (i.e. cars, parking spaces) and the associated consumption. In these conditions, mobility policies significantly affect the productive fabric, and the consequences need to be evaluated.

In this section, we set down the principles to recognize, analyze and evaluate the profound economic effects of mobility policies. We make critical examination of the standard evaluation method to identify the scope of effects and that of the stakeholders, the stakeholder system in a community, and its spatial framework. We clarify how transportation fits into economic production and its impact on identified public finances.

a) The vertical composition of a transport system

A transport service constitutes an economic good produced using a combination of resources: traffic infrastructure, access or parking facilities, a vehicle (except for walking trips), various consumables (energy, etc.), and ancillary services to cover driving, navigating and flow management. The conceptual model consists of a vertical stacking of strata, by decreasing degree of materiality.

The standard cost-benefit assessment method focuses on the interactions between the users and the provider of a service – basically a carriage or conveyance along a path. Such a service entails financial compensation, typically in the form of paying a toll or fee, which amounts to a value flow from the user to the supplier. In the community's account, the individual accounts of all socioeconomic actors are summed, and financial flows within the system are neutralized since user expenditures are offset by the supplier's revenue. The other user expenses are ordinarily counted as sunk costs: energy (except for the fiscal share of fuel prices), maintenance and insurance services, and vehicle depreciation. Put otherwise, the conventional method does not delve in depth into the vertical composition of transport; erroneously, it omits the strata of the vehicle, services other than conveyance or delivery, while on the other hand related expenditures also run high and constitute "tangible values" for suppliers of these goods! The full array of suppliers must all be included within the scope of the evaluation framework, through recognizing the incident value flows.

Let us take the case of automobile transport in France circa 2010: on an interurban motorway itinerary using a passenger car, the user's total cost amounts to about €0.50 per km: this value is typically broken down into 20% toll, 40% vehicle depreciation, 20% fuel and 20% ancillary services, thus revealing that the toll component (traditionally the only one recognized) is merely the tip of the iceberg!

b) Creation and circulation of value flows

Above all else, transportation is a derived economic good whose demand is derived from that for activities. Nonetheless, the user remains an end or final consumer, i.e. positioned at the end of the chain deciding whether or not to consume, and thereby creates value flows towards the immediate suppliers in the chain, who in turn to satisfy their clients produce goods by committing their own resources that they will remunerate financially (i.e. capital, employees) along with intermediate consumption. Value flows are thus spread through the chain to other suppliers: this is the "ripple effect" of economic production activities.

Both the creation and circulation of value flows need to be incorporated into the evaluation framework. Regarding creation by the end client, all value flows are to be integrated, not just for the resources being employed "at the margin" per unit of use, but also encompassing resource acquisition and holding.

As for value flow circulation, the ripple effects from one supplier to the next must be taken into account, as in Life Cycle Analyses when evaluating the consumption of various environmental resources.

Let us get back to the example of automobile transport in France. Over its life cycle, a passenger car is used about 12 years at an average annual use rate of 13,000 to 14,000 km. For a compact model, manufacturing accounts for 40% of the motorist's total costs, vs. 60% use-related costs. Ripple effects can be quantified by multiplier coefficients, on the order of 3.5 for manufacturing and 2.5 for use. Through spreading the value flows, it appears that the final expenditure is due in very large part to wages and associated labor charges: the division of production into specialized stages (vehicle integrators, parts manufacturers, ancillary services, etc.) may dissimulate this fact when examining just a single stage.

By involving both the vertical composition of the transport good and the spreading of value flows into the evaluation framework, the actual scope of transport is restored within the overall economic production and, hence, its true importance can be appreciated. Let us recall that transportation equipment manufacturing is inextricably linked to the economic development of industrialized societies: beginning in the 19th century with maritime and rail transport, and accelerating into the 20th with automobile and airplane transport.

To evaluate ex-ante value flows, a macroeconomic model of both the final demand and the economic/social circuit is appropriate: the simplest form of such a model would be a Leontief-style input-output, which is completely analogous to the matrix model of environmental impacts within the framework of a life cycle analysis (see Leurent and Windisch (2012) for a pertinent illustration).

c) Systemic effects of transport

A "systemic effect" is the consequence of a system's state, with respect to a particular aspect or actor, whether internal or external to the system. At this stage, we have already identified two such systemic effects:

1   Interactions between the end consumer and corresponding suppliers.
2   Ripple effects that take place within the economic production process.

For a given transport system, the following systemic effects also hold:

3   The state of operations is determined by traffic volumes and network loads, with a dual consequence: for operators on the efficiency of their use of resources, and for users in terms of quality of service (travel time, comfort).

4   Users choose from among a wide array of options regarding itinerary, mode of transport and schedule, not only on a one-time basis when organizing individual trips, but also in-depth in their long-run decisions of vehicle ownership or subscription to a mode of public transport.

5   The operational state of transport system influences the basic supply of transport services. Depending on the traffic volumes to accommodate, suppliers are able to achieve economies of scale and benefit from the effects of mass production, especially in the automobile industry or energy production. Public transit operators also know about the "Mohring effect": in order to satisfy higher demand, services are proposed at a greater frequency, which on its own induces additional demand. The same applies to any product of which the suppliers share the benefit of scale economies with the customers, by reducing the price.

6   The in-depth feedback of system state (especially quality of service) on transportation demand: on activity choice and organization, possibly facilitating the individual (in most cases professional) specializations of transport consumers, on activity location and spatio-temporal scheduling, thus on trip-making. These phenomena are already complex for an individual within a household, but even more so for a firm choosing to locate production units and designing the size of its activity in relation to the market areas made accessible thanks to available transport services.

7   Spatial influences: residential location decisions by socioeconomic actors determine global effects: notably the formation of real estate prices. The configurations of individual locations with their respective accessibilities determine the advantages of vicinity for economic production and its output (i.e. agglomeration economies).

8   Environmental impacts, which are now fairly well recognized.

To take these effects into account in the framework of ex-ante evaluation, they must be modeled and simulated. The state-of-the-art can be described as follows: effects (1) through (4) are partially treated, with an emphasis on user incidences; effect (5) is not treated at all, though the Mohring effect seems straightforward to integrate as an endogenous phenomenon into a planning model; effect (6) is only included in certain disaggregated travel demand models, and only for passengers; and lastly, effect (7) is only treated in some integrated models of transport and land use, but such efforts remain for the time being at the experimental level.

This inventory of systemic effects reveals the set of actors involved in the transport system.

d) Actors involved

The set of actors concerned by a transport system mainly encompasses the following types: i) service demanders, users or freight shippers, ii) "direct" suppliers of service components, iii) suppliers with mass production capabilities, iv) employees of various types of suppliers, v–vi) real estate market actors (buyers and sellers), vii–viii) land development actors (buyers and sellers), ix) neighbors and other stakeholders concerned by environmental impacts, and x) public authorities.

Categories i) and ii) for the transport services part, x), also ix) more and more frequently, and at times v) and vi) have appeared in the evaluations conducted up until now. The scope of the scenario (project, scheme, policy) to be evaluated determines the need to complete the list of actor categories that must be addressed explicitly.

For each actor, the impacts (costs, benefits) in a given system state, along with their balance account, must be evaluated. In comparison with a reference state, the test scenario completes a transformation, so its effect on the given actor amounts to the difference in balance accounts between the two states. The market component of costs tends to be a traceable expenditure, thus making it easy to evaluate. The non-market component of costs (i.e. expenses in terms of time or discomfort for the user, a non-internalized externality like environmental impacts or lack of safety) is much more difficult to value. On this topic, typical practices are rather simplistic: the value of time has a strong basis for professional activities or commute trips, but this basis is much weaker for private purposes. Moreover, freight traffic is ordinarily addressed by a simple adaptation of passenger model, whereas the underlying microeconomic rationale is fundamentally different: trip-making is a derived demand for passengers, yet it constitutes the basic activity for a freight hauler. On the benefits side, each actor holds a surplus, which is something monetary for a firm (in the form of revenue) or something valuable with a hedonic connotation for a household (personal preferences, utility). The surplus for each category of actor involved must then be evaluated. In a prudent approach, Venables and Gasoriek (1999) recommended evaluating both the monetary effects (in GDP terms) and well-being effects separately, with the latter including both the market effects and the non-market but individually pleasurable aspects.

e) Public authorities and their financial accounts

Public authorities typically make policy decisions on the transportation system and contribute considerably to financing it. In each system state, they also collect tax revenue based on: monetary exchanges (notably VAT revenue), production (various taxes), real estate and land holdings, energy sources, etc. In some instances, authorities operate transport services and parking facilities, whose specific balance accounts get added to the authority's basic account.

It is critical here to accurately evaluate revenues by their various sources, to the same extent as expenditures, in order to establish an objective balance account, which proves vital when public funding runs low. On the revenue side, tax accounts need to be included, alongside "social" accounts of social institutions which are backed by the public authority: the wages paid by firms provide a source of contributions and withholdings remitted by the employer as well as the employee. Their amounts tend to be quite high: for France in 2010, on the order of 80% of net salary. Mobility policies that exert substantial influence over economic activity must also be evaluated in this respect. Leurent and Windisch (2012) demonstrated that replacing a gasoline-powered car by an electric vehicle generates financial impacts through social bodies exceed fiscal impacts in the financial account of public authorities (on integrating on the social side the cost of unemployment allocations).

f) The notions of territory, community and social solidarity

The economic theory of transportation only pays scant attention to the scope of actors' influence. Yet this scope needs to be recognized and incorporated into the evaluation of mobility policies. In this vein, it is important to accurately define the notion of community: its compilation of actors and spatial limitations, i.e. its own territory. The conjunction of these two aspects determines the territory's domestic level of activity and trade with other territories for each of its impacts: an actor based elsewhere only partially participates towards tax revenues and in no way towards social solidarity – why then should this actor's surplus be counted in an impact evaluation at the community scale?

A policy's industrial, energy, real estate and land use effects must be reexamined according to a territorialism criterion. Real estate rent received by an external actor constitutes an expenditure that goes uncompensated by tax revenue for the community; similarly, the effect on economic production only adds to the surplus for the internal production component, especially as regards transportation vehicles.

g) Example

In an evaluation study of the potential substitution of an internal combustion vehicle (IV) with an electric vehicle (EV), Leurent and Windisch (2012) compared four territorialisation scenarios, making a distinction between the "national" French territory and the outside, as well as the manufacture and usage phases in a car's life cycle. The reference scenario supposed that manufacture and usage were entirely within the country. In the import scenario, the manufacture took place outside the country, but the usage was inside. In the export scenario, manufacture was inside the country and usage outside. Lastly, in the competitive import scenario, an imported electric vehicle replaced an internal combustion vehicle produced in the country, for domestic usage. These scenarios are somewhat of a caricature since national economies are increasingly integrated into the European framework. However, the evaluation of their impact, in both absolute and differential terms, provides an important benchmark. At this stage, the country constitutes the territory that integrates individual interests and is captured in the commercial interaction with the outside, depending on the spatialization scenario.

The domestic public authority (DPA) in a territory implements specific policies, between which they divide and budget their financial resources. The volume of financial flow associated with the life cycle of each car necessarily interest the DPA. In the reference scenario, substituting an IV with an EV had almost no incidence, prior to a purchase bonus. This bonus is a dead expense that the DPA can only sustain for a limited number of vehicles.

In the import scenario, tax on consumption is identical to the base scenario. However, the tax on production in the manufacturing phase is lost for the territory, along with social effects arising from the manufacture. Therefore, the EV loses its principle financial income entries. The financial loss for the DPA is over €8,000 per vehicle before the bonus and €14,000 after the bonus of €5,000 before tax!

However, the worst scenario is a "competitive import", i.e., substituting an IV produced domestically with an imported EV. In this case, manufacturers based outside the territory offer domestic consumers attractive vehicles that convince them to make the change. Prior to the bonus, for the manufacturing phase an imported EV brings €5,000 of financial income (VAT), whereas an IV produced locally brings €14,500, i.e. a loss of €9,500. Including usage, the loss comes to €20,000 before the bonus and €26,000 with the bonus!

The export scenario brings neither VAT (on manufacture and usage) nor social effects and energy surcharges during usage (leaving aside supply of spare parts). Its effects are limited to the manufacturing phase, for which EVs are almost twice as productive as IVs – as long as no bonus is applied, which means making sure that bonuses are attributed to cars used domestically.

Out of all of the scenarios, the export one is the most favourable for territorial public finances, whereas substituting domestically manufactured IVs with imported EVs is the least favourable. In between the two, the basic scenario with manufacture and usage within the territory is slightly positive without the bonus and negative with it; it is less unfavourable than the competitive import scenario.

Once competitive EVs are produced globally, countries that authorize importing them without specific taxation run a high financial risk. The only palliative solution is to produce competitive EVs domestically, and thus enter into technological, industrial and commercial competition.

h) Overview

We have underlined a series of aspects that have already been studied in diverse economic theories (transport, land use, territory, production, trade), but that still need to be integrated into methods for evaluating mobility policies.

The major topic of evaluation is the creation and circulation of value flows between social agents, to be analyzed by paying specific attention to a defined domestic territory as an essential sub-system. Superficial impacts that concern transport users and operators have a well-established form whose treatment in the evaluation does not depend on the community's territory; however, in-depth systemic effects are highly dependent on territorialisation, through the location of agents, the spatial boundary of the community, and its own institutional arrangements (e.g. tax and social regulations).

CONCLUSION

Territorial facilities, as a field of action, constitute a fertile domain for eco-design. We have covered this domain in terms of transport systems and, to a lesser extent, land-use systems, whose complexity increases with spatial scale, from a building block to a major territory, passing by a neighbourhood and an urbanized area. Their planning process involves some valuable design issues!

In these domains, as for everywhere, eco-design is a powerful method for dealing with impacts sensitively, by giving to them the same level of importance as to the function assigned to an object and to the service that fulfils the function. We have shown the need for renewing conventional cost-benefit analysis methods for planning projects. Simulation models also need renewing in order to develop sufficient sensitivity to deal with the relevant stakes. Lastly, an actual design process should match the specific requirements of its object, its contextual dependences, the time scales specific to its components, and internal and external interactions.

REFERENCES

AIPCR (2009) Recommandations aux utilisateurs des systèmes d'évaluation multimodale. Report of Technical Committee 1.1 on Road System Economics. ISBN 2-84060-224-5

COBA (2005) Instruction – cadre relative aux méthodes d'évaluation économique des grands projets d'infrastructures de transport du 25 mars 2004. Ministère français chargé de l'équipement. http://temis.documentation.equipement.gouv.fr/documents/temis/14849/14849_2005.pdf

Leurent, F. (2012) Transport Effects on the Creation and Circulation of Value Flow: principles for policy evaluation. Les effets du transport sur la création et la circulation des flux de valeur: des principes pour évaluer une politique. Routes/Roads n°356, pp. 38–45, October.

Leurent, F. and Windisch, E. (2012) Electric vs. Gasoline Powered Vehicles: the effects on a nation's economic production and public finances. Véhicule Electrique vs. Véhicule à Carburant: effets sur la production économique et sur les finances publiques d'un territoire. Routes/Roads, February 2013.

Leurent, F. and Windisch, E. (2012) Benefits and costs of electric vehicles for the public finances: integrated valuation model and application to France. Proceedings of the ATEC-ITS Congress, February 2012, Versailles. http://hal-enpc.archives-ouvertes.fr/hal-00680987

Saheb, Y. (2008) Analyse critique des quartiers durables. Rapport d'étude OpenExp pour la Fabrique de la Cité, 24 pages.

Venables, A.J. and Gasiorek, M. (1999) The Welfare Implications of Transport Improvements in the Presence of Market Failure. Department of the Environment, Transport and the Regions. London.

WebTAG (2005) Introduction to Transport Analysis TAG Unit 1.1. English Department for Transport. http://www.dft.gov.uk/webtag/documents/overview/pdf/unit1.1.pdf

Part 2

Practicing eco-design

Chapter 5

Urban economics and passenger transportation: concentrate flows, design lines

François Combes, Fabien Leurent & Sheng Li
Paris-East University, City Mobility Transport Laboratory,
Ecole des Ponts ParisTech, Ifsttar, UPEM

INTRODUCTION

a) Transport in the context of buildings and infrastructure

The EEBI Chair is interested in transport for two reasons. On the one hand, buildings are devised to be occupied by individuals who use them to carry out activities but do not remain in them permanently. These individuals must therefore be able to get to the building, which means it must be accessible by road and one or several means of transport. On the other hand, on a larger scale, a bigger set of buildings includes roads that are elements of transport infrastruct ure – their design must fulfil the requirements of the urban fragment concerned, which it must do in interaction with the outside, for internal movements, exchanges with the outside and perhaps also flows in transit.

At the juncture of these two points comes the parking infrastructure, which may be inside the building (partially or even totally in the case of a built car park), or on the street.

b) The designer's position

The bigger the urban fragment, the greater the importance of the transport component, which must be designed in terms of the functions to fulfil, infrastructures, vehicles and operating processes, as well as impacts on the environment, the economy and society, especially local residents, who are exposed to the nuisances of traffic (noise, pollution, etc.) and to the advantages of accessibility.

When planning transport for and in built areas, designers must integrate a variety of functional requirements in line with the nature and motive of the movements to cover, and a variety of local conditions, including in the "locality" not just the interior of the urban fragment but also its interface with the exterior, by the physical interaction constituted by movements.

In the eco-design of a physical system, a fundamental characteristic is the usage model. In the case of a building, the usage model is relatively simple, based on motives and types of occupation (e.g. whether the function is exclusively residential, or tertiary). However, transport usage is inconsistent: for a person in movement, a facility designed for the long term, whether it be an infrastructure, vehicle or public transport service, is only one particular means, optional mode or route, that he or she puts into competition with other options on a frequent basis. This combination of long-term facility and short-term usage decision is a source of tension with a high risk of alteration for the fulfilled function.

For this reason, when planning transportation, the interaction between the supply of transport and the demand for movement plays a crucial role: this interaction is simulated by modelling the usage of supply by demand.

c) Objective: fundamental principles and potential for massive carriage

This chapter sets down the principles for an eco-oriented design of transport systems and services. In a fundamental perspective, design must above all understand its object, in terms of operating requirements and means and physical processes, but also behaviour patterns if the usage is competitive and sensitive to microeconomic conditions. We focus on the diversity of modes of transport and, consequently, on their respective scope. In this respect, the major stake of achieving economic and environmental performance is to carry massive flows to benefit from the economies of scale intrinsic to most modes of transport, and thus amortize investments in physical means and withdrawals of resources, for the highest possible quantity of usage.

We are particularly interested in the public transport of passengers, in which vehicles (from bus to train) and operating processes are used and mutualised by a high volume of passengers – several dozen or hundreds, even thousands, per vehicle run, depending on the capacity of the vehicle.

d) Structure of the chapter

The body of this chapter is organized in four parts followed by a conclusion. We start by taking a more detailed look at transport as a function, including its technical and economic conditions, and the ecological, economic and social stakes of design: adapting to the territory is a strong constraint on several levels, both material and institutional.

We then go on to consider transport's interaction with land use, using an urban economy model to illustrate this interaction and pinpoint the implications for planning services, regulating the urban system, and equipment-funding schemes.

Next, we illustrate the design of a public transport (PT) line in the urban setting, by modelling the respective roles of decision variables for the planner, i.e. space between stations, service frequency, vehicle capacity, and traffic speed. The potential for massively expanding demand along the line determines which mode of transport is most pertinent: bus, tram or subway.

Lastly, we present a more realistic design study, territorialized in the Paris region, to trace out public transport corridors in the form of a ring. This study is based on a professional simulation model.

TRANSPORT DESIGN PRINCIPLES

a) The function of transport, its means and their longevity

The function of transporting a person or good is to move the entity concerned from its original location to its destination, in specified time and carriage conditions. The unit of service provided, also called the functional unit, is the product of a unit of distance multiplied by a unit of quantity, which is a passenger or a unit of freight, i.e. 1 passenger.kilometre or 1 tonne.kilometre.

The service is provided by a "mode of transport", i.e. an assembly of technical means. This will typically involve a circulation infrastructure (also access and parking), a vehicle to contain and move the load, and an operating process, possibly involving agents (driver, transferor, etc.), and always an operating mode (access and usage rules, organization of the service). Some resources are repeatedly used for a long series of utilizations: these are facilities, as opposed to "consumables", like fuel and lubricants (or even tyres or spare parts).

Facilities and the associated consumables are both design objects: in the long term for a vehicle whose technical lifespan goes from several years (car, motorbike) to several dozen years (train, aeroplane, boat), and thus for the associated consumables, since a vehicle is a highly specific object; in the very long term for an infrastructure, which through its construction is located in a dedicated area that it renders artificial and that it occupies exclusively or almost exclusively; in the mid term for a public transport service that must exist long enough to build up a regular clientele to amortize its costs; or in the short term for a specific logistics service.

b) What are the design stakes?

Due to its material conditions, transportation has negative impacts on the local and global environment, including local inhabitants and biodiversity. We will now mention these and then list their social and economic effects, which are mostly positive.

The movement of a vehicle on a supporting infrastructure (e.g. a road) and in a physical medium (e.g. air, water) requires energy, which is consumed more intensively as the speed increases. If this energy comes from a fossil source, then its consumption generates greenhouse gases. If the energy reserve is contained in the vehicle as fuel, then the vehicle releases pollution into the atmosphere throughout its journey. Lastly, the movement of a vehicle and the operation of the motor both emit noise. All of these emissions are more harmful when the journey takes place in a more populous, urban environment, i.e. in the presence of a greater number of local residents who are exposed to them. They also harm biodiversity and the quality of natural environments.

In a permanent way, an infrastructure occupies an area, which it excludes from alternative uses, and its presence breaks up the surrounding area, e.g. species' habitat and/or urban fabric. If the modes are motorized, then the infrastructure constitutes a dangerous place for users as well as humans or animals passing through; its immediate surroundings are particularly exposed to its traffic.

All of these negative impacts constitute what to reduce and avoid when designing a transport system. This concerns vehicles, motors, energy sources, and operating processes, along with infrastructures in terms of materials and different set-ups, decisions to locate, including the layout and sizing of infrastructures. However, the basic stake is the availability of good quality transport services in sufficient quantities, at the places and times when movement is required by users. A society is developed when individuals are able to exercise their specific talents in a productive way by carrying out their activities, thanks to access to the places where these activities are carried out, i.e. professional activities, studies, supplies, access to different services and social, recreational and cultural opportunities, from their home and other locations.

Access to opportunities increases with the spatial scope of movements. The latter is conditioned by the performance of the mode of transport: the average speed from start to finish depends on the nominal speed of the principal section, the conditions of access to the terminus, and the availability of the service if the mode is public rather than private.

To sum up, the fundamental stake of transportation is economic and social. For users, time and cost efficiency for particular requirements is crucial: it determines their choice of means of travel in terms of modal network, route and departure time. Through competition between modes and services, the need for efficiency is communicated to operators, who try to offer effective technical solutions at a moderate price. The winning strategy is to generate economies of scale by standardizing vehicles and processes, creating mass infrastructures and services, and matching services to usage needs in time and space, in order to fill up the capacities installed. Ultimately, it is the territory that benefits from the quality of service and supply, and thus the accessibility of its activities, because these qualities make the territory attractive to potential residents, such as households and businesses. When companies set up in a territory, in addition to providing products and services, they bring revenue through salaries, social payments and taxes that benefit the community.

c) Adapt modes of transport to the geographic, economic and social environment

Designers must identify the needs of a territory and match them with its potential, as well as with the interests of any stakeholders who might implement the projects planned, i.e. territorial planners, operators, constructors, and funders, whose support is needed to accomplish the project.

The physical environment will impose limitations in terms of landscape (higher slopes are easier to negotiate for smaller vehicles and by road), ground (solidity and stability for infrastructures), and hydrology and rainfall (for the infrastructure's solidity and operating processes). If the seismic risk is relatively high, then the modes of transport chosen should be more resilient, capable of working in difficult conditions, and relatively inexpensive to repair.

The inhabitants of a territory determine the transport requirements, by the number and location of individuals, their activities, and therefore the territory's functions (e.g. employment, services, residences). Lifestyle patterns generate demand peaks – higher at home-to-work times – confronted with transport capacities.

In addition, these individuals present an overall sensitivity for transport usage: a certain degree of individualism, and comfort requirements, which can induce people to use a car if the other modes do not offer the required quality (e.g. insufficient stations and lanes for motorbikes and bicycles, lack of in-vehicle comfort and traffic priorities for public transport). Inhabitants' income will also determine their solvency, and the potential to supply them with high-performance modes that are more expensive – a cost that may be borne by the individuals or by the community.

In parallel, inhabitants have a "local resident" tolerance of local impacts and acceptability of modal projects. Legal conditions and social institutions must also be integrated in the analysis and design of transport schemes. The "legal climate" of business determines the security of long-term investments, while civil security influences the service quality of the modes (exposure to the risk of assault in a vehicle, in a station or on a road, or the risk of accident for users and operating agents). In addition, the territory's education system conditions the availability of agents that are sufficiently qualified to operate, build and maintain transport modes.

In this overview of territorial conditions, we have highlighted factors that remain local at a time when other factors, especially energy resources, and many transport materials and equipment, have become globally available, thanks to efficient supply chains and open economic markets, which together ensure that prices remain broadly similar everywhere.

In addition to geographical conditions, transport planning is subject to historical conditions: the territory and its transport system have evolved in close interaction, in a co-development process. The transport system is part of the territory's patrimony, and planning involves devising its maintenance and renewal just as much, if not more, than any developments. Renewal may also provide an opportunity for transformation, by adapting the infrastructure and using different modes. Road infrastructures, in particular, are versatile and allow not just cars and passengers to circulate, but also motorbikes, bicycles and roadway transit vehicles.

In the rest of the chapter, we will concentrate on a specific problem: flow consolidation in an urban environment, in relation to the concentration of activities and buildings. Logically, concentration encourages mass movements and grouping individual trip requirements in pooled resources in terms of infrastructure and/or vehicles. In addition, urban structures result in the relative scarcity of space, rendering land expensive, which stimulates the gathering together of flows on a restricted road network. Thus, a public transport service by line can offer several advantages in urban conditions. Each vehicle can hold several dozen (bus) or hundreds (tram, train) of passengers, and so consolidate flows; if its capacity is sufficiently filled up, then the operating cost per unit of traffic goes down, which makes it more economical and, on a social level, accessible to a greater share of the public. Correlatively, the environmental impacts per unit of traffic are reduced. This reduction is particularly noticeable in urban environments, which contain numerous receptors exposed to local impacts (e.g. pollution, noise).

URBANIZATION AND CONSOLIDATION OF FLOWS

In this section, we recall the economic urban theory that concerns housing, employment and transportation, and we develop a model to simulate the attractiveness of a mass use of transport in comparison to a diffuse mode.

a) Land use, stakeholder projects and regulation

Land use, or the way that human activities use the land, is the distribution in space of these activities and therefore their buildings: housing, production buildings, shops and services, facilities.

Land use generates flows of movement that link one activity to another, since the movements are motivated by the carrying out of activities. Therefore, the number of movements emitted and received by a certain zone during a certain period are proportional to the number of activities in the zone, with a coefficient that depends on the motives and the period.

Reciprocally, the transport conditions, i.e. the point-to-point performance depending on the modal network and the period, confer a specific accessibility quality to each place: a more accessible location is a more desirable to live in, or to place production facilities. The property market gives value to better access by making it more expensive. In turn, property prices, depending on how they are distributed in space, stimulate property development and the artificialization of land.

This mutual interaction between land use and transport is fundamental for a city. It is driven by the major categories of stakeholder that make up the city. On the one hand, residential households choose their home in terms of place, housing quality (primarily the living area) and its spatial location in relation to the activities they carry out, in particular the place of work for parents, and the school for children. Each household with a professional activity generates a repeated trip from home to work. Its localization budget is split between expenditure on housing and the cost of transport. The behaviour of a business is similar, i.e. to carry out certain activities in premises whose size and facilities are acceptable, and with satisfactory access conditions for staff, customers and suppliers.

On the other hand, urban planners integrate the locations of various agents, including households and companies, in order to plan land use and transport in an integrated way. At the same time, individual location and transport suppliers adapt to local conditions: to location requirements, transport requirements, and the economic conditions of these markets. Because the expenditure of those demanding buildings (i.e. households and businesses) is split between location and transport, accessibility advantages (and so transport) are given value primarily by the property market, so that their relative scarcity generates economic rent for the most accessible buildings. Yet this high economic rent results from an overall configuration rather than the creation of an original economic good by a particular agent (who would thus legitimately earn a profit from it). It is therefore up to the territorial government to decide on how to apportion the rent, and to divide the product between the community's needs – in particular needs for land development and transport facilities.

Urban economic theory is very clear in this area. Its applications are highly variable depending on the country and the type of regulation it chooses, whether for transport, the property market or the land market. In Japan, property promoters have to contribute to investments in transportation. In France the employers' contribution to transport (called "versement transport") is an obligatory tax for companies according to the number of employees, which is used to finance the transport system; in addition, companies must compensate employees for half of their public transport expenditure. This funding circuit seems to leave public transport users to pay only a modest share of the associated production costs: but this perspective needs broadening, by integrating into the analysis the economic conditions of the property market (on both the staff and company sides), as well as the salary conditions of the labour market.

b) An urban economic model that makes transport concrete

In urban economic theory, the basic model used to jointly analyze the land use system (including the residential location system and the land system) and the transport system, is the "monocentric model". In this model, each household chooses the surface and position of its location, in relation to its work place and home-to-work travel conditions, in the face of housing prices, which depend on overall household demand. The model makes it possible to establish the "urban equilibrium", i.e. the balance between supply and demand on the housing market, along with the flows and conditions of home-to-work trips.

We devised an original monocentric model to render transport conditions. Usually, these are described only by an abstract cost function that depends on the position of the residence. We modelled two modes of transport, each with realistic conditions and in a competition situation. The first mode is a relatively slow one, available homogenously and isotropically, whose transport cost is reduced to time and proportional to the distance between the user's employment hub and home, with a speed given at c_v (v for private vehicle): let us denote $C_v(r, \theta) = c_v \cdot r$ this cost.

The second is an efficient, rapid, mass mode, but available heterogeneously in the form of radial axes centred on the place of work. By hypothesis, from a given home, a route using this mode includes a main link between the employment hub and an access station i, and thus on a major axis between the origin and position r_i of the station, then a link whose length is $|r - r_i|$ which is still radial but covered in the slow mode, and a circular arc with a length $r\theta$ centred on the hub in order to reach home, still using the slow mode. Note that c_t is the commercial speed of this mode (index t for transit), and g the waiting time in the station. Depending on the polar coordinates (r, θ) of the home in relation to the closest axis, the transport cost for the user is

$$T(r, \theta) = r\theta c_v + \min\{(r - r_i)c_v + (i - 1)g + r_i c_t, (r_{i+1} - r)c_v + ig + r_{i+1}c_t\}.$$

The efficient, mass mode can either represent a network of fast urban roads for cars, or a public transport network on its own site (road or rail).

Taking households to be homogeneous in their preferences and incomes, the balance between supply and demand can be summed up by a single endogenous variable, the level of individual utility, given as u, that each household benefits from. In all residential building places (r, θ), the housing price per surface unit and per occupancy period, $p_{r,\theta}$, is equal to a household's maximum bid to live in this place, which will depend on both u and the minimized cost of transport for both modes.

At the same time, this establishes the housing surface ("lot size") obtained by the household. Strictly speaking, the efficient mode is preferred to the slow mode from point (r, θ) if $T(r, \theta) \leq C_v(r, \theta)$, which constitutes an inequality condition in the plane. The solution is the area of influence of the efficient mode. Along an efficient axis, the areas of influence of neighbouring stations i and $i + 1$ meet together at the following pivotal position:

$$\hat{r}_{i,i+1} = \frac{r_i + r_{i+1}}{2} + \frac{c_t}{2c_v}(r_{i+1} - r_i) + \frac{g}{2c_v}.$$

We have shown that at the urban equilibrium point, the mass mode is preferable in the vicinity of access stations; these stations are associated with peak housing prices and residential density peaks in the number of households per land surface unit, i.e. dips in the lot size per household (if the housing capacity is homogeneous in all ways, cf. Figures 5.1a and 5.1b). The urban form is then a star with knotted arms, centred on the employment hub, with one arm per efficient radial axis, and a knot for each

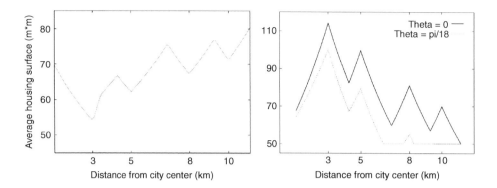

Figure 5.1 Depending on the radius, (a) Size of residence and (b) Land rent.

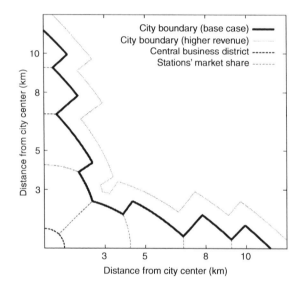

Figure 5.2 Urban form: a star with knotted arms.

station (Figure 5.2). The utility level per household, and consequently the urban form, depends on a set of factors: the number of households, their individual income, their respective preference between residence surface and subsidiary income (after paying housing and transport costs), local property capacities, as well as the cost functions for each mode of transport. For example, a higher income systematically increases the surface of the residence obtained by a household and forces the city to spread.

c) Variations depending the grand socio-political actors' game

The monocentric model, even with more realistic transport conditions, is still an over-simplified version of the urban system, since several of its assumptions are extremely simplistic. For instance, employment is taken to be polarized on a single point, which thus becomes the urban area's centre of gravity: yet in reality, employment is spread over the urban area. In addition, households do not only choose their locations in relation to work; they also take "urban amenities" into consideration, such as education and health facilities, services (shops, leisure), green space, etc. In addition, the housing is taken to be "fluid": the housing capacity (in floor surface) is taken to be dividable into a number of residences of different sizes without transformation or transaction costs, whereas households are reputedly "hyperfluid", i.e. prepared to relocate as soon as a better financial opportunity comes up, with no particular ties, either through habit or progressive adaptation to their previous residence, or ties to the social neighbourhood or local activity network.

Other limitations could be identified. However the fundamental advantage of the monocentric model is that it represents the main elements of the system, its microeconomic agents, and the main goods that they desire (i.e. housing surface, income, spatial accessibility for households), as well as the relationships between these elements. Therefore, the model places each category of agent into relational interplay, with its own interests, decision options and specific economic behaviour.

The model draws conclusions from all of the axioms and hypotheses in order to establish an overall state that is coherent with all of the hypotheses and thus for all of the stakeholders in the system. The hypotheses of fluidity for housing, hyperfluidity and homogeneity for households, and spatial symmetry, can be used to characterize a sufficiently simple state of equilibrium in an analytical way. They could be alleviated using a more realistic, more disaggregated representation, which would be resolved numerically not analytically: cf. integrated land use and transport models.

In the above sub-section, we observed the effect of techno-economic hypotheses on housing capacities and household incomes. At a deeper level, the model can be used to specify stakeholders' respective positions on economic markets, the various transfers, in particular to and from the urban community government: this economic and financial circuit can be used to devise financial schemes to equip the city with local housing capacity as well as transport means.

Thus, the urban equilibrium model can be used to study, *vary and configure the relationships of force between stakeholders, their respective market power, and thus the grand socio-political game between agent categories in the city.*

We have developed alternatives to the urban monocentric model in order to characterize a range of grand socio-political games (Leurent *et al.*, 2010), and then compare their respective effects on urban transformation, as follows: starting with an initial situation featuring only the slow mode of transport, an efficient mode develops with the creation of a structuring network as described above. From the initial situation to the final situation, the cost of transport is decreased for those households with easy access to the efficient mode, and remains constant for the others. The effects on the property market are wholly dependent on the grand socio-political game, with the following alternatives:

1   In the first alternative, the city's government adjusts the local residence tax of each household so that the sum of this tax and the cost of transport remain constant in both situations. As a result, households all maintain their utility and constrained expenditure on housing, residence tax and transport. The residence tax generates

supplementary income for the community for a sum equivalent to the difference in transport costs.

2   In the second alternative, the transport operator adjusts the price of trips on the efficient network and on the circular arcs in slow mode, so that the cost of transport per household remains constant in both situations. As above, households maintain their utility and constrained expenditure, but it is the transport operator that acquires the additional income.

3   The government and transport operator occupy a dominating position in the system – at least for the transformation anticipated here. If they do not exercise these prerogatives, then households' conditions are modified and the housing market may react. Its evolution then depends on one essential condition, i.e. whether or not an alternative housing option exists for a household, and if so, its location in the built-up area (at the city fringe) and in relation to the main axes. In the absence of an alternative option, property owners dominate housing demand, and tenants living in housing that is not well connected to the efficient mode try to improve their utility by reducing their expenditure on transport, bidding to live in housing that is better connected, paying more than the original tenant, in order to supplant and replace it. This principle of outbidding works in a specific way if we consider that households have homogeneous income and preferences, in which case there would be no move, but rather a readjustment of rents so that for each better connected residence, the increase in rent compensates the drop in the cost of transport (at a minimum, or even evolves further, taking advantage of the opportunity for change brought about by the transformation).

4   If housing alternatives exist at the edge of the built-up area, then each household regains market power in the face of property owners. Supposing that they exist in areas well connected by the new mode, then this makes other housing surfaces attractive and results in an ease in housing prices from which all households benefit: their average utility increases and the housing price per surface unit goes down (but rises by lot). Thus, the development of a structured network mainly benefits households, and property owners only partially.

In real life, residential location is highly inert. Most households are not in a position to immediately take advantage of a new housing opportunity. Only a fast-acting minority would take advantage of the boon of improved connections. Among this minority, a significant share would probably be households waiting for a residential transformation to make a demographic transition concrete (e.g. formation of a couple or other family event).

d) Consequences for planning and eco-design

Overall, the urban equilibrium between the demand and supply of housing and transport is primarily controlled by the government and the transport operator – at least in this model, where the specific form of a single employment centre, and the hypotheses of symmetry, mean that local residence tax can be used simply as a regulation instrument, and that the transport price can be simply adapted to match any targeted clientele. For a more complex urban form, specifying economic devices is more difficult and clearly imperfect.

In the absence of a specific political strategy of regulation or price control, the advantages of urban transformation, a priori aimed at households, risk being captured by property owners depending on the scarcity of unoccupied housing (or the difficulty in producing new housing). By playing on another regulation device, property tax, the government could take back from property owners the economic rent that they have gleaned from surplus rents thanks to the boon, rather than due to qualitative improvements of their own.

The consequences are clear for financing an urban transformation: the finance scheme must be devised in line with a policy on transport prices, and according to the regulation modes of the property market, using the devices available, i.e. local residence tax, property tax, establishing conditions for producing housing, and so conditions for land value development and property promotion.

The possibility of funding is a crucial issue when developing an efficient transport network, which consolidates passenger flows on the basis of infrastructure and/or vehicles.

It is up to the community, through its government, to establish the modalities for collecting benefits incurred due to the transformation by the various stakeholders directly concerned (households and the transport operator) or indirectly concerned (property owners, then land owners and property constructors). We have mentioned the regulation devices available (apart from expropriation, which is more radical). In a concrete case, it is important to anticipate the potential by using simulation to evaluate the economic effects on the various stakeholders. The effects on households also have a social scope: communities must take care to make the transformation equitable, compensating the losers by taking from the profits of the winners. To do this, an exchange device is important: this is one fundamental function of currency, which shows once again the importance of integrating the financial aspect into the urban planning design.

Social impacts alone constitute a vast field of research. The transformation considered here raises at least two equity issues: one between tenant households and property owners, and the other between households and the government, which a priori decides on the structuring network's layout, and therefore on who will be well connected and who will remain less well connected. This will go on to determine the efficient mode's market area, and so the natural clientele of modes with better economic performance and better ecological performance. To design a sustainable city, efficient transport means need to be integrated and installed in interaction with the location of households, as part of a joint transport plan and land use plan design, and these modes need to be developed in line with their performance potential.

DESIGNING AN URBAN PUBLIC TRANSPORT LINE

Designing an urban public transport line involves to set up a number of things: which vehicle to use, what stations to serve, what frequency, operation, pricing, etc. These decisions will also determine the costs of the infrastructure, the operating costs, and the quality of the service offered to passengers.

The aim of the models that we briefly present in this section is to balance the costs borne by the operator with the level of service to passengers. Although these models cannot necessarily faithfully reflect local context in detail, they do highlight significant relationships between design choices and their consequences in terms of costs and impact on demand, make it possible to obtain acceptable orders of magnitude, and give interesting sensitivity analysis results.

In this section, we will start by illustrating the spirit of the approach with a very simple example: the choice of optimal frequency on a shuttle. We will then look at how to model production costs and user preferences so as to express an objective function that we will then optimize. Lastly, we will use a slightly more complex example to show how these methods can be used to obtain areas of relevance for different modes like the bus, tram and subway.

a) A simple example

Let us take a shuttle service transporting passengers from point A to point B. The flow of passengers is continuous, and equal to n passengers per hour. The time of a trip from point A to point B is t, and the waiting time at departure is inversely proportional to the frequency f in vehicles per hour from point A. The price of the trip is p. We make the hypothesis that the hourly operating cost, given as C, is proportional to frequency f, at the rate of coefficient c:

$$C = c \cdot f$$

The net profit for the operator is therefore equal to the revenue minus the costs, as follows:

$$\pi = n \cdot p - c \cdot f$$

For the passengers, we consider that the higher the price or trip time, the lower the service quality. We therefore build an indicator called the generalized cost, comprising the price p, and the run time (or the waiting time) multiplied by a coefficient called the time value w_t (or w_a). This coefficient is the amount that a passenger is prepared to pay to decrease his or her travel time by one hour. The service quality can thus be represented, in opposite way, by a generalized cost function G:

$$G = p + \frac{w_a}{2f} + w_t t.$$

The sum of the generalized costs for all of the passengers is simply $n \cdot G$.

The advantage of a generalized cost is that it is expressed as a monetary quantity. For example, if we increase the travel time by one hour, but reduce the transport price by an amount w_t, then passengers are indifferent: the transport service is no more or less attractive in the second situation than in the first.

When we combine the results of the transport operator with the generalized costs of the passengers, we obtain a function of the total cost TC, which takes into account passenger preferences and production costs (profit appears as negative because it is a cost function): $TC = -\pi + n \cdot G$, in other words:

$$TC = cf + \frac{nw_a}{2f} + nw_t t.$$

The objective is to minimize this function because, if is not minimal, that either means that the service quality can be increased at a cost lower than the price paid by the passengers, or that the service quality is too high given how much the passengers are prepared to pay to use it.

By minimizing TC, we obtain the optimized frequency:

$$f^* = \sqrt{\frac{nw_a}{2c}}.$$

The optimal frequency therefore rises with the number of passengers, but not proportionally: demand needs to be multiplied by 4 for the optimal frequency to rise by 2. This frequency goes down with the operating cost c (the latter rises, e.g. over a greater distance). It increases with the coefficient w_a which represents passengers' willingness to pay to reduce the waiting time. Lastly, according to the formula, an error of 10% on one of the parameters is translated by an error of only 3.2% on the optimal frequency: the model is fairly robust.

Note that we can very easily take into account the vehicle capacity constraint K: the optimal frequency becomes simply:

$$f^* = \max\left\{ \sqrt{\frac{nw_a}{2c}},\ \frac{n}{K} \right\} \text{ because } n/K \text{ sets a lower limit for the frequency.}$$

By reinserting f^* into the total cost TC, we obtain the optimal total cost function TC^*:

$$TC^* = \begin{cases} \sqrt{2nw_ac} + w_t t\, n & \text{for } n \leq n_K \\ w_a K/2 + (c/K + w_t t)\, n & \text{for } n \geq n_K \end{cases}$$

where $n_K = K^2 w_a/2c$ is the level of demand from which capacity is saturated at the optimal. This function characterizes the best that it is possible to do with a given mode of transport for a given transport demand. As we shall see below, it is a good tool for determining what mode is suitable for meeting a particular demand.

b) A generalization

We can generalize the above approach in a realistic case, which is a public transport line serving several stations. As we shall see, we need to be more precise in representing the costs, as well as demand: it is no longer sufficient to know the total flow of passengers. We shall also see that this approach can be used to determine the most pertinent transport technique for a particular transport demand (here with the example of a bus, tram and subway).

Modelling the costs

For the type of approach used here, we should not detail the costs any more than necessary for two reasons: first, because the data available do not necessarily allow it; second, because what we want to know is how the total costs change with the variables we are studying.

Let us take the example of frequency: making a first approximation, we can distinguish the infrastructure costs from the operating costs, and make the hypothesis that the former are independent of frequency, whereas the latter increase proportionately with it. We can therefore express and formulate the total cost for a given mode m:

$$C_m = c_{i,m} + c_{o,m} f,$$

where $c_{i,m}$ is the infrastructure cost per hour, and $c_{o,m}$ is the operating cost per hour for a vehicle run.

Assuming that we know the hourly costs of operating a vehicle $c_{v,m}$ and its commercial speed $v_{c,m}$ (v standing for velocity), the following relationship follows:

$$c_{o,m} = \frac{2fLc_{v,m}}{v_{c,m}},$$

where L is the length of the line, which must be covered twice per service cycle for each vehicle.

Modelling the demand

The passenger preferences are modelled as above; however, the volume of demand needs to be treated in greater detail, including details of the starting points and destinations. Strictly speaking, three parameters are necessary:

- n the number of trips per time period;
- x_M the critical load, i.e., the passenger load on the busiest segment of the network;
- d the average trip length.

The impact of waiting time on passengers will be proportional to n; the impact of commercial speed on passengers' trip time will depend on d; lastly, x_M will determine whether or not the vehicle capacity constraint is saturated

Optimal frequencies, and total associated costs

We can quite easily adapt the results of the preceding section to this new context. The optimal frequency becomes (omitting index m to ensure legibility):

$$f^* = \max\left\{ \frac{1}{2}\sqrt{\frac{nw_a v_c}{c_v L}}, \frac{x_M}{K} \right\},$$

whereas the new cost function is, noting $n_K = 4c_v L x_M^2 / w_a v_c K^2$:

$$TC^* = \begin{cases} 2\sqrt{\frac{nw_a c_v L}{v_c}} + w_t n\, d/v_c + c_i L & \text{for } n \leq n_K \\ nw_a K/2x_M + 2c_v Lx_M/v_c K + w_t nd/v_c + c_i L & \text{for } n \geq n_K \end{cases}$$

c) Area of relevance of different modes of transport

To illustrate the type of results that can be obtained using these models, let us take the case of a 15 km-long public transport line. We assume the average length of a trip on this line to be 3 km. In addition, we take the critical load ratio over total demand

Table 5.1 Hypothetical characteristics on three modes of public transport.

	Bus	Tram	Subway
Infrastructure cost (€/km·h)	0	150	1500
Operating cost (€/veh·h)	80	200	500
Commercial speed (km/h)	12	16	25
Capacity (pass/veh)	70	250	600

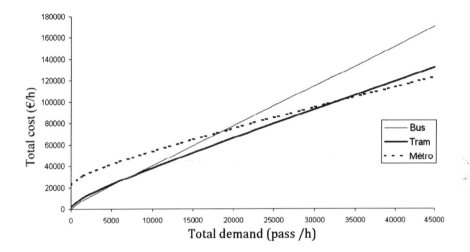

Figure 5.3 Total cost function associated with the three modes: bus, tram and subway.

x_M/n to be constant and equal to ¼ (which corresponds to a fairly evenly distributed demand; this ratio would be equal to ½ for a line where all passengers get on or off at the same station). The following parameters are assumed as shown in Table 5.1.

Figure 5.3 shows the evolution of the total cost TC^* for each of the three modes, assuming a line length of 15 km and average trip length of 3 km.

The area of relevance for each mode corresponds to the zone in which its total associated cost is lower than that of the two others. We obtain fairly reasonable orders of magnitude, with the bus recommended for demand under 7,000 trips per hour, the tram for demand under 30,000 trips per hour, and then the subway.

The convexity in the cost curve at low demand shows that the frequency effect goes up when demand increases, permitting economies of scale. These economies are easier to obtain for low-capacity modes, but are almost exhausted once the capacity constraint is reached. This effect is slower with higher capacity modes, but these are more expensive, mainly due to infrastructure costs.

d) Overview and discussion

The above example shows the advantage of this approach, i.e. providing orders of magnitude to make design choices, with easy utilization models, and requiring little data. These models are sensitive both to production costs and the quality of service for passengers.

We could fairly easily extend the above method to take into account the spacing between stops. The impact on costs is twofold: first, more stops implies more infrastructure costs and maintenance costs, and the latter can vary widely between modes. Second, more frequent stops reduce the commercial speed, which implies that to provide a given frequency, the operator needs more vehicles circulating at the same time. Clearly, adding a variable can make the model impossible to resolve analytically, but it would be totally usable numerically.

Note that several important points have been omitted in the above approach. First, the issue of anticipating demand is absent; here it is not only assumed as known, but as independent of the quality of service provided. Design models that render demand endogenous would be worth using here, but their development is currently one of the difficult issues of transport engineering. Second, a number of impacts have also been omitted: externalities like noise or pollution could quite easily be integrated; however, other subjects, like the opportunity for urban renewal created by a public transport project, lie outside the scope of these tools and need to be tackled differently.

APPLICATION: DESIGNING HIGH-QUALITY SERVICE BUS LINES IN GREATER PARIS

In this part, the methods discussed above are illustrated using a real-life case, which is a project study in greater Paris (the Ile-de-France region in France). At the time of the case, 2010, the public transport supply is limited for passengers wanting to make long, non-radial trips. As the transport demand for these trips is not as concentrated as in the radial directions, it may be not so relevant to implement circular suburban subway lines.

An alternative is to set up bus rapid transit (BRT) lines, travelling at high speed between stations placed far apart. The advantages are many: they would be inexpensive, would use the existing motorway infrastructure, their vehicles and agents could be redeployed, even the whole line could be suppressed if the traffic observed did not meet expected levels. Lastly, they would be a competitive alternative to the car, and thus ensure a certain level of demand.

The study summed up here (Li, 2011) comprises several stages. We shall leave aside the problems of route and feasibility studies of the stations, and constituting scenarios; we shall first take a brief look at traffic forecast methods; we shall then concentrate on the financial and socio-economic situations.

a) BRT project: characteristics of lines, layout, size of fleet

The BRT's main quality target is speed and high-quality service. We therefore assume that vehicles travel on motorways, in a bus lane, at a commercial speed of around 65 km/h. The frequency is 12 vehicles per hour (for an average waiting time of 2.5 minutes), with stations spaced from 5 km to 8 km apart. All passengers are seated, and the vehicles have a capacity of 110 passengers. The price is fixed at €0.1/km.

The line routes (see Figure 5.4, which also shows traffic forecasts) include two circular routes, the first of which is relatively close to Paris and includes stops at La Défense, Val de Fontenay, Orly airport and Versailles; the second is further out and

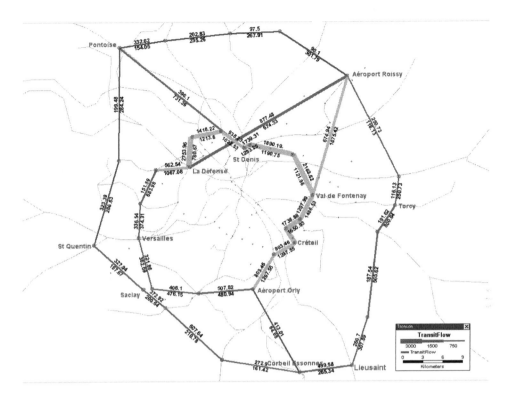

Figure 5.4 Layout and traffic forecasts for BRT lines.

stops include Pontoise, Roissy CDG airport and Corbeil-Essonne. There are several direct links between the two circular routes, especially between the two airports. The total length of the lines is 459 km; given the frequency sought and the anticipated commercial speed, the fleet required is 180 buses.

b) Modelling costs and demand

To establish the predictive financial results for the operator of these lines, we need to estimate the infrastructure and operation costs, and also anticipate demand. The cost estimate encompasses building works generated by the project (mainly new stations), purchases of vehicles, and expenditure on operations. Provided that the estimated unit costs linked to these expenditure items are correct, there should be no particular problem.

Modelling demand is somewhat more complex. Here, we have assumed that the spatial structure of trips is maintained, and therefore that the trip flow matrix by origin-destination pair will not change. However, we assume that passengers can change mode, which involves modelling two things: the modal choice itself, and the choice of route.

Table 5.2 Financial versus socio-economic balance of the BRT scheme.

Agent	Item	Amount (€M/yr)	Total (€M/yr)
Operator	Infrastructure	−23.6	−46.8
	Rolling stock	−3.3	
	Operating	−24.2	
	Revenue	4.3	
Passengers	Pass. improvement on public transport	48.6	50.6
	Pass. improvement on private car	2	
Externalities	Reduced congestion for private cars	2.1	3.2
	Road safety	0.6	
	Pollution	0.4	
	GHG emissions	0.1	
Socio-eco result			7

For each of these two stages, passengers' preferences regarding travel alternatives are represented by generalized costs that comprise the price of the trip and different service quality variables: run time, feeder time, waiting time, transfers. A weight is associated with each of these components. For the modelling carried out here, the money value of in-vehicle time is set to be €12/h, while waiting time is taken as 1.5 times more tedious than in-vehicle time, and feeder time twice as tedious.

The route choice stage involves, for each mode, finding the route with the lowest generalized costs. This stage takes traffic congestion into account. The modal choice stage is based on a standard discrete choice model, the multinomial logit model, which takes generalized costs and produces modal shares for each origin-destination pair. The traffic forecasts obtained from these two stages are illustrated in Figure 5.4.

c) Financial results and socio-economic analysis

From demand estimates result revenue estimates, which can be used to establish the financial and socio-economic sheets. The financial balance takes into account the operator's situation, i.e. the costs (of the infrastructure, vehicles and operation) and revenues. The socio-economic balance features supplementary items, which include: improved service quality for passengers, reduced congestion on the roads, better road safety, and reduced pollution and greenhouse gases. The Table 5.2 shows both results for one year.

The socio-economic result of the project is therefore positive, even without taking into account externalities, which would improve the results still further. However, the financial results are totally different: revenues are insufficient to cover the operator's costs. Supplying the service at the price considered in the simulation would require substantial subsidies.

d) Overview

This part only gives a brief outline of a comprehensive study that comprises different scenarios and sensitivity analyses, including of the operator's receipts in relation to the price. More targeted impact studies have also been carried out, including how reserving a special bus lane on motorways would affect road traffic.

More generally, the possibilities offered by this type of approach are very broad. Unfortunately, it is impossible to remain within the analytic frameworks given in the previous sections, which necessarily means working by defining scenarios and sensitivity analyses, which is much slower and precludes the exhaustive exploration of a great number of options.

An interaction between the two types of model can be fruitful for this type of project study: a simple analytical model can be used to obtain orders of magnitude, while a more comprehensive model provides more realistic data on traffic, costs and receipts. We could set up a to-and-fro system between these two types of model, until we obtain a coherent solution with the two approaches.

CONCLUSION

In this chapter, we started off by presenting the physical characteristics of transport means, then the economic interplay of urban stakeholders for housing and transport of households. We then gathered the technical and economic characteristics to highlight the designer's sphere of action, and the optimal design of a public transport line that optimizes modes and service frequency. The search for an optimal solution led us to the notion of a mode's "area of relevance" compared to other modes, before making an implementation decision. Lastly, we explored an economical solution involving public transport on circular corridors in the outskirts of greater Paris, based on a territorialized simulation model. This application takes all means of transport within a built-up area, and sets them into local competition, as well as an interaction of complementariness, to undertake a trip using several means to cover space (intermodal concept).

Two key messages emerge that concern on the one hand *the intensity of demand*: the intensity of human presence in a city makes mass flows possible; and on the other hand, the *technical capacity of supply*: the hierarchy of means of transport according to their area of technico-economic relevance, in terms of both technical characteristics and territorial conditions.

In a specific territory, or in other words, "on the field", the intensity of demand and technical capacity come together, and confront each other. They set limitations for each other: for a given demand, supply must not be more expensive to develop; and for a given supply, demand must not be in excess – at the risk of saturating and potentially breaking down the system performance. Ultimately, a built-up area's financial capacity is confronted with the development cost of additional urban transport capacity: if the all-dimensional size of the built-up area involves substantial means (e.g. high-frequency rail links), then we need to re-think not just the transport system, but also the very form of the built-up area, the total size of its population and jobs (spreading could constitute a development option), its internal configuration in the geographic area, and the financial circuits of its economic interplay.

To devise supply that matches demand, is ideally located and of an ideal size, i.e. neither too much nor too little, it is useful to avail oneself of a simulation model that is sensitive to demand and supply characteristics. The need for a design tool led the Chair to initiate the development of the original model presented in the following chapter: to simulate the current operating state of the supply of transportation and the demand for movement. This will need to be completed with a model of the city's mid- to long-term operations, including the respective dynamics of demand and supply, both for property and transport, and in particular the formation of housing and land values, along with the production costs of housing and transport.

REFERENCES

Kilani, M., Leurent, F. and de Palma, A. (2010). Monocentric City with Discrete Transit Stations. Transportation Research Record Volume 2144, Pages 36–43. DOI 10.3141/2144-05.

Leurent, F., Li, S. and Kilani, M. (2010). Patterns of land-use equilibrium for city's response to planning schemes: an analytical framework and a case of transit development in the Paris area. Conférence Politiques publiques et organisation industrielle dans la ville, en l'honneur de M. Fujita. Lille University 3, 24–25 June. 28 pp.

Li, S. (2011). Summary of thesis studies. Working paper Enpc. 7p.

Chapter 6

The eco-design of parking systems: systemic analysis and simulation tool

Fabien Leurent, Houda Boujnah & Alexis Poulhès
Paris-East University, City Mobility Transport Laboratory, Ecole des Ponts ParisTech,
Ifsttar, UPEM

INTRODUCTION

a) Context: a parking system as an object of design

Parking is an integral function of mechanized mobility, to *keep in waiting,* and *maintain available* vehicles that are used to connect the places where its user(s) carries out his or her activities. Reciprocally, for a building accommodating an economic or social activity (e.g. shop, service), the availability of parking places close by provides access opportunities for "clients" from further afield and thus determines a *recruitment area* for potential customers. For this reason, both customers and "suppliers" of activities are interested in parking and the availability of places for two- or four-wheeled vehicles. In urban environments, these places are *mobility facilities* that may be private (in a private building), semi-public (e.g. shop car park) or public (on a road or in a special car park), but are always located on a territory. When space is scarce, facilities are more likely to be built constructions; in densely populated environments, "parking silos" constructed underground or as a superstructure can be implemented to provide large capacities.

Parking is a design object on several scales: from the micro-local layout of a car park comprising a few places, to a parking plan and policy for an entire town, passing by the design of a large car park and by the scale of the neighbourhood, which is the most determining, since neighbourhoods are local catchment areas for activity buildings and parking places, used by inhabitants going from home to their vehicle, as well as workers and visitors.

The Eco-design Chair investigated parking on two scales of, respectively, the neighbourhood and all or part of a built-up area, in order to study how supply matches demand at local level, and the possibilities of transfer between places. Indeed, demand that is not satisfied locally, but that persists, is transferred to a neighbouring area that it will "parasitize": local parking policy must be designed taking into account the basins of pedestrian accessibility. And, to understand local demand for parking, customers should be analyzed according to their motive for coming and where they have come from.

b) Objective and organization of the chapter

In this chapter, our aim is to establish the basic eco-design principles for a parking system: firstly, we clearly set out the system, i.e. how it is composed, how it works, and specific effects like the formation of vehicle stocks and transfer flows; we then confront intuitive planning with economic benchmarks in order to identify the planning principles; lastly, we present the spatialized model of parking supply and demand, ParkCap, devised to simulate plans and policies in urban areas.

The rest of the chapter is divided into six parts. We start by considering parking from an eco-design angle, describing the functions it fulfils, the actors concerned and the eco-design challenges, in a general way based on significant technical conditions and with the perspective of renewing technologies and services (section 2). We then focus on the system effects, which we identify and qualitatively characterize (section 3). The economic benchmarks are presented in the next part (section 4), followed by the planning principles (section 5). We then look at modelling supply and demand, setting out the physical and economic principles (section 6) and taking a typical case to show the conditions and results of an application (section 7). After the conclusion, a detailed annex provides a systemic analysis of parking in terms of supply, demand and regulation in a territory.

AN ECO-DESIGN APPROACH TO PARKING

a) Function, devices and services

At a basic level, parking is the *function* of accommodating a temporarily stationary vehicle in a specific place. In real time, the place has a state of occupancy: either occupied or vacant. For users, the need to park stems from the need to make journeys with their vehicles, and behind that the need to carry out an activity in a location close to the parking place. In some cases, the nature of the activity imposes the use of a vehicle and makes parking indispensible, e.g. accompanying a dependent person, or purchasing supplies from a supermarket.

The *material device* that provides the function is a place, i.e. a marked-out, stationary space that is functionalized, and tends to be more artificial and anthropic (i.e. marked out, constructed and used) when the geographic environment is denser; if necessary, associated with operating facilities like a payment point or electricity supply point, or an access barrier. If the vehicle is a bicycle, a stand or railing can be used to chain up the vehicle and keep it vertical. Generally, each place results from a particular built element: a section of road, or a special building (above or below ground) or a specific car park (above ground, underground or elevated). The type usually determines the holding status and relates to an operating mode.

A parking *service* involves providing places for those clients who request them for a desired duration. Self-produced parking is when individual demanders equip themselves with places that are thus private. Public parking is devised for others, and makes available a set of places in the form of a "lot" depending on the location: inside a location, any place would be suitable for clients. The service may be offered for free with or without a time limitation, or for a fee. The quality of the service covers the facility of access and use, the pleasantness and security of the activity location targeted by the client, and also the availability of a place, and the possibility of anticipating with certainty how a journey will unfold.

Figure 6.1   Parking lots: (a) private mode and (b) public mode.

b) Concerned parties

The parties concerned by parking can be divided into specific types: individuals, companies or authorities, property promoters, planners, operators.

Individuals who hold vehicles must have a parking place near to their homes: this facility is easier for individuals living in private houses rather than apartments, but in any case, the property market gives value to parking places, and attaches a price. The availability of a car park interacts with the choice of residence and the use of a vehicle, a usage that is all the more convenient when the parking place is close to the place of residence.

The parking possibilities at the location of an activity also influence the use of vehicles: for instance, when parking is available at a work place, it is a determining factor for using the car as travel mode from home to work. For *companies*, the availability of parking is important for not just their employees, but for their customers and suppliers. It influences the location of establishments – in particular major stores. Given the influence that parking availability has on property prices, property promoters accordingly size up the parking capacity in their projects. A significant reduction of places in a programme could discourage customers that would be ready to pay for investment in facilities.

The demand for parking varies in space depending on the qualities of access provided by different modes of transport. A location that is very well connected by public transport or soft modes of transport requires less parking per unit of activity. It is up to the spatial *planner* to locally balance the intensity of land use by activities, with a parking capacity, in line with modal accessibility; capacity includes private places, and semi-public and public parking, with the latter devised to be mutualised between different successive uses. Restricting private capacity will have an impact on the use of public capacity, and vice versa. Local shortfalls affect neighbouring areas with a spread of the parking occupancy. These transfers cause additional traffic, which increases environmental impacts in line with the level of congestion.

Once parking capacity is installed at a location, it needs to be managed: the size and intensity of demand can justify calling on a specialized *operator* (company or state controlled), remunerated by sales receipts or a local authority subsidy. In dense urban zones, commercially operating places along the street, which are inexpensive to install and charge for, should achieve financial profitability. Levying a charge for parking reduces occupancy duration, improves availability and reduces the time users spend searching for a place – in exchange for a fee.

c) Techno-economic domain under renewal

Nowadays, mobility is managed according to an integrating *multimodal policy*. Parking can be moved to a relay car park at an intermodal station connecting to efficient public transport. In dense zones, the road capacity is divided between types of vehicle to encourage two-wheeled or specific-use vehicles (e.g. deliveries, disabled people).

In our *information era*, parking is managed more subtly: users guided to available places in real time (using message panels or internet); more systematic fines; time modulation of tariffs, etc. The development of centralized information systems in parallel with the reallocation of some types of parking capacity have resulted in different forms of *vehicle sharing*: vehicle-sharing systems for two-wheelers and four-wheelers, car sharing and car pooling (i.e. ride sharing), which in principle reduce the time that vehicles spend parked. Another solution is placing rental vehicles in a public car park at users' request to ease the time constraint of an appointment between renters and customers (VINCI Park innovation).

Lastly, some *vehicle evolutions* concern parking, e.g. the parking aid system; small, easy-to-park cars; charging equipment for electric vehicles, and soon, self-parking of the vehicle without its user. This will probably be preceded by "co-parking", i.e. pooling of private places between individuals (people or companies), similar to car pooling, with direct matching using information and communication technologies (specialized applications that users access permanently via smartphones or other mobile computers).

d) Decision issues: the place of eco-design

The parties concerned can contribute to making parking environmentally viable, depending on their role and situation.

- Users, by equipping themselves with small vehicles, save space on the public roadside. By using a mode other than the car to access a saturated zone, or by parking on the outskirts of this zone, they avoid additional car travel.
- Car manufacturers, by facilitating parking manoeuvres with a specific device.
- Parking operators, by restricting the occupancy rate to improve the flow of access; by managing air quality in confined underground car parks; by dimensioning car parks sufficiently for easy manoeuvring.
- Traffic operators, by combating unauthorized parking that blocks traffic and provokes congestion.
- Planners, by locally adjusting parking capacity and activity intensities, locating places in the immediate vicinity of places required, and inversely avoiding locating certain activities in places where parking would inhibit traffic (typically close to junctions). And also by managing multimodal mobility through incentive

instruments rather than restrictions (slowing down, discomfort, queues): making places available in relay parks, intermodal integration in terms of physical layout, information and tariff levels.

PARKING SYSTEM PHYSICS

This part sets out the notion of a parking system and its concrete aspects in terms of stakeholders, technical resources and the areas concerned. In addition, we characterize the occupancy of local parking capacity as a physical process that operates in space, by propagation, and that interacts with traffic conditions, and with environmental resources.

a) Parking as a subsystem

We have already identified the stakeholders in a parking system, in their roles of direct demanders (individuals) or indirect demanders (companies), suppliers (operators), providers (property suppliers) and regulators (planners). This set of roles is typical of an economic supply and demand system.

On a basic level, an elementary parking usage concretely involves a previously vacant place, a vehicle and its user. Their one-off composition is made in internal interaction with the circumstances of the user's activity (e.g. motive, duration), with the journey sequence (e.g. main route in his or her own vehicle, final journey using another mode): the parking system interacts on a deep level with the activity establishments system and with the multimodal traffic system. It is preferable to design them together as subsystems integrated into a main system of activities and mobility (cf. annex), including uses and facilities.

To delineate this kind of system in space, all of the places that can be connected by users need to be integrated: neighbouring places, or long-distance transfers between a place of destination or a relay park, etc. and the initial location of the users, with their own parking conditions at the origin place of the trip.

b) Aggregated effects

In real time, a user's demand for a place is confronted with the occupancy stock rather than other basic demands. In other words, users interact with the state of the system as characterized by local aggregates: the state variables are macroscopic despite being micro-local.

The phenomenon of public parking occupancy is a macroscopic physical process, which operates in space by local diffusion (local transfer of a usage) and potentially by remote propagation (more distant transfer by anticipation, which can be based on real-time information).

The state of parking affects the state of traffic on local roads: when users cannot find a place, they continue to circulate until they do, or block a road or access to a building by stopping there to wait for a place; their search takes time and causes wear and tear to the vehicle. The increased usage leads to an increase in energy consumption and pollution emission, and wear and tear of resources (vehicle, infrastructure).

It constitutes additional circulation, which impedes the traffic of other users, wasting their time and increasing their environmental impact.

Along with the influence of parking on the traffic state through the traffic load, another influence can affect the flowing capacity: unauthorized parking on the roadside temporarily hinders a traffic lane, and potentially leads to a traffic jam.

ECONOMIC BENCHMARKS

The parking system is both physical and economic, because of the stakeholders and their actual roles. In real time, users act micro-economically to choose their parking place and route: this contributes to the physical operation of the system. In addition, on a macroscopic level, aggregating individual situations, the state of the system presents characters of an economic kind that affect its users, operators and the regulator, as well as other stakeholders in the activities and mobility system.

The aim of this part is to provide several elements of an economic nature as benchmarks, i.e. what is the value or cost, for a particular stakeholder, of a certain situation or decision? We successively look at the investment in a parking place (for an operator), then the diversion between places (for a user), next the collective cost of a lacking place, and lastly the cost of obstructing the traffic.

a) Investment cost of a place

A parking place for a car occupies a ground surface of around 10 m², not counting the access. On the surface (roadside or private area), investment in a place includes preparing a terrain (waterproofing, site servicing) and especially the ownership of the terrain; the cost of land can range from 1 to 100 or more depending on the geographic environment: from a few dozen euro per m² in a village to several thousand in a large urban area in France in 2013. In a silo, the surface area per place must include the share of access, and the investment cost of the construction will be as much as the cost of occupying the land (up to 50 thousand euro per place at La Défense, Paris). Since facilities are solid, a lifespan of around 20 years is credible: an investment of 10 thousand euro represents around 500 euro in amortization per year (leaving aside interest on capital), to be related to, say, 200 business days per year, giving a daily value of 2 to 3 euro. Added to this comes the operating cost, of which the order of magnitude is similar. The economic profitability of the service requires receipts of 3 to 5 euro per place per day.

b) Individual cost of diverting from one place to another

Let us now look at the user to evaluate the cost of parking one's vehicle further afield when a place is not available in the desired location. A transfer carried out at distance d requires travelling once in the car, at vehicle speed v, and twice by foot respectively before and after the activity, at walking speed w. The transfer costs a time $t = d/v + 2d/w$, plus energy consumption of, let us say $c \cdot d$ not counting other usage costs.

Let us take the values of $d = 300$ m, $v = 15$ km/h and $w = 4$ km/h: assuming consumption of 10 litres at 100 km at a price of €1.5/litre, the unit cost is $c = €0.15$/km.

Energy consumption thus costs €0.05, which is negligible compared to the time $t = 10$ min which is worth $g = \alpha \cdot t = €1.5$ of generalized cost to the user, assuming a time value of $\alpha = €10/h$ which is standard in France at 2010 conditions.

In this microeconomic model, users feel the effects of the time spent much more than the energy consumption. However, this transfer, when repeated every business day for 20 years, consumes 120 litres of fuel, to be compared with the energy used to construct a place.

c) Collective cost of a lacking place

Let us now look at the lack of a place in a given location, during a time period Δh: assuming that during this time there is always a car driver waiting for a place, and that one driver is progressively replaced by another. For all of the users concerned, the lack of a place costs $\gamma \cdot \alpha \cdot \Delta h$, where $\gamma \approx 2$ a factor of tediousness of waiting, which is psychologically uncomfortable. A shortfall lasting eight consecutive hours in a saturated parking zone would thus cost €160 under current French conditions. For 200 days per year and a lifespan of 20 years, the corresponding value would come to 640 thousand euro, which would justify investing in not one, but several, places! In other words, using a basic calculation, we have illustrated the opportunity for investing in parking facilities where demand is constantly present.

d) The cost of a traffic obstruction

Lastly, let us examine the simple case of obstructing traffic: taking a case of unauthorized parking that blocks a street with a single traffic lane for Δh, for a traffic demand of x veh/h much lower than the nominal capacity given as k.

In a bottleneck model familiar to road network operators, an obstruction disrupts traffic during $\Delta h + h'$, where h' is the duration required to clear the tailback built up during Δh: therefore $(k - x) \cdot h' = x \cdot \Delta h$. The time wasted is $T = \frac{1}{2}x(\Delta h + h')\Delta h$. For the values $k = 800$ veh/h, $x = 200$ veh/h and $\Delta h = 5$ minutes, the time wasted comes to $52'$, at a total cost of $G = \gamma \cdot \alpha \cdot T = €16$ including tediousness. It is justifiable to make the emitter pay for the hindrance caused, with an appropriate parking fine.

e) Recapitulation

We have associated economic costs with investment in facilities and usage situations. The values constitute the opportunity costs for managing parking: not using an available space to supply parking places results in losing the possible tariff revenue, and additional costs for potential users. Not controlling unauthorized parking in real time comes at a cost for all road users.

Lastly, in the cost balance, energy consumption weighs significantly less than the consumption of time, at least in users' individual decisions.

The overall conclusion is that the time aspect drives user behaviour; a lack of parking places can influence modal choices; however, as users are willing to pay to save time, an investment policy to satisfy demand could prove profitable, with beneficial industrial impacts on a social level, to be balanced with impacts on local residents. The environmental balance is a priori not obvious, since additional traffic due to diversion between places and slower general roadway traffic increase energy consumption and pollution emissions.

LOCAL PLANNING PRINCIPLES

Sustainable mobility has become a widespread political claim, yet sometimes to be paid lip service only. The naïve application of simplistic rules can be appropriate in some cases, but risks being counterproductive in others, often due to transfer phenomena, or unauthorized parking, which risk taking more away from a community than it hoped to gain through its policy.

This section puts forward some design elements for planning local parking, contributing to bottom-up design that is closer to local needs and possibilities. We successively look at a private establishment, then a road section and a road junction, then a public car park or relay car park, and end with the notion of a neighbourhood.

a) Parking for a private establishment

One of the simple, challengeable precepts is restricting the number of places per building according to its accommodation capacity, with on one hand the flooring values, and on the other the ceilings … it is best to let the property promoter choose, and oblige him to make a financial contribution in line with the number of places and thus develop the modes of access to the site together. It is also a good idea to impose a property tax per parking place to improve the flow of possession and contribute to the needs of the neighbourhood.

When land space is scarce, car parking places should be incorporated into buildings, on the inside rather than alongside. Places should be provided for different modes of vehicle, not just cars, but motorbikes and bicycles. Some places should be equipped with an electricity supply, whose infrastructure should immediately be included in large buildings.

It is also worth anticipating vehicle pooling by building occupants, reserving an area to make it easier to create a user club. Lastly, vehicle access between the building and the road should as far as possible avoid hindering the operation of the pavement and the road.

b) Standard road section

By definition, the basic function of a road is to be available for circulating traffic: therefore its viability is its primary quality. For parking along the side of such a road, sufficient availability should be maintained to anticipate search routes and unauthorized parking, and avoid reducing the viability.

Parking capacity should be in proportion to access requirements, and so the intensity of local activities, depending on the modal alternatives. The following basic functions should be covered: public transport stops, short deliveries, occasional brief visits (e.g. to use a cash machine). Parking for self-service vehicles should be conditional on a rotation rate, while shared vehicles should be done so with local establishments. It is important to maintain the viability of pavements for pedestrians and the integrity of vehicles (e.g. padlock points for two-wheelers). Creating places for two-wheelers is a good supply policy for developing these modes, along with bike lanes and respect for pedestrian routes.

c) Planning a junction

A road junction is a sensitive area mutualised by multiple flows in a limited space. Its overall viability interacts with the viability of each of its roads: it both depends on them and determines them. An accumulation of vehicles within the junction should be avoided – apart from zones earmarked for short waits in a turning movement[1].

Given the above, a junction does not marry well with parking. Parking should be avoided close to junctions, even for public transport vehicles, whose manoeuvres disturb the general flow. At an even earlier planning stage, premises located close to the junction should be reserved to activities that do not require parking (whether authorized or unauthorized).

d) Public car parks and relay car parks

A public car park is a large group of places in a given location, accessible to all users provided they respect specific rules and pay a fee. Most car parks are located in dense urban zones whose activities they serve: above ground, or in a silo if density is very high.

A relay car park is a public car park associated with public transport – a railway station or a station of bus rapid transit or of car-sharing. These are often free in suburban areas for regional liaisons, but fee-paying for inter-regional or international liaisons.

A public car park serves a variety of activities that generate a continuous flow of customers during the day, in the evening for leisure activities, or at night for local residents who do not have a private parking place. These car parks open up flows in the vicinity: an installation above ground diversifies access for pedestrians and prevents them from making detours, but takes up significant ground space.

A relay park's main purpose is to provide customers with access to trains. To reduce the journey for pedestrians between parking places and the platform, car parks can be spread over several floors. Relay parks contribute to making a station multi-modal: it is preferable for car parks to accommodate cars and two-wheelers, which should have the advantage of places close to platforms and easier access from the outside.

Along with multi-modality, another trend for train stations is multi-functionality, whereby a group of shops and services are located close to or on another floor of the station for the traveller to use in addition to his or her voyage. This kind of development interacts with the relay park, on the one hand by competing with its location and need for surface area; and on the other hand more advantageously by developing a clientele

---

[1] For left turns, an Indonesian-style scheme is appropriate: out of three lanes by axis crossing the junction, one receives traffic leaving the junction, while the two others accommodate the flow entering it. The lane on the side continues with the movement or the right turn and the middle lane can stock vehicles waiting to turn left. (assuming driving on the right hand side of the road)

attracted by the additional functions, over and above rail journeys, at diverse times, which spreads occupancy, and with an increased disposition to pay to park.

e) What kind of parking suits a neighbourhood?

The parking possibilities in a neighbourhood constitute a major component of its opening up to the outside through individual mechanized modes that allow mid- to long-distance connections (on an urban scale). In addition, local parking operations, places available close to activity destinations, and traffic conditions for accessing these places all play a part in the local operation of the neighbourhood as a fabric of social and economic activities.

We recommend the following planning principles:

1 Detailed coverage: spread overall capacity throughout the neighbourhood by ensuring detailed coverage.
2 User pays: regular parking users should have a private place for their vehicle and so invest in their own facilities. This goes for residents and companies' agents and customers.
3 Multimodal transport service to go with the road system: parking facilities are a transport infrastructure that should be coupled with circulation routes for the same mode. Two-wheelers integrate into a neighbourhood more successfully than cars, and the neighbourhood layout should encourage them.
4 Sizing in line with the intensity of activities in the neighbourhood and its multi-modal transport service: a priori the public transport service saves on parking for visitors and residents who choose not to equip themselves with a private vehicle.
5 Operation modes that avoid internal or external diversion, reduce search traffic, eradicate unauthorized parking, to ensure that traffic is more efficient for those travelling and in terms of environmental impacts.

In particular, neighbourhoods must not allow their demand for parking to infringe on their neighbours, except when it is advantageous to the latter. A relay park is an example of this kind of set-up, and can involve close or remote neighbourhoods.

The city of Bogota (Colombia) is an example of a modern transport organization design, with successful decentralized parking. The city's major road network is a star centred on the town centre, made up of wide, fast roads. Most of these axes include a couple of lanes dedicated to each direction of Bus Rapid Transit services, belonging to the so-called TransMilenio mode, which is highly efficient as it carries up to 30,000 passengers per hour per lane in its direction. On the main trunk roads, roadside parking is unauthorized, with high fines at peak travel times. Each building much possess its own parking solution: apartment blocks, houses, shops and services. For private or professional visitors, mini public car parks exist on ground level, offering several dozen places payable by the minute and densely spaced to cover the whole territory. In these conditions, potential users see the car as a mode with access stations; public buses are in efficient competition with cars thanks to their hierarchical organization and the way they feed into major axes where they travel at high speed and frequency. Parking does not overflow into the next neighbourhood, parking facilities are integrated into the urban infrastructure, and their private operation participates in local economic activity.

PARKCAP: A SUPPLY-DEMAND MODEL FOR PARKING IN A TERRITORY

The ParkCap model focuses on supply and demand for traffic and parking in interaction with an urban territory. It is devised to simulate mobility policies. Version 1 of the model, which we developed in work for the Chair, deals with a critical period: the morning rush hour, when residents leave their homes and workers arrive to park their vehicle.

Traffic is processed by a static model that assigns traffic to paths on a road network, but gives details of the sub-paths made to search for a place: if necessary, these sub-paths include loops. Parking is modelled by "lot", i.e. groups of places according to a place and management mode. Each lot is demanded by a certain number of candidates, and this determines the probability of immediate parking. In case of failure, users transfer to neighbouring lots according to probabilities that depend on the destination and access conditions. Starting with a target parking lot, the final sub-path is evaluated on the basis of the expected cost of parking and pedestrian access to the destination. Before his trip, each user chooses a target lot and a route from his or her starting point to the target lot, in a rational manner. The traffic and parking system shows a joint state of traffic equilibrium.

a) Application context and objective of the model

Mechanized modes of transport provide a way of efficiently covering space, but occupy more space, in terms of traffic and parked vehicles. In dense urban environments, parking places are a limited resource whose division between demanders needs to be organized and regulated: size of lots, time restrictions, pricing, etc. Parking revenues can be used to develop supply and invest in public car parks off the road, either above ground, underground or in a superstructure.

One parking constraint is the price and time spent searching for a place, and this constraint transforms car trip conditions. It can induce travellers to transfer to a more remote place that costs less or is more immediately available, or to use other modes like two wheelers or public transport. Parking policy is recognized as a major component in a multimodal mobility policy in built-up areas. It includes not just managing places according to location and time, but also sizing in terms of urban requirements (residential, commercial, tertiary, industrial, etc.). In sum, parking systems interact with traffic systems and activity location systems (cf. figure in annex).

The ParkCap model, for "Parking Capacity", jointly simulates traffic and parking for trips on a road network. The model concerns a limited time period, usually one hour, in a quasi-static regime. It is mainly devised for the morning rush hour, when residents are leaving their night-time locations and workers are looking for a parking place close to their work.

The model is applied to an urban area that is sufficiently coherent in terms of car trips and parking capacity, i.e. so that any exchanges with the outside are rare or well channelled. It processes all places and journeys, distinguishing "segment" demands according to the layout of private places (at work or home), destination and solvency. It establishes local macroscopic conditions for occupying lots, trip times, and time spent searching from a target lot, along with individual conditions per trip.

b) Scientific and technical aspects

The interaction between all of the supply and all of the demand is made up of a set of relations and thus the conditions between elements, which are the lots and the journeys. A system state is a supply-demand equilibrium of the traffic when all of the conditions are jointly satisfied.

The ParkCap model establishes an equilibrium state by conciliating physical laws for the occupancy of lots and the circulation of vehicles, with microeconomic laws for users' behaviour when choosing a route and searching for a place.

By destination, each lot constitutes a parking option. By origin-destination pair, a travel option is the combination of a main path from its origin to an "initial target lot", and a final path from this target lot to the destination, to search for a place and achieve ultimate access to the destination. By trip, the main path and the target lot are chosen in line with the destination and the quality of the service, both on the network and in the lots identified by place and management mode. The occupancy of each lot depends on local capacity and demand, in relation with demand on a wider scale and with neighbouring capacities.

These are the *physical laws* integrated into the model:

1   By road section, the individual run time is an increasing function of the flow of vehicles during the period.
2   By parking lot, if the number of requests during the period exceeds the number of places on offer, then the lot is saturated. The ratio of available capacity compared with the number of requests, truncated to the unit, is the probability of success per request.
3   In case of success, the ultimate access to the destination is a sub-path by foot, let us say a pedestrian leg. In case of failure, the diversion to another lot results in an additional journey and thus a time and cost overrun.
4   Per travel option, the ex-ante expected travel time is the sum of the run times throughout the path up to the target lot, and the expected time between the target lot and the destination.

The *microeconomic laws* are the following:

1   Users evaluate travel options through an ex-ante generalized cost in which the physical times per item are weighted by a specific "value of time" that also integrates tediousness, added to the monetary cost (traffic and parking expenditure).
2   At the trip origin, users choose the ex-ante option of minimum cost.
3   If users do not obtain a place in a lot, they transfer to alternative neighbouring lots according to probabilities that decrease with the expected cost between the lot and the destination.

The distinction between the search for a place and the main leg of a car trip can be used to distinguish two scales of territory as experienced and used by the car driver: the local area, better known because actually used to carry out the activity at the destination, versus the area to be run through. This distinction naturally satisfies the respective functional specialization of parking and traffic, each with a specific infrastructure and operation.

The model deals with these specific features by distinguishing subsystems, each of which is the object of a sub-model. The connection between these two scales is attained by the notion of a zone of lots associated with a particular destination. This kind of lot zone is different from a public transport station due to the sophistication of the search journeys, which can include loops.

In computing terms, these original notions give rise to an additional layer of objects compared to standard components in a model of traffic assignment to a road network (i.e. the layer of node/arc items and the path layer). Processing them also requires multiplying the algorithmic complexity in time and space: the factor is the square of the average number of lots per segment of demand.

CLASSROOM INSTANCE

The typical case developed by Leurent and Boujnah (2012) shows a frequently requested central lot, surrounded by a first ring of less frequently requested lots and a second ring of non-saturated lots (Figure 6.2). The state of equilibrium is particularly dependent on the volume of trips whose destination is the central lot: a rise in this volume increases occupancy from the centre towards the edge (Figure 6.3), as well as search traffic (Figure 6.4) and the expected cost per trip towards each destination (Figure 6.5).

A more concrete application to the Cité Descartes in Marne la Vallée is reported elsewhere (Leurent and Boujnah, 2014).

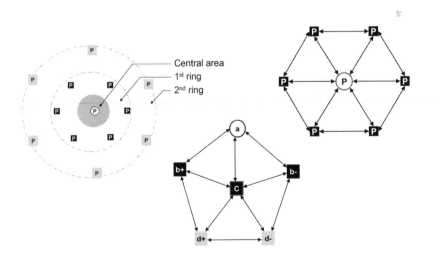

Figure 6.2 (a) General layout, (b) Local transfer network from the centre, and (c) Local transfer network from a destination in the first ring.

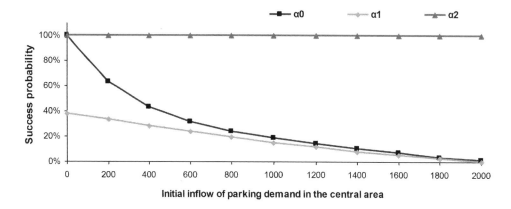

Figure 6.3 Success probability of a request, by type of lot, according to volume of demand for the centre.

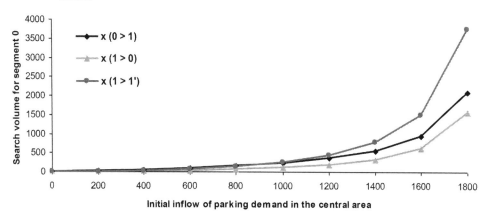

Figure 6.4 Search traffic by type of arc on the network, according to volume of demand for the centre.

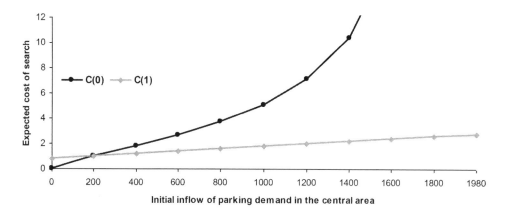

Figure 6.5 Individual cost of search, by destination, according to volume of demand for the centre.

CONCLUSION

In this chapter, we addressed parking as a design object, and in particular an eco-design object. We have set out the different aspects of parking, and the different technical components concerned as well as the stakeholders and concrete modalities. We have presented the economic benchmarks, planning principles, and a simulation model that is sensitive to supply and demand in a territorial area.

In addition, the chapter on land use and mobility in the Cité Descartes deals with parking in this area, in a reference situation and according to a project scenario.

The planning principles proposed for parking, thanks to their analysis, could be debated and improved upon; the same goes for the economic benchmarks, to be completed by parking-specific indicators of social impacts.

The simulation model is a design tool to be developed: its first generation concerns a single mode of vehicles and a single time period. The avenues for development include, on the one hand a dynamic extension, in order to simulate the evolution over one day, in particular during the afternoon, often the peak parking time; and on the other hand, inclusion into a multimodal framework, with different individual or collective modes, both public and private, which are jointly available in a territory and result in complementary features and healthy competition.

ACKNOWLEDGEMENTS

The first phase of the EEBI Chair resulted in fertile interaction with VINCI Park, a major international group of parking operators, and with Sareco, a consultancy firm specializing in parking studies. We are grateful to the managers at VINCI Park, Serge Clémente and Lydia Babaci-Victor, for sharing their invaluable experience of relations between clients and operators in different countries, and for their explanations of market innovations in parking services. At Sareco, Eric Gantelet and Jean Delcroix shared their experience of designing local parking to match local needs. We sincerely thank all of them.

REFERENCES

Leurent, F. and Boujnah, H. (2011) Une analyse offre-demande du stationnement. Application à l'agglomération parisienne, In: Proceedings of the ATEC-ITS congress, Versailles, France, 2–3 February.

Leurent, F. and Boujnah, H. (2012) Traffic Equilibrium in a Network Model of Parking and Route Choice, with Search Circuits and Cruising Flows. In *Elsevier Procedia – Social and Behavioral Sciences*, Volume 54, 4 October 2012, Pages 808–821. Available online at: http://www.sciencedirect.com/science/article/pii/S1877042812042590

Leurent, F. and Boujnah, H. (2014) A user equilibrium, traffic assignment model of network route and parking lot choice, with search circuits and cruising flows. Forthcoming in *Transportation Research Part C*.

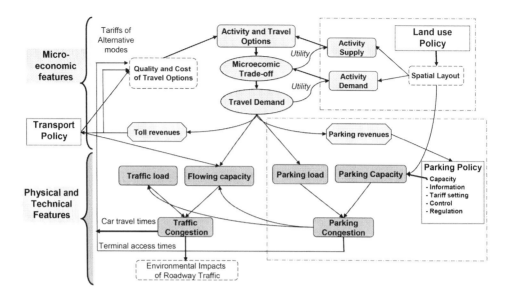

Figure 6.6 Physiology of parking (Leurent and Boujnah, 2011).

APPENDIX: AN ANALYSIS FRAMEWORK FOR A PARKING SYSTEM

This appendix presents an abstract and generic analysis of a parking system, distinguishing between a physical supply subsystem, a demand subsystem and regulation subsystems for parking, multimodal transport and spatial planning (Figure 6.6).

A.1   Physical supply: practical aspects

Within a given area, the physical supply of parking consists of all the parking spaces that are available at a given time, with their respective means of operation and quality of service. This supply is integrated with a multimodal supply of transport, and interacts with highway capacity and other modes of transport.

A.1.1   The different types of spaces and their respective conditions of use

A parking space is characterized by its location, its mode of operation and its availability for a potential customer according to its occupancy status. The mode of operation includes the access conditions, and, in some cases, a space may be reserved for private use (home, firms) or a specific functional purpose (commercial car park, parking spaces reserved for deliveries, etc.), and specific conditions may apply to its use: no charge up to a given duration, or with a charge and a surveillance and punishment system to deal with violations.

The mode of access brings us to the role of the player who owns and operates the space: this may be a household or group of households (co-ownership), a firm for its employees, clients or suppliers, a commercial parking operator, or the non-commercial manager of the road system and off-street parking spaces. This stakeholder has to bear long term purchase or rental costs and the maintenance expenses.

Finally, the use of a space by a vehicle at a given moment is exclusive: a space which is occupied by one vehicle cannot be taken by another.

A.1.2 Capacity and parking load

For each location and mode of operation, the number of parking spaces constitutes the theoretical parking capacity. The number of vehicles occupying them at a given instant constitutes the parking load. The difference between the theoretical parking capacity and the load constitutes the residual capacity which expresses the actual availability for a potential client of that type of place at that location and at that instant.

A.1.3 Quality of service and parking options

The availability, ease of use (required manoeuvres) and the terms of use at a given time determine the quality of service and the price for a potential client who qualifies for a space (i.e. who meets the access conditions), on the basis of his or her desired duration of occupation.

For a given location at a given time, the mode of operation provides a driver who qualifies for a space with a parking option, which is characterized by its quality of service and a price. The occupancy rate, which is the ratio between the load and the capacity, is the degree of congestion for the option. In the case of a widely accessible mode of operation, in particular on-street or car park parking, a high occupancy rate jeopardizes availability and generates transaction costs in order to find a space, consider a series of possible sites, obtain information and then remember the location that has been identified and walking, not just to the end destination, but back again to recover the vehicle.

A.1.4 Combining options, and transfer phenomena

The respective qualities of service and prices of the options oblige clients to perform trade-offs, potentially leading to transfers between different parking modes and parking locations. Transfers are governed by the spatial principle of proximity to the destination and the economic principle of cost minimization and lead to the spatial propagation of parking stocks when these stocks attain the capacity.

The local options for a mode therefore combine to form a single option whose availability is aggregated and whose transaction costs are fairly random.

A.1.5 Interaction with road traffic

In the case of every car trip, a journey has one parking operation at the origin and another at the destination. Together, these three components determine the quality of service delivered to the car as a transport mode. A vehicle's ability to move allows it to search for and access a parking space.

When few parking spaces are available, the search process is lengthened and becomes a specific type of car traffic that reduces the traffic capacity for other types of traffic proportionally. This leads to a slight reduction in highway capacity. A more serious threat is the large reduction in capacity that occurs when a vehicle is parked in a traffic lane, temporarily blocking it.

A.1.6   Linkage with other transport modes

In a car parking situation, the parking component also includes walking, which is an indispensable complementary mode for reaching one's vehicle. Public transport, particularly rail, is also suitable for intermodal links. The creation of park-and-ride facilities, possibly with integrated fares, facilitates such connections.

The other type of linkage is competition between modes, i.e. between the car and alternative modes, such as walking for short distances, two-wheelers for short and medium distances and public transport for medium and long distances. The terms of competition are determined by the qualities of service, fares and transaction costs of the different modes.

A.2   Demand: microeconomic and spatial aspects

Demand is made up of all the demanders, i.e. potential clients and users, who wish to park their vehicle in order to perform an activity under certain spatial and temporal conditions. Demanders make decisions in situations that involve choices at several levels: for a given parking situation, in the larger framework of a trip or an activity schedule, or in the long term in connection with strategic decisions to do with vehicle ownership or locational issues. Each situation is influenced by the demander's activity purposes and specific characteristics.

A.2.1   The choice of a parking situation

There is an obvious hierarchy between different modes of parking based on the quality of service they provide. Reserved parking spaces (at home or at work) come above dedicated parking spaces (when paying a business call or shopping) which in turn come above public parking spaces (public roads and car parks). When reserved or dedicated parking is available it automatically becomes the preferred mode.

The choice is less straightforward when only public parking spaces are available, for example when reserved or dedicated parking spaces are too distant and therefore too costly. In this case the choice will depend on the quality of service, the price and the transaction costs of the different options. When parking costs are proportional to the duration of the activity, as is the case with pay parking, users may decide to shorten their activity.

A.2.2   The choice of a travel option

With a given range of facilities for the demander, the travel options relate to the route, the departure time and the transport mode. In principle, when demanders travel by car, they select the route and the parking space together, modifying the route in order to find a parking space or access a reserved or dedicated space. The selected route also depends on the traffic conditions.

The departure time can also be chosen on the basis of the traffic and parking conditions and how they vary according to the time of day. In particular, arriving at the destination earlier or later may change the parking conditions, in terms of both price and, above all, availability.

For a person with a car, the possible transport modes include a car with parking at the destination (the case of the user driver), walking, a privately owned or self-service motorized or non-motorized two-wheeler (if available), public transport (if there are any services locally) and, perhaps, an intermodal combination of a car and public transport, or even a taxi (public or private) which obviates the need to park or drive.

The available parking options influence the choices that are made with regard to use of a car to reach the destination or a park and ride site. Moreover, the parking conditions at the trip origin influence the other modes that are chosen – users remain responsible for their vehicle even if they do not use it. The same applies, incidentally, to privately-owned or shared two-wheelers.

A.2.3   Activity purpose and choice of destination

In the same way that parking charges can influence the duration of an activity, the price and the quality of service can influence the choice of the location for an activity when this can be changed, as is the case with shopping or some types of services and leisure activities.

Other activity purposes, such as work or personal business, are more spatially and temporally constrained but may provide compensation which influences parking: a reserved parking space, the reimbursement of a user's costs by their employer.

The choice of a means of transport may be influenced by logistical constraints that are inherent to the activity purpose: transporting bulky objects, or activities that involve a group that travels together, with specific costs and a greater preference for privacy which can represent an advantage for the car and encourage the group to park as near as possible to their destination.

More broadly, the availability of a car facilitates activity chaining, and easier parking makes car use, and hence activity chaining, easier.

A.2.4   The specific features of the demander and long-term choices

Each demander has specific exposure to parking conditions, because of their rights to access and use private parking spaces, because as residents they have the right to use public parking, because of the size of their vehicle which may enable them to use a small parking space, or because of their manoeuvring skills or ability to walk, etc.

Demanders are also characterized by their preferences, the trade-offs they make between quality of service and transaction costs. Such trade-offs are influenced by the demander's financial income and time budget, which varies, in particular, between those who work and those who do not. The preference for reliability and certainty influence the decision to have access to a private parking space at one's home.

By owning a car, possessing a parking space (as an owner or tenant) and even choosing a certain type of home (and living environment), demanders gain a set of facilities that determine their short term choices as regards parking, transport mode and activities.

Conversely, when making long term decisions about acquiring access to a certain facility, demanders take a set of short-term consequences (benefits and disadvantages) into account.

A.2.5   Parking pressure is determined by geographical factors

For each location and mode of parking, individual demands to occupy spaces for a certain duration create parking pressure, whose unit is the product of a space and a unit of time. This pressure interacts with capacity and at a given instant is exerted over a certain area.

Some destinations that offer a larger number of activities are more attractive to individuals demanding activities. The quantitative relationship between local opportunities and demanders, hence between land use and expressed demand, is complex as it depends on the spatial configurations of the destinations in relation to trip origins.

Depending on the performance of the transport networks, a destination draws in traffic from an attraction zone, which is defined by accessibility and competition from other destinations. The traffic it attracts is multimodal, making use of all the modes that are available, according to their specific quality and the origin-destination distance. The range of available transport modes thus influences the parking demand generated by the supply of activities.

At night, the supply of activities is essentially produced by households wishing to stay at home to rest. Parking pressure is therefore at the trip origin for the residential purpose.

Last, both at night and during the day, the total stock of vehicles owned determines the total need for parking and hence the sum of local parking space occupation for each mode. It depends both on the size of the population and the economic capacity of the agents (households and firms) and the need for activities which is the social and economic source of demand.

A.3   Operation, regulation and planning

The economic structure of the parking market is as follows: atomized demand i.e. demand which is distributed between a large number of demanders each with only a slight influence; segmented supply that is divided between three sectors, namely, reserved, dedicated and public. Locally, the public parking sector is either a monopoly or an oligopoly, with or without coordination between the different locations. The three sectors are in weak competition in the short term, the main linkage between them being local parking pressure rather than demander choice (as a result of the hierarchy between different modes).

The collective regulation of parking involves the public parking sector in the short term, and all the sectors in the long term. It is one component of multimodal transport policy, and it is also determined by spatial planning.

A.3.1   Operation and regulation of the reserved sector

A reserved parking space is managed individually by the person who holds the right to it, who uses it and makes sure nobody else does. In the long term, an owner or tenant decides whether or not to exercise the right to be a permit holder.

The reserved sector increases the overall capacity of a zone, but only permit holders can use the spaces, on an individualistic basis. In areas with low population densities, the roadside parking space in front of an individual's dwelling, particularly in front of their drive, provides an intermediate way of parking one's car, as, in principle, these are reserved for residents.

Collective regulation of the reserved sector mainly takes place in the long term, during construction of the building, which includes measures to provide access to the road system. In the short term, regulation can involve such access measures and the road in front of the building. It would be possible to levy a specific charge for each parking space, based on the external impacts of use of the vehicle on the road system, but to our knowledge a measure of this type has never been applied.

A.3.2   Operation and regulation of the dedicated sector

The dedicated sector resembles the reserved sector in that permit holders have access to an area with a number of spaces, but differs as regards the number of parking spaces involved. This larger capacity is intended for a more varied group of users, which means it interacts more strongly with the capacity of the public parking sector. The set of spaces exists within a delimited area and once it is built it is difficult to extend it upwards. Planning measures are thus restricted to the initial sizing of the facility.

The aim of day-to-day management is to ensure that parking spaces are available, under capacity constraint: to limit pressure the manager can limit the time of stay and, more rarely, charge for parking (in the case of commercial car parks, and above all, park-and-ride sites).

Collective regulation operates in the same way as for the reserved sector, with greater emphasis on access to the road system and, the possibility of limiting access flows to the car park under normal traffic conditions.

A.3.3   Operation and regulation of the public parking sector

This sector covers parking spaces that are available to the public which are not allocated for any function and which may be on-street or in an off-street car park. In certain cases this type of car park may be commercial i.e. run by an operator for commercial purposes. It is necessary to single out those public parking spaces that are reserved for a given categories of vehicle: deliveries, taxis, shared vehicles, electric vehicles, persons of reduced mobility.

The management of the public parking sector is more influenced by public opinion and more dynamic than the other sectors. The capacity that is available on the road system can be modified, within limits imposed by the layout of the area. Limiting time-of-stay and charging for parking are widely used measures. Privileges may be granted to residents to encourage them to limit their car use – to the detriment of daytime capacity for more productive purposes.

A.3.4   The contrasting goals of a parking policy

Two major goals for parking policy are maintaining free flow for access and capacity constraint. These goals are pursued separately or together by a managing authority, by applying a set of measures.

When capacity is lower than the unregulated demand, in order to maintain free access flow it is necessary to introduce regulation through constraints or pricing or by limiting time-of-stay. For this system to work, effective monitoring and enforcement are required. These are also required in order to prevent illegal parking.

When capacity constraints are applied without measures to ensure free flow, demand produces queues, thereby regulating itself via transaction costs. But this provides no revenue and therefore no economic resources, while consuming valuable resources – the user's time, and wear, energy consumption and pollution due to vehicle use.

Ideally, local parking policies should be coordinated to ensure spatial coherence, with clear signals for users.

A.3.5 Multimodal transport policy

The goal of a multimodal transport policy is to guarantee the accessibility of activities, the mobility of persons and goods with a good quality of service and affordable prices, while limiting negative impacts whether environmental or social (inequity, segregation). Controlling parking, in the short term in the case of public parking and in the long term for the other types, provides a way of channelling demand for activities towards the locations and transport modes that satisfy a criterion that takes account of all the issues, in particular technical and economic efficiency.

To achieve this, a spatially coherent policy is required that is attuned to the geographical structure of the area in question.

A.3.6 Land use and spatial planning

Spatial planning determines how public space is organized and allocated and influences the location of social and economic agents. These are independent entities and it is their decisions that are responsible for assigning functions to buildings, for the supply and demand for activities, and, finally, for the physical configuration of the area in question.

In a similar way, most planning projects are designed by the private sector, i.e. spatial developers. The spatial planner supervises these projects, imposing built-surface ratio limits, land prices, the preservation of space, and, in some cases, the transport conditions including the parking conditions, at the design stage prior to implementation. The spatial planner also regulates parking capacity and the access distances to activities. It also controls the functional allocation of land and built-surface ratios, and thereby regulates the supply of activities and, hence, the demand for activities and therefore parking.

Chapter 7

Eco-operation of road traffic

Vincent Aguiléra
Paris-East University, City Mobility Transport Laboratory, Ecole des Ponts ParisTech,
IFSTTAR, UPEMLV

INTRODUCTION

Road transport is the source of numerous environmental nuisances and/or negative economic externalities, e.g. fine-particle air pollution, time lost in jams and accidents. Some of these effects can be dealt with, wholly or partially, by considering a vehicle or infrastructure subsystem in isolation. Thus, car manufacturers are constantly improving the safety and energy efficiency of cars. Similarly, both the design and maintenance of road infrastructure benefit from progress made in road engineering research (e.g. size of roads, design of intersections) and road security (e.g. crash barriers, concrete central dividers, surface drainage, etc.). However, some typical road traffic phenomena, particularly traffic congestion, are essentially systemic, in that they naturally result from interactions between system components, and more or less from the individual characteristics of each component. A traffic jam forms when the density of the flow of vehicles present at a given place and time exceeds the infrastructure's local capacity to clear the flow. This local capacity overshoot (in time and place) may be exceptional and unpredictable when it is itself the result of an exceptional, unpredictable event (e.g. accident, storm). It may also be recurrent and thus predictable. This is the case, for example, for congestion phenomena observed on road networks in major cities during morning and evening rush hours, or on inter-urban networks during major summertime movements.

This chapter, entitled *eco-operation of road traffic*, presents a set of studies undertaken as part of the Eco-design Chair's work. All of the studies look at recurrent congestion phenomena, by modelling the dynamics of interactions between transport supply and demand, and from this point of view the operation of the networks. We can therefore take as constant both the road network itself (i.e. no new infrastructure) and the spatial structure of the activities at the origin of the journeys. For example, we can assume that the locations of jobs and housing are invariable in the case of metropolitan networks. Within this framework, the overall volume of demand is constant. The only thing that can vary is its distribution in space (users are assumed to be free to choose their route) and in time (users are assumed to be free to choose their departure time and under imposed constraints at the arrival time). Supply is also likely to vary over time: the price of tolls, maximum authorized velocity, route information, and capacity are all infrastructure characteristics that can change over the course of a day, and consequently constitute levers for action to modify the distribution, in space and time, of network users, by influencing the choices of departure time and route.

This chapter is organized into three sections followed by a conclusion. The first section looks at the congestion phenomenon. It starts with a very simple model of a stream of vehicles in a lane to illustrate the fundamental notions of traffic engineering, including the notions of capacity and quality of service. We then stipulate the notions of supply and demand on road networks, leading to a table at the start of the second section summarizing the alternatives available to operators for managing traffic on their networks. Each variable identified is accompanied by a concrete illustration of the operational systems used to manage traffic. We focus in particular on systems with dynamic capacity management, dynamic modification of authorized velocity, user information, and variable pricing. In the third section, we present the research carried out at the LVMT with support from the Eco-design Chair, with an aim to facilitate design and evaluate innovative operating strategies.

CONGESTION PHENOMENON

Obviously, only a limited number of vehicles can simultaneously occupy a given section. When a section is at maximum occupancy, bumper to bumper, we can easily imagine that the velocity of the flow of traffic can only be nil. Traffic engineering studies carried out since the 1930s have identified relationships between variables that describe the flow of vehicles, typically the occupancy rate, velocity and flow. These relationships, called the fundamental diagram in traffic engineering, are set out in a), based on a simple flow model and experimental measurements. The notions of capacity and quality of service are then introduced in b) and c) respectively. Section d) extends the analysis framework to the scale of a whole network, introducing the notions of supply, demand and choice of route. The content of this first section is aimed to identify, at the start of the second section, the main alternatives available to operators to limit the impacts of recurrent congestion phenomena.

a) Fundamental diagram

Let us take a lane comprising n vehicles following each other one after the other without being able to overtake on a section of length L. The vehicle with index i moves at velocity v_i. Note that d_i is the distance that separates vehicle i from the preceding vehicle, with index $i + 1$. To avoid any collisions in the lane, the controller of vehicle i (whether it is a human or an automaton matters little here[1]) must maintain an inter-vehicle distance of d_i above a critical value whose value has a lower limit of quantity $\tau_i \cdot v_i$, where τ_i is the reaction time of the controller of vehicle i. If the kilometric density of vehicles $\rho = n/L$ is low, the controller of each vehicle is free to choose its velocity. This is no longer so when the kilometric density of vehicles is high. In this case, the inter-vehicle distances d_i are constrained. Since τ_i are constant, the individual vehicle velocities v_i become constrained. Over and above the critical density, the interplay

---

[1] Automated devices for longitudinal vehicle control (velocity and inter-distance) are available on the market. For example, ACC (Automated Cruise Control) combines standard speed control with a radar that can estimate a vehicle's relative distances and velocities and thus modify the regulator's speed limit.

of mutual inconveniences impairs individual velocities and overall quality of service. From now on, we will take the characteristic size of vehicle density to be occupation rate k rather than the kilometric density ρ. For a section of length L occupied by n vehicles, we obtain $\rho = n/L$ and $k = \sum l_i/L$. Taking $\bar{l} = \sum l_i/n$ as the average vehicle length, we obtain a simple relationship between the density and the density: $k = \rho\bar{l}$. The occupancy rate variable has two advantages. Firstly, it is a dimensionless variable that varies in the range of $[0; 1]$. Secondly, the sensors that are currently most frequently used to measure road traffic are electromagnetic loops inserted into the road surface. These loops essentially measure the presence/absence time of metallic masses passing over them, which directly corresponds to the density of the road surface.

Below a critical value k_c of road surface occupancy, vehicles' individual velocities are free. If we assume that they are on average equal to the maximum authorized velocity v_{max}, we can expect that the flow rate q of the traffic flow, a product of density and flow velocity, varies linearly with the occupancy rate. Indeed, in these conditions, and equating the flow velocity v with the spatial average of vehicle velocities $\bar{v}$, we obtain:

$$q = \rho v = \rho\bar{v} = \rho v_{max} = \frac{v_{max}}{\bar{l}} k$$

Beyond the critical occupancy rate, the velocity of the flow varies depending on the occupancy, but the flow always varies linearly with the occupancy. In this case, we obtain:

$$q = \rho v = \frac{v}{\bar{l} + \bar{\tau}v} = \frac{1}{\tau}(1 - k)$$

This simple model is consistent, on average, with experimental observations. The curves on Figure 7.1 represent the projection in the planes kv and kq of velocity, flow and occupancy measurements. These measurements are made by electromagnetic traffic loop sensors. The section considered is around 500 m long and comprises 6 traffic lanes. It is located east of Emeryville, on interstate highway I-80, in the region of San Francisco, California. As part of the NGSIM project, data from 8 stations equipped with one sensor per lane spread longitudinally and homogeneously along the section were recorded for ten consecutive days. Every 30 seconds, a sensor measured the velocity of the traffic flow, the occupancy rate, and the vehicle flow. The points represented in Figure 7.1 are, for each density value from 1% to 70%, in 1% steps, the average of all of the recordings of velocity and flow observed for this occupancy rate. For both flow and velocity, we observe that these average points are organized in a regular curve. For the flow, the curve increases linearly for low-density values ($k < 10\%$). It also decreases linearly for high-density values ($k > 20\%$). The flow reaches its maximum for a critical density k_c at around 17%.

The curve that emerges in the density-flow plane is called the fundamental diagram. It dates from experimental research done by Greenshields in the 1930s and is now the basis of traffic engineering. Thus, hydrodynamic traffic models, which model traffic flows as a continuous fluid and aim to resolve a system of differential equations connecting flow, velocity and occupancy, often use the fundamental diagram to express a functional relationship between occupancy and flow. This remains controversial,

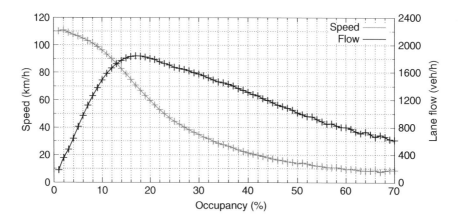

Figure 7.1 Experimentally observed relationships betweem density and velocity (red curve), and density and flow (blue curve). Data NGSIM-I80.

and justly so. The fundamental diagram observes that a functional relationship exists between occupancy rate and the average values of instantaneous flows. It is risky to deduce that a functional relationship exists between occupancy rate and instantaneous flows[2].

We can nevertheless deduce from the fundamental diagram two fundamental notions for operating road networks, i.e. capacity, discussed below in b), and quality of service, discussed in c).

b) Capacity

We can see on the occupancy rate/flow graph in Figure 7.1 that the average instantaneous flow has a maximum value of close to 2000 veh/h/lane. This value is a characteristic volume of motorway traffic that can be found in most traffic engineering reference books, in France and elsewhere. To refine, we could take 1800 veh/h/lane for an inter-city motorway and 2200 veh/h/lane for an urban motorway. Where does this magical constant come from? A quick look at the simple model set out above shows that the variable that essentially determines the size of the fundamental diagram is the average reaction time of drivers. In fact, the model's critical flow is given by:

$$q = \frac{1}{\tau}(1 - k_c) = \frac{v_{max}}{l}k_c$$

For an average reaction time of 1.5 seconds, an average vehicle length of 5 metres and a maximum velocity of 113 km/h (70 mph), we attain a critical flow of around 2200 veh/h. The statistical analysis of measurements on the section from the NGSIM-180

---

[2]The proceedings of the Greenshields Symposium (Khüne 2011), organized in 2008 by the Transportation Research Board for the 75th anniversary of the fundamental diagram, give a good idea of the diverse approaches to traffic theory.

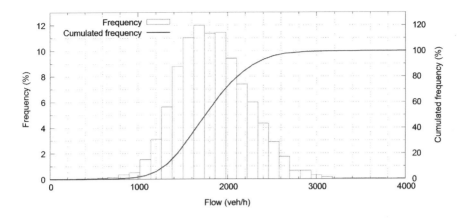

Figure 7.2 Distribution of instantaneous flows for a density of 17%. Data NGSIM-I80.

data set confirms that the flows measured rarely exceed this critical flow. Figure 7.2 represents the distribution, in frequency and cumulated frequency, of flow values observed on the NGSIM-180 data set, for an occupancy rate of 17%. We can see that the flow can reach very high values, admittedly rarely, of up to 4000 veh/h. The median value, at slightly below 1800 veh/h, is lower than the average value, which is slightly higher than 1900 veh/h. The model's critical flow, 2200 veh/h, is located in the last two deciles.

This justifies the concept of a section's traffic capacity, understood to be a "rarely exceeded flow value", while underlining the practical difficulties of definition and measurement. Leurent (2005) has critically reviewed several empirical definitions of capacity (as used in road engineering in practice), a rigorous measurement method based on a probabilistic model, and complementary bibliographical avenues on this notion of capacity, which is more complex than a simple model implies. We should also point out that the notion of capacity "internalizes" numerous factors. Some are constant and very directly linked to the road, e.g. number and geometry of lanes, surface quality, roadside developments. Others are variable and/or dependent on external factors, e.g. traffic composition, such as the percentage of heavy-goods vehicles, climatic conditions and hours of sunlight. The neighbouring environment can also have a very direct impact on capacity by influencing drivers' average reaction time, which we have seen is a variable that essentially sizes the fundamental diagram. For example, on motorways, experience shows that an accident in one traffic direction distinctly modifies traffic conditions in the opposite direction, and these modifications can be put down to the simple variation in drivers' average reaction time.

c) Quality of service

Looking once again at the occupancy rate/flow graph in Figure 7.1, we can easily verify that, for a given flow value below the critical flow, two operating points are possible. One corresponds to free traffic, with an occupancy rate below the critical occupancy rate and high traffic velocity. The other corresponds to congested traffic, with an occupancy rate above the critical flow, and low traffic velocity. Beyond the critical occupancy rate, the further the flow moves away, in higher values, from the fundamental diagram, the more instable the traffic flow becomes. It enters into an "over-critical" state, in which any disruption is likely to propagate further back in the traffic flow and amplify. The back end of a traffic jam then moves upstream faster than the front end does, causing violent waves of braking that propagate through the network and provoke long-term drops in capacity. It is in the operator's interest, both to make operations easier and to offer the best possible service to users of the infrastructure, to maintain flow and density below the critical threshold, and in any case to characterize the level of service for users.

Different levels of service exist depending on the country. We use definitions of six levels for inter-city motorways, rated A to F, taken from the American Highway Capacity Manual. The advantage of these definitions is that they associate an occupancy rate with a user's particular experience and consequences for the operator in case of an incident. They can be used to understand the fundamental diagram in Figure 7.1.

- *Level of service A*. Occupancy rate of around 4%. Vehicle velocities are very close to the speed limit. Drivers are able to change lane as needed. The other vehicles have little or no influence on driving. Driving comfort is at a maximum. No jams occur in case of significant disruption (accident).
- *Level of service B*. Occupancy rate of around 7%. Vehicle velocities are close to the speed limit. Changing lanes requires greater care. Driving comfort is at a high level. No long-term jams occur in case of significant disruption.
- *Level of service C*. Occupancy rate of around 11%. Changing lanes requires particular care. Inexperienced drivers feel uncomfortable. Speeds start to decrease. Jams are likely in case of significant disruption.
- *Level of service D*. Occupancy rate of around 17%. The flow is on the edge of stability. Changing lanes becomes difficult. The comfort of all drivers decreases. The level of vigilance required is high. A slight disruption can provoke a traffic jam. Brake waves can start downstream.
- *Level of service E*. Occupancy rate of around 25%. Capacity is reached. Velocities remain relatively high, but flow is very instable. Changing lane is difficult. Drivers experience significant discomfort. Constant caution is required. Any disruption downstream, even minimal, is transferred upstream and tends to amplify.
- *Level of service F*. Occupancy rate exceeds 30%. Frequent total stops of different durations are required. This level of service is encountered when demand structurally exceeds capacity.

d) Supply, demand and route choice on a network

A road network generally offers users several routes for making a journey from a starting point to a destination. Each route can be seen as a sequence of homogeneous sections, each of which is characterized by a number of attributes, and each of which can influence a user's choice of route. Thus capacity, authorized speed, road surface quality, landscape quality, and operating modes (e.g. toll) are attributes that, when put together during a journey, can explain the mechanism of users' route choices. Some attributes are independent of the class of vehicle, for instance, capacity and landscape quality. Others depend on the class of vehicle, for instance, authorized speed and toll charges, which are not the same for regular vehicles and heavy goods vehicles. Some attributes vary depending on the route choices made by other users. This is the case for journey time, quality of service, and more generally all factors that depend on density. In addition, networks accommodate heterogeneous users. These users can be differentiated not just according to the class of vehicle they use, but also according to their own preferences and/or constraints, i.e. starting point/destination of journey, departure time, constraints on arrival time, preferences in accentuating sections and routes.

Travel demand on the network is made up of the sum of individual journeys, each of which is characterized by a class of vehicle, a starting point, a destination, a departure time, etc. For each individual journey, the network supply is made up of all of the possible journeys between the starting point and the destination. The interaction between supply and demand acts in two stages. The first stage, for journeys considered in isolation and all else being equal, involves comparing the routes available, in line with users' preferences and/or constraints. In the second stage, with fixed route choices, flows are distributed over the network by starting point-destination-departure time and run for the length of the routes. In this way, because of congestion, local conditions of flow in the sections can modify certain attributes, e.g. journey time and comfort, thus invalidating some initial route choices. The supply-demand equilibrium is characterized by the coherence between the state of the network (which results from the distribution of demand on the routes) and the distribution of demand over the routes (which results from the state of the network).

TRAFFIC MANAGEMENT

In the light of the above, this section starts with a list of the main alternatives available to operators to limit the impacts, on traffic[3], of recurrent congestion phenomena, e.g. capacity, velocity, time structure of demand and route choices. We then present the operational systems currently in use.

Capacity. Congestion is a phenomenon whose origin, on a network scale, is occasional. At a given point in time, at an intersection or on a section, traffic demand exceeds capacity. Appropriate management of intersection capacity can be used to optimize usage. This point is covered in a). The specific case of access control is treated in b). Section capacity can also be dynamically adapted: dynamic allocation systems for lanes are presented in c).

Velocity. The maximum authorized velocity is also an interesting control device. As seen in section 1, this parameter controls the slope of the free part of the fundamental diagram. Reducing this velocity moves the critical density to the right and attenuates the wave break. Speed regulation throughout a journey is the subject of d).

Time structure of demand and route choices. We saw in d) that demand, at any point in the network, results from the sum of route choices made by users, and traffic conditions along the length of the routes. By informing users, operators can modify certain user choices, in particular routes and departure times. User information devices

[3]The evaluation of other impacts, especially environmental ones, is discussed in the third section.

are covered in e). Lastly, price signals can be effective. The issue of congestion charges is the subject of f).

a) Intersection management

When different streams of vehicles come together at an intersection, its flow capacity is significantly reduced. Cyclically attributing capacity to antagonistic streams can avoid conflicts and improve the capacity usage rate. Traffic lights can be used to separate in time the admission of different streams of conflicting vehicles. In urban environments, traffic lights can be used for objectives other than simply regulating traffic made up of individual vehicles, e.g. protecting pedestrian crossings, giving priority to public transport lines. The capacity allocated to an entrance is, at an initial approximation, equal to the section's capacity, multiplied by the ratio between the green time allocated to this entrance and the duration of the lights' cycle. This capacity is optimum when the cycle is calculated so that each entrance is allocated with a green time proportional to the flow of demand. The following strategies are frequently used.

The *fixed time* strategy is the simplest. Signal programmes are pre-calculated using historical data, then inserted into the light control systems. In general, for a given junction, several signal programmes are used, each corresponding to a typical period in the day, such as morning or evening rush hour, quiet periods. In *adaptive* control strategies, traffic lights adapt to real-life traffic conditions. Controlling a junction is thus technically more complex. Traffic sensors are added to the light controller (usually traffic loops) that can estimate demand at each entrance and then choose a signal programme that matches supply closest to demand. Lastly, in a *coordinated* strategy, rather than installing signal programmes independently for each intersection, they are installed to take into account a set of neighbouring intersections, generally along a main road (e.g. green band strategy).

b) Access control

The principle of access control involves preventing the occupancy rate of a main road from reaching critical level, by reducing the flow from side roads as needed. Side roads are thus used as temporary storage zones. The aim is to maintain the main road in a stable operating regime, below the critical density threshold. In practice, access control is usually done using traffic lights on side roads. The supply flow is converted into the green phase duration. Isolated access control employs data from sensors located near the access, in particular data on the flow of the main road before the access, and the density of the side road after the access. The supply flow on the side road is calculated as the residual capacity on the main road. Ideally, real-time data should be input into access control. When such data are not available, static control using historical data may be employed. Access control tends to encourage flow on the main road. When accesses are regulated independently from each other, those located further down are penalized *de facto*. Multi-access control can be used to ensure a degree of fairness to users.

c) Dynamic lane allocation

Continuing with the idea of locally adapting supply to demand for movement, the allocation of lanes to a particular traffic direction can be altered. Several types of dynamic lane allocation exist.

Reversible lanes. In some cases, central lane(s) can be used for traffic travelling in one direction or another. This dynamic allocation is done using lane allocation signals. Numerous bridges and tunnels are equipped with this apparatus. See Wolshon & Lambert (2006) for the state of the art.

Using the hard shoulder. The hard shoulder is transformed into an additional lane the use of which is authorized during heavy traffic and when security conditions permit. For the rest of the time, this auxiliary lane is reserved for emergencies and must remain free for vehicles that have broken down or been involved in an accident, and for allowing emergency vehicles to pass. In most countries that employ this solution, users are informed by overhead illuminated signs when the lane is open to traffic, as for reversible lanes. Specific radars usually monitor breaches. A different solution, using a mobile barrier system, has been set up in France on the shared A4-A86 road section. The system has so far not displayed beneficial operational properties. The barriers are fragile, resulting in long periods of unavailability.

Reserving lanes for vehicles with high occupancy. The operational implementation of lanes reserved to vehicles transporting several passengers (HOV, High Occupancy Vehicles) was used extensively in the United States in the 1980s. Given the low number of HOVs using these lanes, the concept of HOT (High Occupancy Tool) quickly emerged. In this case, all vehicles are authorized to use reserved lanes, but those that do not carry the minimum number of passengers (either 2 or 3) have to pay a toll. In France, an early experiment authorizing bus traffic on the hard shoulder was carried out in 2007 close to Grenoble on the A48.

d) Velocity control

Dynamically modifying speed limits in line with traffic conditions is not a new concept: the highway code sets out different limits depending on weather conditions; lowering the speed limit is common practice in urban areas during air pollution peaks. However, lowering the authorized speed in order to control traffic is a relatively recent idea. Velocity control aims to spread traffic congestion over the sections before a main corridor. In France, speed restrictions are most frequently applied in the summer on the A7 and A9 roads (ASF network). These two motorways on average accommodate 75,000 veh/day, with peaks of 165,000 veh/day in summertime. The system, which was devised and set up by ASF, was originally installed between Vienne and Orange on a 250 km stretch. It is based on a traffic flow model that anticipates the risk of a jam 30 minutes in advance. Users are informed in real time through messages posted on variable message panels (VMPs) and the motorway radio station. The results, as evaluated by the operator, appear to be highly successful, i.e. 90% of drivers respect the 110 km/h limit, flows increase by 15% to 20% during rush hours, drop of 20% to 30% in congestion size and numbers of accidents.

e) Guiding users through information

The main means of informing road users is fixed signalling, which includes direction signalling. This medium provides static information. Historically, the first medium used to transmit dynamic road information was the radio, which still works very well. In the 1980s, dynamic signalling systems started to emerge: VMPs. At the same time, the radio developed a standard, the RDS-TMC, which transmits numerical data either to drivers (posted on an on-board screen), or to on-board navigation systems. Until the 2000s, road network operators produced most of the user information. Since then, the development of wireless communication has led to huge changes. Increasingly, users can now get access to a range of information sources that extend much further than a single operator.

Direction signalling is devised to guide users in prime comfort and security conditions, day and night, in urban and inter-urban environments, from their starting point to their destination. The hierarchy of networks and the "unambiguous" design of direction signalling thus encourage a single route for each starting point-destination pair, called the operating route. In congested circumstances, the operating route is not necessarily the best solution for users. It is therefore necessary to provide information on possible alternatives. In parallel with routes conveying major seasonal migrations, alternative routes, called *Itinéraires Bis* are often set up to absorb some of the traffic from the main routes. The alternative routes are marked out in a specific and permanent manner, except for at the entrance point, which is only opened in case of need. When open, users are encouraged to use alternative routes via standard means (e.g. radio, VMPs).

Dynamic route signalling is used on urban networks, often sufficiently meshed, instrumented and congested so that alternative routes can emerge during rush hours. This measure is applied on the expressway network in the Paris region. It is based on the SIRIUS real-time traffic control system.

f) Congestion charging

In congestion situations, individual users are subject to a waiting cost, and also cause others to endure a waiting cost. The marginal social cost of congestion is the waiting cost that a user, through his or her simple presence, causes all of the other users to endure. Let us imagine a queue comprising n users of a service of which the process time per user is constant and equal to τ seconds. The total private cost, which is the sum of the individual waiting costs, is $\tau n(n + 1)/2$, and the average marginal private cost is $(n/2)\tau$. If an additional dishonest user edges her way to the head of the queue, she causes each user behind her to lose τ seconds. Since n users are behind, her marginal social congestion cost comes to $n\tau$ seconds. If this same additional user takes his place at the end of a queue, she does not cause any other users to endure an additional delay: his marginal social congestion cost comes to 0. On average, with this example, the marginal private cost and the marginal social cost are equal, and are both worth $(n/2)\tau$. As we saw in section 1.1, in the road domain, speed decreases with density. As a result, journey time increases with density. Following the rationale of the economist Prud'homme (1999), the congestion phenomenon is taken into account by:

Modelling the marginal congestion cost by a curve that increases in line with traffic volume.

Distinguishing the marginal private congestion cost from the marginal social congestion cost, and taking the function of the rising marginal social cost to be faster than the function of the marginal private cost.

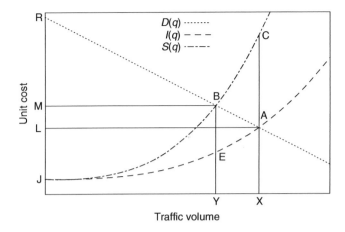

Figure 7.3 Economic modelling of congestion (Prud'homme, 1999).

Now let us consider the diagram in Figure 7.3. The x axis represents the volume of traffic q and the y axis shows the unit costs. The curve $D(q)$ is the demand curve. It represents the use of the road according to the unit cost of using a vehicle. The main component of this cost is the time required to cover a unit of length. The curve $I(q)$ is the supply curve. It represents the private usage cost borne by a user. This cost includes a fixed part, comprising mileage expenses for using the vehicle, and a variable part that depends on the journey time. When users are few in number, this cost tends towards a lower limit given as J. In fact, for low densities, velocity is at a maximum, and consequently journey time is at a minimum. When demand q rises, velocity drops and the journey time increases. An equilibrium is reached at the intersection point A between $D(q)$ and $I(q)$. In A, marginal users gain a benefit from the infrastructure that is equal to their private cost. However, this equilibrium is not optimal. The marginal social congestion cost is represented by the curve $S(q)$. The equilibrium reached at B, the intersection point between $D(q)$ and $S(q)$, is the optimum operation point for all users. This optimum situation can be reached by imposing a congestion charge, equal to EB, so that the private individual cost has the same value as the social individual cost. In this graph the social congestion cost is equal to the area of the triangle ABC. A few examples of congestion charges are given below.

In France, congestion charges are rarely applied in inter-urban situations, except for some bridges, with a summer/winter variation (e.g. Millau viaduct), and in peri-urban areas on several sections (A14, duplex A86). The CCZ (Congestion Charge Zone) in London is an example of charging by band. Access is chargeable from 7 am to 6 pm on working weekdays and vehicles are monitored by automatic number plate recognition. State route SR-91 (Orange County, California), operated by Cofiroute USA, has been extended from 4 to 7 lanes in each direction since 1995. It was the first road to apply a congestion charge in the United States. On central a 16 km section, the three additional lanes in each direction, called Express Lanes, are equipped with open road toll barriers. Vehicles must be fitted with transponders. One of these lanes is an HOV (High Occupancy Vehicle) lane. It is reserved for vehicles with three or more passengers, and also open to motorbikes and zero emission vehicles (electric or hybrid). The toll varies depending on the time of day and the direction travelled.

DESIGN AND EVALUATION OF OPERATING STRATEGIES

For road transport users, recurrent congestion is synonym with losing time, consuming surplus fuel, and prolonged exposure to factors with long-term health effects (i.e. stress, fine particles). The unpredictability of traffic conditions means that users are obliged to allocate a greater time budget to ensure that they arrive at their destination at the chosen time. Vehicles' mutual inconveniences in congested traffic increases the sensation of insecurity, reduces driver comfort, and increases the risk of accident. Lastly, the congestion phenomenon amplifies the environmental nuisances of road transport, in particular noise and emissions of atmospheric pollution and greenhouse gases.

Examples of operational tools for managing traffic were presented in the previous section. These tools can be used to act at different spatial and temporal scales on journey supply and/or demand: locally on capacity and journey length using dynamic speed control; at network scale, modifying departure times and/or route choices by supplying information to users and/or congestion charges. The combined effects of this range of tools are not easy to apprehend, whether the focus is on traffic effects or other criteria in general. In addition, the social acceptability of certain measures is evolving (e.g. speed control, congestion charges) at a time when information technologies are opening up new potential to users and operators.

Engineering innovative operation strategies requires a new generation of decision-making tools that can realistically model supply/demand interactions on networks and provide multi-criteria evaluations of these strategies.

The next part of this section presents recent studies carried out at the LVMT with this aim. The fundamental mechanism common to these studies is that of allocating traffic to the user optimum. A reminder is given in a). We then present in b) an allocation model used to take into account the dynamic information possessed by a portion of users. Section c) gives an analysis example based on the notion of a marginal congestion cost extended to a network-scale dynamic case, in order to determine where and when it is pertinent to act on demand, before congestion occurs. The example used is the Rhône Valley during a summer weekend. Taking this same example, section d) presents a scenario combining time-related toll modifications and choice of departure time. Lastly, section e) looks at pollution emission models.

a) Principle of assigning traffic

Let us take a simple network comprising a starting point o, a destination d, and two arcs, 1 and 2, leading from o to d. We can imagine, for example, that o and d are located on either side of an urban area, that arc 1 represents a bypass route, and that arc 2 represents a route passing through the town centre. The capacity and free journey time of arc 2 are assumed to be lower than for arc 1. The curves in Figure 7.4 represent the evolution in time of the journey according to the flow on each arc: in red, arc 1;

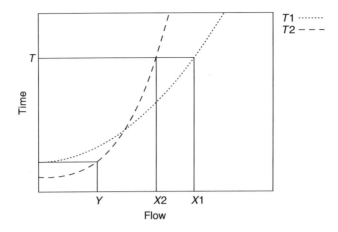

Figure 7.4 Principle of assigning traffic to user equilibrium.

in blue, arc 2. Since arc 2 has a lower capacity than arc 1, its journey time function according to flow increases faster than that of arc 2. The flow of demand between o and d is given as q. A user beginning at the starting point can choose between the two routes. Demand q is thus expressed as the sum of flows on each of the arcs: $q = x_1 + x_2$.

Let us assume that the users are rational and fully informed: they choose the route that minimizes the time (and more generally the cost) they spend on the route. If demand is low (below value Y in Figure 7.4), they all choose route 2. For $q \leq Y$, the flows assigned to the routes are therefore $x_1 = 0$ and $x_2 = q$. As soon as $q > Y$, route 1 becomes competitive: if demand remained totally assigned to arc 2, route 2 would have a longer route time. If a fraction of the demand is transferred to route 1, we find flows are assigned to routes to achieve equal journey times. This is the case shown in Figure 7.4 for demand $q = X1 + X2$.

We have illustrated by example the principle, initially formulated by Wardrop, that bases the assignment of traffic on the user optimum: if a route is taken between a starting point and a destination, then it must be at a minimum cost. The example is based on two arcs. Generally, the problem of assignment can be expressed and resolved at the scale of a network, between several starting points and several destinations. When the dynamic of congestion is only taken into account through flow-time functions, as in the example, we talk of static assignment. When the time dimension of the traffic flow is taken into account, we call it dynamic assignment. Static assignment is expressed as a convex minimization programme. Standard optimization methods may be adapted to take into account the transport network's support graph. The power of current computing means that it is used commonly by engineering firms, including for major metropolitan networks. Dynamic assignment is now on the point of being operational, even though important research questions remain, including in particular the number and stability of equilibrium states. The LVMT has developed a model named Ladta (for Lumped Analytical Dynamic Traffic Assignment) of which the implementation, the LTK (for Ladta ToolKit) was devised to distribute calculations on a grid. This parallelization means that, using greater computing resources, we can attain process-ing times similar to those of static assignment. Examples of application of LTK are given in sections 3.3 and after.

b) User information

This section takes another look at static assignment, and sets out a way of modelling dynamic information for users, using the analysis framework proposed by Leurent & NGuyen (2010). We consider once more the two-arc network seen in section 3.1, this time assuming that demand q, which is constant, to be expressed as the sum of a population of users not equipped with a dynamic information system, of volume y, and an equipped user population, of volume $z = q - y$. Each day, on each arc, a random disturbance occurs. It is assimilated here to a supplementary volume of traffic, given as w. The volumes on the arcs are thus written as:

$$x_a = y_a + z_a + w_a \quad a = 1, 2$$

We assume that in the long term, the user equilibrium is reached, so that the average journey times balance out:

$$E[t_1(x_1)] = E[t_1(x_2)] \tag{7.1}$$

We assume that non-equipped users are informed of traffic conditions on each route through experience acquired over time, in the long term, and that their distribution on the routes does not vary from one day to the next, so that:

$$y_a = E[x_a] - (E[z_a] + E[w_a]) \quad a = 1, 2 \tag{7.2}$$

Let us now assume that we know the distribution of non-equipped users on the routes (i.e. y_1 and y_2) as well as, for a given day, the disturbances w_1 and w_2. We assume that, each day, equipped users are distributed over the routes so that the volumes $x_1 = y_1 + z_1 + w_1$ and $x_2 = y_2 + z_2 + w_2$ result from a static user equilibrium assign-ment. On the graph in Figure 7.4, we look for z_1 and z_2 so that we can minimize the journey times of equipped users, no longer beginning with the starting point, but with the respective x-axis points $y_1 + w_1$ and $y_2 + w_2$. For each day, we deduce x_a, and $E[x_a]$ and $E[z_a]$. An equilibrium is reached when x_a confirm (1) and $E[x_a]$ and $E[z_a]$ confirm (2).

By linearizing (7.1) in the form $\tau_1 + \gamma_1 E[x_1] = \tau_2 + \gamma_2 E[x_2]$, Leurent & NGuyen propose analytical solutions in some cases, and a numerical method using a two-level optimization algorithm, relaxing the volumes of non-equipped users y_1 and y_2. Sev-eral questions are covered, based on users' rate of equipment, in particular the utility (individual, social) of information and interactions between congestion charges and information.

c) Marginal social congestion cost: dynamic analysis at network scale

The notion of marginal social congestion cost was presented in f), in a static context, and in the perspective of congestion charges. The example of a queue was taken to calculate the marginal costs of average, private and social congestion. Looking back at this example, it is clear that the marginal costs vary over time, and in an opposite manner. The marginal private congestion cost increases with the order of appearance in the queue, while the marginal social congestion cost decreases with the order of appearance in the queue. If we take into account the spatial and temporal dynamics of traffic flow on the network, the dynamic analysis of the marginal congestion cost can be used to quantify the interactions between flows from starting points to destinations, in time and space. For the example, we imagine a network in Y, with a queue of vehicles at the start of each branch at the same time. Depending on the journey time of each branch, and depending on the length of each queue, several interaction configurations are possible at the convergence point. If the journey times are very different and the length of the first queue to arrive is sufficiently short, vehicles in the two queues will follow each other in time at the level of the convergence point without interacting. If, however, the journey time and length of the queues are comparable, then congestion may occur at the level of the convergence. In the dynamic case, the marginal social cost of a vehicle depends in particular on the order in which the vehicles in each queue arrive at the convergence point. In Aguiléra & Leurent (2010), we formalized the marginal social cost of dynamic congestion for a queue with variable capacity, as well as for a starting point-destination pair, and defined indicators that can be used to decode the spatial and temporal interactions of traffic flows. The following outlines a case in the Rhône Valley.

The Rhône Valley accommodates a significant share of trans-European North-South traffic. This is seen each year during the major summer migrations, when holiday makers from northern Europe (i.e. Belgium, Netherlands, Germany and the UK) take to the French road network to reach (or return from) tourist zones in southern countries (i.e. Italy and Spain). These transborder flows combine with national traffic, a significant share of which comes from the Paris region. The main bottleneck is on the A7 motorway, in particular in the section between the south of Lyon and Orange, for a length of around 200 km. We started off by calculating a dynamic user equilibrium assignment on the entire French road network. The required data (for supply, network characteristics, and for demand, a dynamic starting point-destination matrix) were supplied by the company Cofiroute, in partnership with two other VINCI Autoroutes companies, APRR and ASF. The demand matrix comes from electronic ticketing data. It comprises 628 starting point-destination pairs for the date 14 July 2007. The traffic conditions, resulting from a dynamic user equilibrium assignment, are represented in Figure 7.5 at four different points in the day: 00.30, 06.00, 12.00 and 18.00. We were able to verify, by comparing with data from sensors at 200 measuring stations spread over the network, that the simulated flows provide a realistic representation, for the whole day, of the flows measured. Two zones are characterized by high congestion levels during a significant proportion of the day: one is the Parisian region, a continual source of high traffic volume; and the other is the Rhône Valley, where North-South flows come together. For each arc of the network, and for every instant of the day, the assignment can be used to disaggregate flow by origin, destination, and departure time from the starting point.

The criticality indicator $\chi_{od}(h)$ measures in min/km the marginal social cost of a starting point-destination pair for a departure from the starting point at time h. The maps in Figure 7.6 show the evolution of this indicator during the day, at 0.00, 06.00, 12.00 and 18.00. The orange lines correspond to the starting point-destination pairs

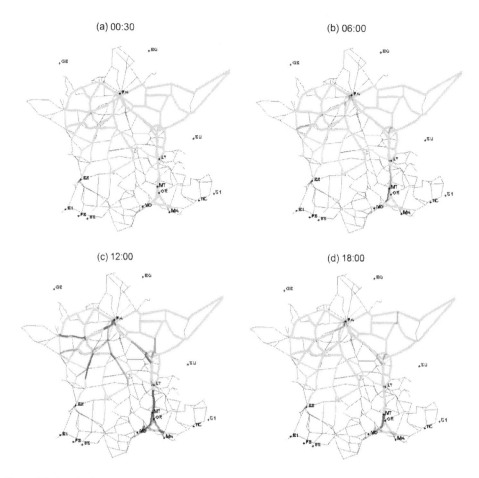

Figure 7.5 Level of congestion on the arcs, resulting from dynamic assignment, depending on the time of day. a) 00.30. b) 06.00. c) 12.00. d) 18.00. Green shows fluid traffic. Orange shows dense traffic. Red shows congested traffic.

that have a positive criticality indicator below 1 min/km; the red lines correspond to the starting point-destination pairs with a criticality indicator above 1 min/km. These maps can be used to determine where and when measures aimed to limit congestion can be applied. For example, on Figure 7.6a, North of Lyon, five starting points, of which two are close together and the other near to Paris, have a positive criticality indicator at 0.00. These starting points emit traffic flows that, when they come together several hours later in the same place, will create significant congestion. The congestion map (Figure 7.5) confirms that at 6.00, congestion points appear north of Lyon and Dijon.

d) Modulation of prices and departure time choice

Badly coordinated departure times can lead to a capacity overload when flows converge at the same instant in the same place. Toll price is a variable that is likely to modify

a) 00:30   b) 06:00

c) 12:00   d) 18:00

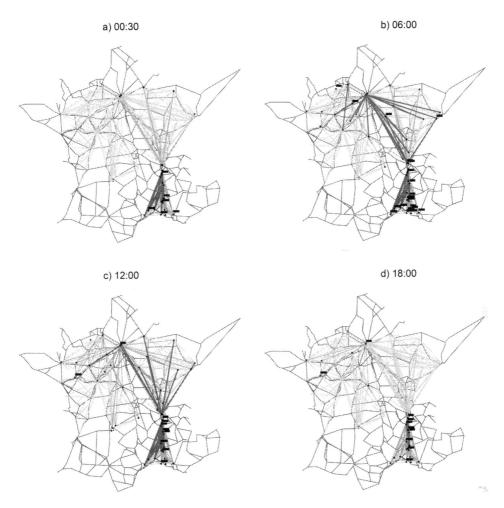

Figure 7.6 Critical SD pairs. Orange shows pairs $\chi(h) > 0$. Red shows pairs $\chi(h) > 1$ min/km.

the choice of users' departure time. This section presents the principle of modelling the choice of departure time, and then the application of an extension of this model to inter-urban network cases.

–   Principle of modelling the choice of departure time

The first studies on the choice of departure time date from Vickrey (1969). We take a volume of n users who would all like to arrive at their destination at the same time b^*. For each user, the total cost of the journey, given as c is expressed according to the time of departure from the starting point, given as h, by:

$$c(h) = \alpha t(h) + \beta d^-(h) + \gamma d^+(h)$$

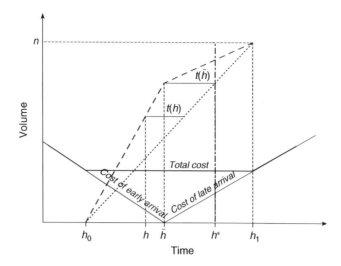

Figure 7.7 Choice of departure time, Vickrey model.

where:

$t(h)$ is the journey time for departure at the instant h.

α is the unit cost of the journey time.

$d(h) = h + t(h) - h^*$ is the function of delay for a departure at the instant h. This is the difference between the expected arrival time h^* and the actual arrival time $h + t(h)$.

$d^-(h)$ which has a value of 0 if $d(h) \geq 0$, and $-d(h)$ otherwise, measures progress to the arrival point.

β is the unit cost of progress to the arrival point.

$d^+(h)$ which has a value of 0 if $d(h) \leq 0$, and $d(h)$ otherwise, measures the delay at the arrival point.

γ is the unit cost of delay at the arrival point.

The problem is to determine the distribution of users over departure times so that each minimizes his total cost, in a hypothesis where the capacity constraint between the starting point and the destination is modelled by a queue whose exit flow is fixed. The problem and its solution are shown in Figure 7.7.

The n users who want to arrive at time h^* are represented by the green segment. The red segment represents the progress of volume at the queue exit, in relation to time. The slope of this segment is the maximum flow at the queue exit. We try to determine, on the one hand, the time interval between departures $[h_0; h_1]$ so that all departure times in $[h_0; h_1]$ are isocosts, and so that the departure times outside $[h_0; h_1]$ have a greater cost than the times inside them; and on the other hand, the distribution of users inside $[h_0; h_1]$, in other words the function of input volume.

Let us assume that a departure time exists in $[h_0; h_1]$, given as $\bar{h}$, which makes it possible to arrive at the expected time h^*. By definition, the delay $d(\bar{h})$ is nil. The journey time $t(\bar{h})$ is thus necessarily at a maximum. Between h_0 and $\bar{h}$, arrivals are early. The cost of earliness has to be compensated by a reduction in the journey time.

Similarly, between $\bar{h}$ and h_1, arrivals are late. The cost of lateness must be compensated by a reduction in the journey time. Because these variations, whether early or late, are linear, it is easy to determine $[h_0; h_1]$, $t(\bar{h})$ and $\bar{h}$ according to the problem data, which determines the volume profile of entrances, represented in blue on the graph in Figure 7.7. The problem has been extended to deal with urban networks, with several starting points and several destinations. Most of these extensions (e.g. see Heydecker & Addison, 2005 or Palma & Lindsey, 2006) consider normal distributions of arrival times at the destination, and a cost model for the user that basically follows Vickrey's model.

– Modulation of prices and choice of departure time: application to the Rhone Valley

The basic model and its extensions mentioned above are not suitable for inter-urban networks, especially during summer migrations. In this case, users differ significantly, both in terms of their expected arrival times (some want to arrive in the morning, others in the evening, and in any case, the hypothesis of a normal distribution around a target time, which is justified for peri-urban networks, no longer holds), and in terms of their evaluation of a delay at arrival (in some cases, lateness is not an option, e.g. if a ferry needs to be caught). Lastly, a non-negligible proportion of users are potentially susceptible to change the day of departure (to the day before or after), especially if toll prices are more attractive.

In Aguiléra & Wagner (2009), we proposed an algorithm for resolving the problem of choosing a departure time on a network, so that we could take into account any volume profile to express the desired time of arrival, along with more complex functions for evaluating the cost of delay than those generally used in published studies. This algorithm was entered into the LTK, and we applied it to the case of the Rhone Valley to study a scenario of toll modulation based on the time of day.

The essential parameters of the time modulation scenario studied are shown in Figure 7.8. The coefficient of time modulation used is show in Figure 7.8a. The time interval [0; 24] corresponds to 14 July 2007. The toll increases from 5.00 in the morning by 12.5%. From 9.00 to 13.00 it is at its maximum rate, with a 25% increase. It then gradually goes down to reach a lower threshold at 65% of the nominal value during the night and the start of the following morning. It gets back to its nominal value at 10.00 the next morning. The sections where the toll price has been modulated are shown in red on Figure 7.8b. Two examples of the function for evaluating delays are shown in Figure 7.8c. The class of users VL2, in red, clearly prefers earliness to lateness, and retains the expected arrival target time. Class VL1, in blue, bears earliness and lateness with comparable costs, if the delay does not exceed 6 hours in absolute value. Beyond this limit, the costs of earliness and lateness become infinite in a 12-hour band on either side. They return to finite values for the preceding day and following day.

The results obtained are shown in Figure 7.9. The left-hand map (Figure 7.9a) illustrates congestion on the Rhone corridor in the simulated reference situation, with no price modulation, on 14 July 2007 at 11.00. The right-hand map (Figure 7.9b) gives the same information with modulation. The curves on Figure 7.9c show the total time flow at the departure from all starting-points: in green, without modulation; in red, with modulation. The peak demand on 14 July is delayed, cut down, and partially moved to the next day.

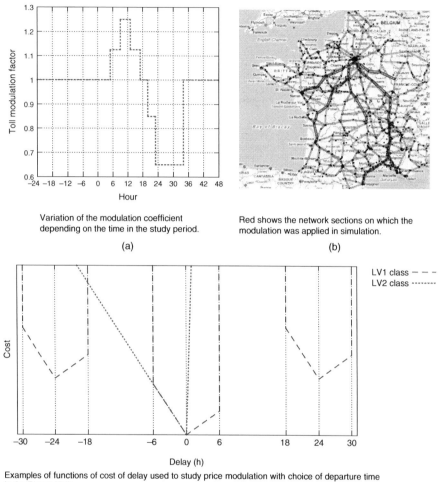

Variation of the modulation coefficient depending on the time in the study period.

(a)

Red shows the network sections on which the modulation was applied in simulation.

(b)

LV1 class — — —
LV2 class ········

Examples of functions of cost of delay used to study price modulation with choice of departure time in the Rhone Valley.

(c)

Figure 7.8 Time-based modulation of toll prices.

e) Emission of pollution

Road transport represents a significant share of the overall balance of atmospheric pollution and greenhouse gas emissions. The potential or proven health impacts are proportionately higher than for other sectors because the toxic emissions are mostly produced in urban and peri-urban areas, close to inhabitants. The main atmospheric pollutants in the road sector subject to regulation are:

- *Nitrogen oxides* (NO_x): the road transport sector emits the most NO_x in France, with 57% of emissions in 2011. Nevertheless, NO_x volumes emitted by road transport have stabilized since 1993 been regularly decreasing since 1996, with

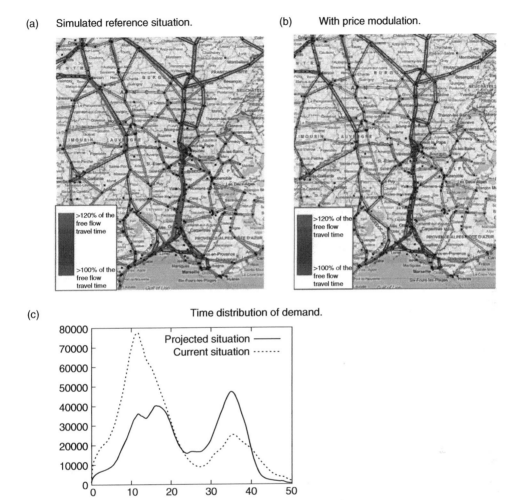

(a) Simulated reference situation.

(b) With price modulation.

(c) Time distribution of demand.

Figure 7.9 Levels of congestion in the Rhone Valley at 11.00 on 14 July 2007, with and without price modulation.

the arrival of catalyzed vehicles. Of the different NO_x, nitrogen dioxide (NO_2) is a pollutant whose concentration in the air is regulated by air quality standards. NO_x either directly, or indirectly (as precursors to the formation of tropospheric ozone), have immediate direct impacts through irritation of the respiratory tract and eyes.

- *Particulates* (PM): the particulates emitted by transport include those emitted from the operation of a motor through exhaust fumes, and those linked to abrasion of the road, tyres and breaks. These are local pollutants whose concentration in the air is regulated. They are classed into several categories according to diameter, in particular: PM10 (particulates with a diameter under $10\,\mu$m); PM2.5 (particulates with a diameter under $2.5\,\mu$m); and PM1 (particulates with a diameter under $1\,\mu$m). In 2011, road transport was responsible in mainland France for 18% of particulate emissions, ranking fourth in the emitting sectors. Exhaust fumes contribute significantly to emissions of the finest particulates, including diesel motors, which mostly emit PM2.5 and PM1. PM1 only comes from combustion. In 2012, the International Agency for Research on Cancer classed exhaust fumes from diesel engines as carcinogenic. With 60% of its cars running on diesel fuel, France is significantly concerned by this situation.

- *Carbon dioxide* (CO_2). CO_2 is the main greenhouse gas emitted by road transport. The sector contributes around one third of total emissions, making it the highest-emitting sector. CO_2 emissions from vehicles are not currently regulated, unlike NO_x and PM. In 2009, the European Union adopted a regulation aiming at an average emission level of CO_2 per car manufacturer of $120\,$g/km for new fleet by 2015.

Traffic emissions respond to three main effects. Firstly, the traffic effect: the greater the traffic, measured in veh.km, the higher the emissions. Secondly, the fleet effect: emissions are closely related to the fuel and technologies used in vehicles' engines. The progressive creation of increasingly restrictive standards has led to considerable progress for some pollutants in Europe. Thirdly, the velocity effect: the standard shape of a curve of pollutant emissions according to speed is a U-shape, with high values for slow and fast speeds.

– Effect of traffic and fleet on emissions

We can start by looking at the joint progress of traffic and road traffic emissions, shown in Figure 7.10 for 2001–2012. During this period, although traffic increased, the overall volume of road traffic emissions went down constantly and significantly, continuing a downward trend that started in the mid-1990s. From 2001 to 2010, according to figures published by SETRA, the total volume of traffic went from $415\,$G.veh.km to $490\,$G.veh.km, a relative rise of 18%. The significant decrease in 2008 can mainly be explained by the impact of the economic crisis on HGV traffic. At the same time, NO_x emissions dropped by over 35%, and PM1 emissions by almost 50%.

These decreases can be explained by the progressive equipping of vehicles with catalytic convertors, the application of increasingly restrictive emission thresholds (EU standards), and the renewal of the fleet. The progress in PM emissions differs according to their size. The most significant drop is for PM1, which only comes from exhaust fumes. The larger the size of the particulate, the less significant the drop. Some of the gains obtained from exhaust fumes are lost due the fact that particulate emissions from abrasion increase with traffic.

– Effect of velocity and fleet on emissions

Independently from the volume of traffic and the composition of the vehicle fleet, the average velocity of traffic also influences road traffic emissions. Figure 7.11 shows the PM emissions of an average vehicle in France in 2007, along with projections carried out by SETRA for 2015, 2020 and 2025. The model used by SETRA to carry out these

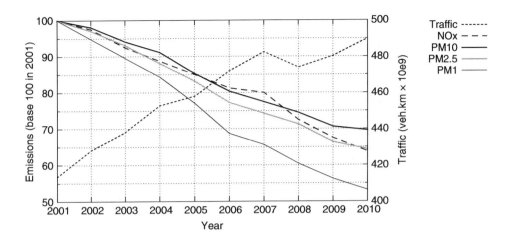

Figure 7.10 Comparative progress of traffic and road traffic emissions over the period 2001–2010. Sources: SETRA for traffic data; CITEPA for emissions.

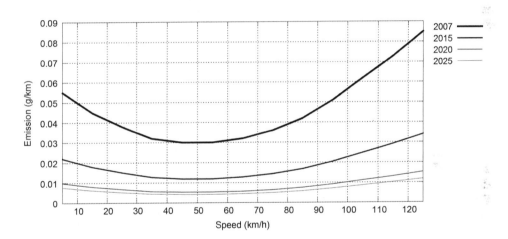

Figure 7.11 Projection of PM emissions of an average car in France according to average velocity. Source SETRA.

projections is the European standard COPERT IV. COPERT defines emission factors for the main pollutants that depend on average velocity, without clearly stipulating what this notion of average velocity means. In general, depending on the average velocity, emission factors present a characteristic U-shaped curve, with a minimum velocity of around 60 km/h.

– Modelling emissions

Calculating emissions on a road network is often done by applying an average velocity model to the flow/velocity data of a network's arcs. In France, this is the case for example for ADEME's IMPACT software, based on COPERT emission factors.

(a) 04.00   (b) 08.00

(c) 12.00   (d) 18.00

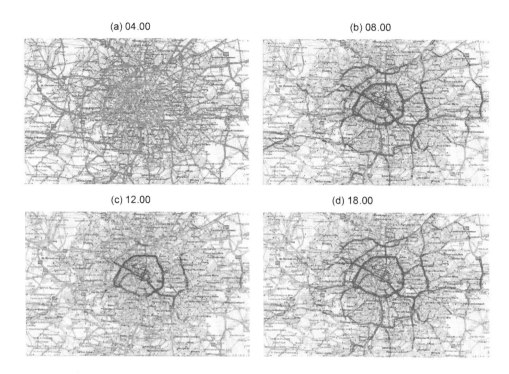

Figure 7.12 Emissions of carbon monoxide calculated from traffic volumes and velocities resulting from a dynamic assignment on the greater Paris road network.

In (Aguiléra & Tordeux, 2013), we showed that, in a congestion situation, this calculation method is not consistent with the emissions calculated by a microscopic model of emissions, which inputs data on individual vehicle trajectories at a temporal resolution of 1 second. Both models produce highly comparable emission curves in free and saturated traffic conditions. The gap widens close to the critical flow: the average velocity model does not correctly take into account the dynamics of congested traffic, and underestimates congestion emissions by around 20%. We proposed a method for disaggregating the average velocity of the flow on distributions of vehicle velocities. By applying the average velocity model to these distributions, we can take better account of the dynamics of congested traffic. The method has a practical advantage in that it does not require additional data.

In addition, we saw in 3.5.1 how the effect of the vehicle fleet influences emission calculations. If we want to model traffic emissions in a more detailed way, in time and space, we need to know the composition of the rolling fleet for each arc and at each instant. We can thus rebuild an image of the rolling fleet on an arc at an instant, by propagating the fleet to the starting points all along the assigned routes. In the case of an urban area, we can see the advantage of this type of process if we imagine how users' vehicle stocks can differ depending on the town.

The maps in Figure 7.12 illustrate a case using the emission calculation module included in LTK. They represent the spatial and temporal distribution of carbon monoxide emissions corresponding to journeys within greater Paris carried out in private vehicles, for a demand corresponding to an average day based on data taken from the Enquête Globale Transport 2001 in the Ile de France (greater Paris) region and the area's road authority, DiRIF.

CONCLUSION

This chapter presented a set of studies carried out in the frame of the Eco-Design chair. The studies all focused on recurrent congestion phenomena by modelling the dynamics of interactions between transport supply and demand, and thus on the operation of networks. The first section looked at the phenomenon of physical congestion and identified the notions of capacity and quality of service. The second section presented a range of operational instruments that operators can use to limit congestion's impact on the quality of service to users. The third section illustrated through examples how research can contribute to the emergence of a new generation of decision-making aids for engineering innovative operating strategies: through realistic modelling of supply/demand interactions on a network, leading to *ex-ante* and *ex-post* evaluations relating to several criteria.

Most of the discussion focused on the recurrent phenomena of congestion involving private vehicles on urban and inter-urban networks. Within this restricted framework, a great number of important themes relating to road network operations were not mentioned, including: accidentology, energy efficiency issues, goods transportation and heavy vehicle traffic, public transport, noise, the reliability of journey times, and the social acceptability of different measures.

In fact, road transport provides a significant share of the transportation of workers and goods. The issue of operating road networks thus lies at the heart of production system operations. The volume of road transport supply is constrained by the joint impact of shrinking financial resources and the maintenance costs of the existing network; as a result, the engineering of network operations is moving towards demand management. In doing so, it is moving away from simple capacity questions, in an approach that necessarily integrates varied issues (e.g. spatial planning, organization of production systems, behaviour of economic agents, etc.) made more complex by the increasing use of information and communications technologies.

REFERENCES

Aguiléra, V. and Tordeux, A. (2013). *A new kind of fundamental diagram, with an application to road traffic emission modelling.* Due for publication in the Journal of Advanced Transportation.

Aguiléra, V. and Leurent, F. (2010). *The Network Cost of Congestion: Analysis and Computation of Marginal Social Cost Disaggregated by O-D Pair and Departure Time.* In Proc. of the 89th Transportation Research Board Annual Meeting, Washington D.C.

Aguiléra, V. and Leurent, F. (2009). *On large size problems of dynamic network assignment and traffic equilibrium: computational principles and application to Paris road network.* Transportation Research Record: Journal of the Transportation Research Board. Vol. 2132, pp. 122–132.

Aguiléra, V. and Wagner, N. (2009). *A Departure Time Choice Model for Dynamic Assignment on Interurban Networks*. Proceedings of the European Transport Conference, Leeuwen, The Netherlands.

Buisson, C. and Lesort, J.-B. (2010). *Comprendre le trafic routier – Méthodes et calculs*. Editions du CERTU. Collection Références, June 2010.

de Palma, A. and Lindsey, R. (2006). *Modelling and Evaluation of Road Pricing in Paris*. Transport Policy (13), pp. 115–126.

Kühne R., ed. (2011). *75 Years of the Fundamental Diagram for Traffic Flow Theory – Proceedings of the Greenshields Symposium*. Transportation Research Circular, number E-C149.

Heydecker, B. G. and Addison, J. D. (2005). *Analysis of Dynamic Traffic Equilibrium with Departure Time Choice*. Transportation Science 39, pp. 39–57.

Leurent, F. (2005). *La capacité d'écoulement du trafic – un modèle désagrégé et des méthodes de mesure*. Les collections de l'INRETS. Rapport N°265.

Leurent, F. and NGuyen T. P. (2010). *Network Optimization with Dynamic Traffic Information and Tolling*. In Proc. of the 12th World Conference on Transportation Research, Lisbon, Portugal.

Prud'homme R. (1999). *Les coûts de la congestion dans la région parisienne*. Revue d'Economie Politique, 109(4), pp. 425–441.

Vickrey, W. (1969). *Congestion theory and transport investment*. American Economic Review, 59, pp. 251–261.

Wolshon, B. and Lambert, L. (2006). *Reversible Lane Systems: Synthesis of Practice*. Journal of Transportation Engineering 132(12): 933–944.

Chapter 8

A model of passenger traffic in public transport, sensitive to capacity constraints

Ektoras Chandakas, Fabien Leurent & Alexis Poulhès
Paris-East University, City Mobility Transport Laboratory, Ecole des Ponts
ParisTech, Ifsttar, UPEM

INTRODUCTION

a) Background to network planning

In an urban area, travellers use different modes of transport to make their trips. Public transport (PT) is a way of concentrating trips and carrying massive passenger flows, involving a range of solutions: buses, trams, subways and trains, in ascending order of capacity. Planning a network involves establishing the layout of lines, along with modes and services on each line. The stake is to adjust service performance to match user requirements, this by geographical liaison, and balancing potential costs and benefits. If planning is to be relevant for a territory's needs, it is essential to know the passenger flows and their effects on network performance.

Several technical guides provide designers with indications, such as a report in French by Certu (2003) and manuals in English (TRB, 2003; Vuchic, 2006). In urban areas that are large and already equipped with a complex network, adapting existing lines and developing new means of transport will depend on the territory (i.e. location of residents and jobs) in relation to network effects: the load of passengers on a line is influenced by the other lines in the network via connecting stations; diagnosis and forecast require a system simulation, using a model assigning traffic to routes on a network.

b) Problem statement: equip design with a congestion-sensitive model

Engineer-designers of road networks are well aware of the impact that congestion has on traffic flows: on a route where the traffic load is heavy in relation to flowing capacity, the quality of service deteriorates and the route becomes less attractive to users, who turn to another option if they can. In other words, the usage model is sensitive to congestion. The same is true for public transport (PT) networks, in which a greater diversity of capacity constraints and congestion phenomena affect travellers and/or service vehicles. On PT networks, capacity constraints act not just on linear sections, but also on accesses, vehicles, platforms and connecting transfers.

These congestion effects are highly non-linear (the order of complexity often reflects the number of diverse flows interacting) and when all flows increase there is a risk that critical points will saturate, with disastrous chained consequences on the services that stop at stations, and ultimately the establishment of a broadly sub-optimal operating regime in terms of both service quality and capacity available to users. The biggest misuse would be if a critical point induced the sub-optimal operation of a large portion of the network, without being identified nor resolved! For designers, it is crucial to identify the critical points in order to understand the concrete problem, and to avail themselves of a sensitive model in order to devise possible solutions.

To sum up, in a metropolis where rail transport is the most massive mode of transport, it is fundamental to clearly set out the congestion phenomena in order to successfully simulate route choices, traffic loads and service quality, in close interdependence.

c) Objective and structure of the chapter

This chapter presents an innovative model for assigning passenger flows on a PT network, named CapTA (for Capacitated Transit Assignment), and its application to greater Paris. The model is sensitive to the configuration of transport means, their performance, the starting point-destination structure of flows, and the behaviour of route choices; in addition to these standard aspects, it is sensitive to diverse traffic phenomena. We clearly set out the load of passengers waiting on a platform, vehicle capacity, the quality of service in a vehicle in line with the capacity of seated places, and the impact of boarding and alighting flows on a line's operation and frequency. Its application to the greater Paris network illustrates the CapTA model's behaviour and shows the concrete importance of these effects during rush hours.

The rest of the chapter is organized into five parts. We start by setting out the principles of the CapTA model. Then, we describe the line model: line system and capacity constraints. The next two Parts are devoted to presenting a simulation of the CapTA model on the public transport network in greater Paris and the detailed results of this simulation on the regional express line, RER A. To conclude, we mention current and potential developments for our model.

PRINCIPLE CHARACTERISTICS OF THE CAPTA MODEL

In this section, we cover the basic characteristics of the CapTA model in three sections. Firstly, we provide hypotheses concerning modelling transport supply and demand. Secondly, we present the bilayer, duplex representation in our model. Lastly, we describe the assignment process and state the traffic equilibrium on the network. In this chapter, we emphasis the systemic aspects and describe the relationships between the model's components. For additional details and a comprehensive mathematical formulation, see the following documents and articles on this research project: Leurent *et al.*, 2011; Leurent *et al.*, 2012; Leurent and Chandakas, 2012; Chandakas, 2014.

a) Modelling supply and demand

A public transport network comprises lines with one or several transit services (often called "missions"), operated by vehicle runs along line-haul routes, and pedestrian arcs that model travellers' movement by foot. The pedestrian arcs are used for feeding, transferring between lines and diffusing travellers. A transit service is identified by a specific route and the series of stations stopped at along the line. It employs a single type of vehicle (if necessary by homogenization) and is therefore associated with precise capacity features.

The study territory is divided into demand zones. On each zone a precise point, called centroid, is purported to represent one endpoint of every trip that takes its origin or destination in the zone. The transport demand is modelled as a set of agents that carry out journeys trips from an origin u to a destination v. For each origin-destination pair (u, v) on the network, we can identify a flow of travellers, denoted as q_{uv}.

A passenger is taken to be an agent whose economic behaviour is rational. Each path on the network is characterized by a generalized cost. It is constructed in a sequential manner based on the generalized costs of network components: nodes, arcs and services. The generalized cost of a component, through a multiplicative coefficient, takes into account the physical times of components according to the specific tediousness experienced along them by a passenger, or a class of passenger. Each agent chooses a path, or a combination of paths, on the network in order to minimize the trip cost to himself on his origin-destination pair. In the CapTA model, we assume that at each point of path choice, passengers establish an optimal strategy for reaching their destination, in line with the model by Spiess and Florian (1989).

b) Representation on two layers, in a duplex

A fundamental characteristic of the CapTA model is its representation on two, superimposed levels, i.e. in a duplex: the upper layer (network scale) and the lower layer (line scale); the representation is adapted to the precise requirements of the effects modelled on each layer.

The upper layer describes routes for passenger trips for all origin-destination pairs on the network. Passengers can therefore take pedestrian arcs and line "legs" (Figure 8.1b). Pedestrian arcs are taken for feeding into the network, transferring lines, and passenger diffusion. A line leg represents a path from one boarding station to one alighting station along the line, i.e. a line sub-path. The cost of the leg results from the average cost of the run along the sub-path, and the average waiting time of the services that make up the leg: these components come from the line model described below. To sum up, on the upper layer, we derive the flow of passengers on the network's arcs, from the costs of these components and the principle of cost minimization.

The lower layer individually deals with each operating line (bus or train) and represents the local vehicle traffic and passenger traffic on it. At this level, a line is represented by an independent sub-network (see second subsection in next Part) made up of boarding, alighting, station stay and inter-station arcs (Figure 8.1a). The line model derives the effect of capacity constraints on the performance of a line and the leg conditions.

We can establish a two-way relationship between the layers per line ℓ: from bottom up, from the vector of passenger volume per vehicle by line leg (denoted as $\mathbf{x}_\ell$) the line model gives the vectors that characterize the cost of the legs: $\mathbf{g}_\ell$ of the average generalized leg run cost, $\mathbf{w}_\ell$ of the average combined waiting cost and φ_ℓ the combined service frequency. In other words, to the upper layer, the line model amounts to a relation between flow and cost (including frequency) in a vector form: $\mathbf{x} \mapsto (\mathbf{g}, \varphi)$.

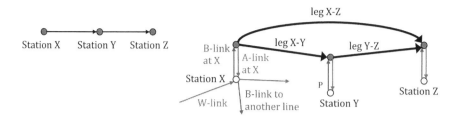

Figure 8.1 From service route (a) to leg-based network model (b).

Conversely, from top down, based on the arc and leg attributes, the upper layer model yields the passenger flow vector by upper layer network arc, i.e. $(\mathbf{g}, \varphi) \mapsto \mathbf{x}$.

c) The process of traffic assignment and equilibrium

On the upper level, a stationary equilibrium of passenger traffic is defined as the conjunction of: (i) the hyperpath choice at minimum cost, (ii) the assignment of passenger flow on the paths of the upper layer, (iii) flow conservation at every node of the upper layer, and (iv) the dependence of the generalized cost, g, waiting time, w, and frequency of services, φ, on the flow vector x of the upper layer. The traffic equilibrium is formulated as a fixed point problem of the flow vector on the upper layer per destination node. The regularization procedures (described in Leurent *et al.*, 2012 and Chandakas, 2013) have been devised to guarantee the continuity of the function of cost and the existence and uniqueness of the solution of the line model.

The issue of traffic equilibrium on a network can be resolved by an iterative algorithm based on the method of successive averages. This algorithm comprises a series of distinct stages: establishing the cost of the components; the composition of the auxiliary state, with the search for an optimal strategy towards a destination; the convex combination of the preceding state and the auxiliary state to yield the next traffic state of the network; and the evaluation of the convergence. The CapTA model's most characteristic feature is the use of a line model during the stage of establishing the cost, enabling a more detailed identification of the capacity effects. The line model is described below.

LINE MODEL

In the CapTA model, each subsystem is dealt with using a specific sub-model. In this section, we shall describe first the line subsystem which gives rise to the line model. The modular architecture of the CapTA model enables us to integrate the representation of the two layers in a coherent way and to derive the local effects of capacity phenomena using local models. These models are presented in the second subsection.

a) Line system

A line constitutes a basic subsystem of the public transport system (Leurent, 2011). The line operates independently from the rest of the system, in particular when it circulates with separated right-of-way. In this context, each direction of the line is operated

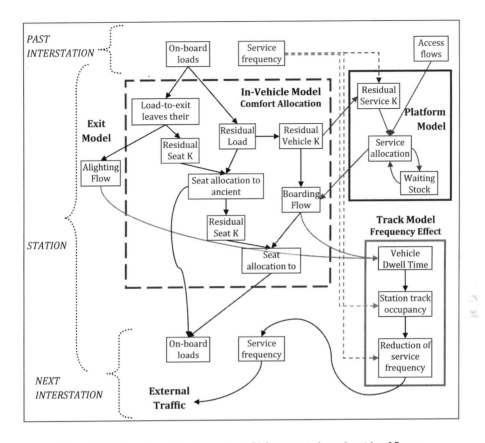

Figure 8.2 Overview of the line system (K for capacity) on the side of flows.

separately and different lines seldom share run sections. Lastly, between the lines there are weak interactions induced by passengers who are feeding into or transferring lines. By line, the line model deals with the capacity effects that are internal to the subsystem and the travel cost of each entry-exit station pair on the line. Consequently, a line is reduced to a sub-set of services in which the vehicles cannot overtake each other because they use a shared infrastructure (line lane and platform track).

A line is represented by an acyclic sub-network that is arborescent in a single traffic direction. This network is made up of subsets of vehicle run arcs (inter-station and station stay arcs) and subsets of passenger pedestrian arcs (boarding and alighting arcs). A line is used by one or several services z. Each service has different characteristics (e.g. service policy, frequency, vehicles).

The topological order of the arcs and nodes along a line and its services is useful not just to represent the service and line legs, but in particular to establish the chronological order of traffic operations. Figure 8.2 illustrates the line system and the process of operations. Five successive operations are involved around a station:

(I)   Passengers moving in vehicles from the preceding station to the current one;
(II) Passengers alighting at the current station;

(III)   Passengers waiting on the platform and boarding into vehicles that have available capacity and that directly stop at their exit station;
(IV)   Dwelling and occupation time by vehicle on the platform track, which determines the performance and frequency of services;
(V)   Interaction of vehicles with general traffic on inter-station arcs.

Each local process is addressed by a specific local model. The results of a local model can be used as inputs to the next local models. The line model coordinates the local operations using two sequential models as follows:

- A physical model to establish the vehicle load and traffic flow. It derives the physical interactions of the passenger flows and the service vehicles on the line's components depending on the capacity constraints. It proceeds in the forward topological order of stations and uses local models.
- An economic model to evaluate the cost of an individual leg by entry-exit station pair along the line, depending on the outcomes of the physical model. It proceeds in reverse topological order and the cost of travel on a service leg is calculated by reverse accumulation of the cost of the leg's components.

b) Modelling capacity constraints

Along a line, the sequential models use specific local models to establish the effect of capacity phenomena on the components of passengers' trips and vehicle traffic. Each local model deals with one out of the following capacity constraints: the seat capacity in a vehicle, the total capacity of a vehicle, the vehicle capacity of platform tracks. They are described below.

- Vehicle comfort model
  The vehicle comfort model is an adaptation of the seat capacity model developed in Leurent (2012). A vehicle contains several states of comfort, denoted as r: passengers can travel seated or standing. However, standing is considered as a more penalizing state. Consequently, for a vehicle running for service z the volume of standing passengers y^r_{zi} at each station i is compared to the available capacity in seated places, k^r_{zi}. The chronological order of operations can be used to establish the priority for assigning seated places: seated passengers getting off at the station make seated places available and a two-phase competition (passengers on board have priority over boarding passengers) gives the probability $p = \min\{1; k/y\}$ of occupying a seated place at a given stage. Lastly, the cost of a leg is evaluated according to the probability of sitting down at each stage along the leg.
- Platform storage and waiting model
  The platform storage and waiting model developed by Leurent and Chandakas (2012) derives the effect of the total vehicle capacity constraint on the local choice of vehicle and on the average time spent waiting on the platform. Thus, the vector $[x_{is}]_{s>i}$ of passenger flows entering at station i is faced to the residual capacity vector $[k^R_z]_{z \in i}$ of services z that stop at the station. The model is based on the explicit description of passengers waiting for a vehicle per exit station s, whose number constitutes the stock of passengers σ_{is}. For a vehicle from service z, the

stock of passengers who are candidates for boarding, n_{zi}, is the sum of stock σ_{is} for all the stations s served by the vehicle. Consequently, a stock per service n_{zi} contends with the available boarding capacity k'_{zi} and gives an immediate boarding probability of $\pi_{zi} = \min\{1, k/n\}$. The average waiting time on the platform per exit station s is calculated using a traffic bottleneck model.

– Platform track occupation model

The platform track occupation model introduced in Leurent $et\ al.$ (2011) considers platform tracks as a scarce resource. A vehicle circulating on a line blocks a section of the line for a period determined by the occupation time, including the time to dwell passengers and the operational time for door opening and closing as well as vehicle deceleration and reacceleration, plus the operational margins required for safety. A vehicle's dwelling time depends on the boarding and alighting flows, thus establishing an interaction between flows of passengers and vehicles. A long dwelling time per vehicle, summed over the vehicles scheduled during the reference period, could result in the total occupation of a platform track for a period that exceeds the reference period. Thus, the frequency of all the services that use this lane is adapted in a proportionately reverse manner. This adaptation is propagated down the line and reduces the service's capacity.

APPLICATION: SIMULATION OF THE GREATER PARIS NETWORK

Greater Paris has a dense public transport network that offers a wide range of road and rail options. With 8.5 million daily journeys (OMNIL, 2012) made on public transport, the system is under constant pressure. Matching supply on the network to the demand for transport is therefore a crucial issue.

This study does not claim to offer a diagnosis of the Paris region's PT system. Although it identifies some of the sticking points on the greater Paris network, its main aim is to illustrate the original capacities of the CapTA model and its behaviour. We start by describing transport supply and demand (subsection a) and the model's variants and parameters (subsection b), which we used to carry out the simulation for the Ile-de-France region. We then examine the assignment of passenger flows on the network (subsection c) according to two alternatives model specifications, and the effect of capacity constraints on the lines' operation (subsection d). Lastly, we evaluate the impact of capacity constraints on passengers (subsection e), before concluding with a discussion of the salient features in this simulation (subsection f).

a) Demand and supply of public transport in the Paris region (Ile-de-France)

A simulation involves assigning a demand for transport to a network according to the supply. We apply the simulation to the morning rush hour in 2008, as described by DRIEA (the State agency for regional and inter-borough department infrastructure and development). More precisely, Ile-de-France is split into 1305 Traffic Assignment Zones. The trip demand between the zones is described by a 1305 × 1305 matrix of origin-destination (O-D) flows, and totals 1.15 million trips for one hour. We want

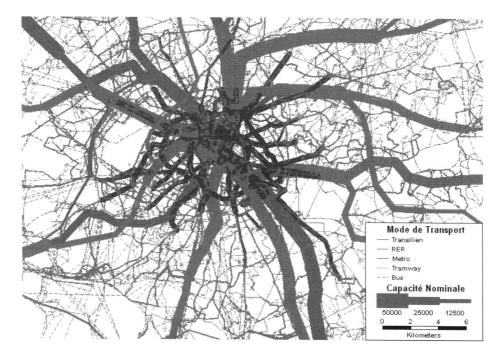

Figure 8.3 Total capacity of lines per mode of transport (central part of the urbanized area).

to highlight the model's behaviour and capacities, and so we use the O-D matrix homogeneously enlarged by 30%.

The public transport network in greater Paris is characterized by a wide range of modes of transport and types of service. Buses, trams, subways, regional express trains (RER) and railway lines run through the region offering different services in terms of frequency, commercial speed and capacity. The transport supply we consider corresponds to the annual service in 2008, as described in DRIEA data, with several necessary modifications. Concerning vehicle characteristics, we defined 17 types of vehicle on guided transport systems and 7 types of bus and coach, which we then associated with services (Chandakas, 2012). Figure 8.3 illustrates the capacity of transport lines per mode of transport.

Lastly, the service network is transformed into a calculation network. The process involves creating line legs that correspond to entry and exit station pairs along a line (see second subsection in second part). In total, the calculation network comprises 160,000 nodes and 307,700 arcs and service and line legs (of which 30,000 are line legs).

b) Specifying the model's sensitivity

The main aim of this study is to identify how capacity constraints affect the CapTA model's behaviour. This involves comparing two variants of the model, either with or without account of capacity constraints. Common to the model's alternatives are the parameters linked to the generalized time. To calculate the generalized time of a trip, we multiply the physical time inherent to each component (according to a passenger's specific physical state) by a coefficient that indicates the discomfort. If we take 1 minute of being seated in-vehicle as a reference, the multiplicative coefficients for waiting on the platform, feeding and transfer come to 2.

A set of parameters chosen for the local models and capacity constraints included in the CapTA model can be used to specify each model variant. The two variants used in this study are defined below:

- UC: This is the baseline alternative with no account of capacity constraints. Most concrete planning studies are still carried out in this way. In this case, passengers are concentrated on the most efficient routes, in terms of nominal performance (scheduled run time and frequency, no in-vehicle discomfort). We can expect the structural lines of the network to be heavily loaded – eventually beyond their nominal capacity.
- CVCW: This variant includes all capacity constraints. We distinguish guided transport modes from road transport modes. For the former, the constraints represented pertain to total vehicle capacity (platform and waiting model), seated capacity (vehicle comfort model) and line capacity (platform occupation model). For the latter, the constraints concern the occupation of seated places and in-vehicle comfort. Concerning the in-vehicle comfort model for all modes, the discomfort coefficient varies depending on the density of standing passengers, ranging from 1.2 for the first standing passenger, to 2 for 4 people/m^2, which corresponds to maximum density. Taking capacity constraints into account, passengers can choose between different lines, avoiding overcrowded lines by opting for longer routes.

On the UC variant an equilibrium state is computed in one iteration only, whereas on the CVCW variant the determination of an equilibrium involves a series of iterations. The simulation's convergence level is evaluated as the average gap in passenger flows on the arcs. An acceptable level of convergence is reached after 50 iterations, with an average gap that is reduced to 1‰ of the initial value.

c) Assignment results: passenger flows onto the network

To system planners, the principal result of a traffic assignment model is the flow of passengers on the arcs of the network. The variant without capacity constraints (UC) corresponds to the choice of optimal hyperpaths without taking into account the effects of capacity constraints on passengers' route choice. The simulation results indicate that some structural lines are heavily loaded, notably so the RER A (east-west) and RER B (north-south) lines: the acronym RER stands for Regional Express Railways. More precisely, on RER A, westbound flows on the most loaded sections reach 100,000 passengers per hour, for a supplied capacity of 58,000 passengers: a ratio of 1.7 flow compared to capacity. This corresponds to an excess of 42,000 passengers who – taking into account capacity constraints – must choose an alternative route to get to their destination. In total, flow exceeds nominal capacity on 58 sections out of the 1,750 sections of guided modes.

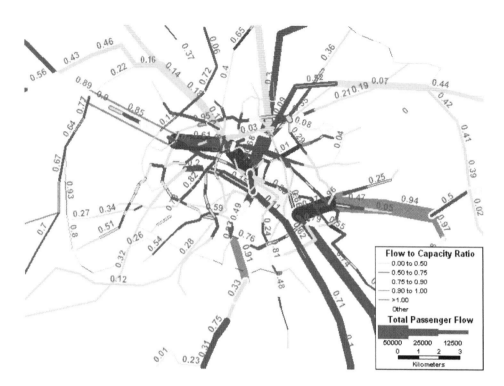

Figure 8.4 Capacitated passenger flow and flow ratio compared to nominal capacity on the CWCW variant.

In the capacitated variant CWCV, passengers choose their route according to the generalized cost and local availability of lines, resulting from various local capacity constraints: the total capacity of the vehicle and the transit service, the occupation of seated places, and the track platforms' vehicle capacity. Figure 8.4 shows passenger flows on the network (by line thickness) and the flow ratio compared to capacity (by colour) for the modes of guided transport on the CWCV variant of the model. Sections with a flow up to 75% of nominal capacity are shown in light and dark green. The sections in yellow (or orange) present a ratio of flow compared to capacity of between 75% and 90% (or 90% and 100%). Lastly, sections where hourly flows exceed the line's nominal capacity are shown in purple. In the CWCV variant, these sections are few (9 out of 1750 sections in total) and the maximum ratio observed is only 3.75% above unit. Clearly, taking capacity constraints into account affects route choices. Passengers who face long waiting times and low levels of comfort opt to transfer to alternative routes in order to reduce their perceived cost, as evaluated by the platform storage and waiting model and the seat capacity model.

d) Operating services under capacity constraints

The CapTA model (on the CWCV variant) is unusual in its sensitivity to how passenger flows affect the operation conditions of a service, followed by the lines' performance.

The local model deals with this vehicle capacity constraint by reducing the service frequency. The simulation of the greater Paris network leads to two remarks on this local model:

- We cannot faithfully represent this concrete phenomenon in an assignment without capacity constraints (UC variant). By including capacity constraints (as in CWCV), some sections are relieved and passenger flows are transferred to alternative lines, thus modifying the flow structure. Consequently, the sticking points in the respective results of the two variants (UC and CWCV) are not identical and the network's effects influence the location of flow transfers and their size.
- Frequency adaptation is primarily triggered by an overflow of the service exchange capacity, since the dwelling time of vehicles depends on the product of boarding and alighting flows by elementary times that depend on the vehicle exchange capacity. This exchange capacity involves the number and width of the vehicle doors, which serve as passenger channels for boarding and alighting. On a given line, a frequency reduction may take place upstream from the most loaded sections, and thus reduce the capacity available on sections where it is most strongly required.

The simulation of the greater Paris network suggests that a dense network with massive flows of passengers undergoes a frequency reduction on some of the structuring lines. Figure 8.5 shows the drop in frequency simulated by the CWCV variant of the CapTA model. We can see that the frequency of the Métro's M1 line (east-west) drops from 34 veh/hour (nominal frequency) to 27.4 veh/hour eastbound and 31 veh/hour westbound. Similarly, line M14 (automatic) undergoes a reduction in nominal frequency (40 veh/h) of 6% to 15% depending on the direction, while the frequency of the westward RER A decreases from 30 veh/h to 27.7 veh/h, or 7.7%. This reduction in frequency has significant secondary impacts: 7.8% less capacity on the line, or 4,500 passengers and 1,400 seated places during the rush hour. The lost capacity is the equivalent of 1.7 double-decker trains.

e) Consequences for users

The average generalized cost on the network includes waiting time (WT) and in-vehicle transport time (IVTT), as well as time spent walking to transfer between two lines and times to feed into the network (at the origin or destination of a trip). Table 8.1 summarizes the average generalized time of variants of the CapTA model and details the components. Let us analyze the composition of the average generalized time (GT) of the CWCV variant. The waiting time corresponds to waiting time on the first line and waiting time included in transfers. The average perceived waiting time comes to 29.1% of the total time. At 41.5% of the total time, in-vehicle time constitutes the largest component of generalized time, whereas walking time during transfers comes to 5.6%, and feeding time constitutes 23.8% of generalized time.

It comes out that the generalized time of the CWCV alternative increases by 11.2% compared to the UC alternative. Among the components of generalized time, in-vehicle time increases the most (23%). In fact, ¾ of the increase in generalized time can be attributed to in-vehicle discomfort. On the other hand, the increase in waiting time is

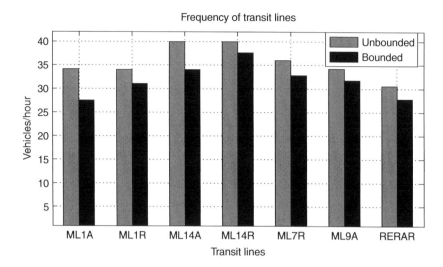

Figure 8.5 Comparison between nominal frequency and adapted frequency at the terminus of several structuring lines on the greater Paris network.

Table 8.1 Average generalized time (in minutes) on the greater Paris network.

Model variant	Optimal generalized time	Actual travel time	Perceived waiting time	Perceived in-vehicle time	Perceived Transfer Time	Perceived Feeding Time	Average No. of Transfers per trip
UC	61.56	40.63	18.79	23.10	3.96	15.71	1.42
CWCV	68.45	41.70	19.90	28.40	3.88	16.27	1.35
%diff	11.2%	2.63%	5.9%	23%	−2.04%	3.6%	−5.13%

limited, which indicates that the lack in total service capacity applies to a subset of sticking points which is fairly limited on the network scale.

f) Discussion

The greater Paris public transport system, which includes 13 rail transport lines, 14 subway lines, 4 tram lines, and several hundred bus services with duplicated lines on the central sector, provides an ideal field to test an assignment model with capacity constraints. From the simulation results, it comes out that significant flows of passengers are concentrated on structuring north-south lines, such as RER B and subway line M13, and east-west, like RER A and subway lines M1 and M14. By integrating capacity constraints, flows can be spread out to alternative routes in a more realistic way. Flows on the lines only occasionally exceed the nominal capacity.

The passenger flow on the network influences lines' performance, and especially their service frequency. This frequency decreases by as much as 19% for some lines and directions, with secondary effects caused by the drop in downstream capacity. For passengers, capacity constraints contribute to increasing the average waiting time by up to 14% and 22% for subway lines M1 and M11 respectively. This moderate increase corresponds however to 230 additional hours of passenger waiting on line M1 only during one morning rush hour. In terms of quality of service in vehicles, the average standing time for a passenger depends on the structure of demand and the topology of the line. The longest times are mainly on long radial lines, such as subway line M8, with individual values of 5.5 and 7.5 minutes depending on the direction.

FOCUS ON LINE A OF THE RER (REGIONAL EXPRESS RAILWAY)

The CapTA model acts on two superimposed levels: at the upper layer of network route choice, the passenger flow is assigned to routes and lines depending on the passenger cost of these items; on the lower layer, the model evaluates the cost of entry-exit station pairs along the line according to the passenger flow by leg that are determined on the upper layer.

Let us from now restrict our analysis of the simulation results to the RER A line, which is the busiest in the greater Paris network, often accommodating over one million travellers per day. RER A comprises 46 stations spread over 5 branches and the central trunk. Two branches go eastwards (Marne-la-Vallée north-east and Boissy-St-Léger south-east) and three go westwards (Cergy-le-Haut and Poissy north-west and St-Germain-en-Laye south-west), converging on the central section. As a result, eastern and western suburbs are connected to central Paris and the La Défense business area.

We shall describe more precisely the initial supply and demand data linked to RER A (in subsection a). In subsection b, we look at the results of the reduced frequency model, and in subsection c we present the behaviour of the passenger stock on the line's central trunk, as it comes out from the local model of platform storage and waiting. We then highlight the consequences of integrating capacity constraints into the generalized time of journeys on the line (subsection d), and end with a discussion of the salient points of the line model (subsection e).

a) Supply and demand on RER A

The sub-network of the RER A line comprises 1,200 service nodes and 182 initial arcs, transformed into 1,702 line legs. Demand, resulting from the CWCV assignment (cf. subsection b in previous Part) is made up of 107,000 eastbound passengers and 141,000 westbound passengers; the total is 248,000 passengers in both directions for one hour in the morning rush.

We focus the analysis of the results in the direction of the morning rush, i.e. from central Paris and eastern areas towards La Défense. During the most intense hour of the morning rush, transport supply comprises 18 trains on the Marne-la-Vallée (MLV) branch and 12 trains from Boissy-St-Léger (BOI), which converge before the Vincennes station (cf. Figure 8.6). Consequently, 30 trains per hour are scheduled to travel on the central trunk to La Défense. Out of them 4 trains per hour terminate their service at La Défense, while the remaining 26 diverge on the three branches, including 5 towards Poissy (POI), 5 towards Cergy (CRG) and 16 towards St-Germain-en-Laye (STG).

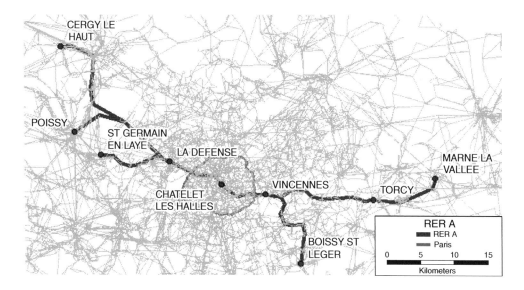

Figure 8.6 RER A (in red) in relation to central Paris (green ring) and the La Défense business district.

Table 8.2 Types of vehicle on RER A and their characteristics.

Type of Vehicle	Seated Places per train	Total train Capacity	Number of doors per train side	Passenger channels per train side
MS61	600	1888	36	72
MI84	432	1760	32	64
MI2N	1056	2580	30	90

The westbound line comprises 15 different services, each of which is characterized by a service frequency, a subset of stations where it stops and a type of vehicle. The characteristics of the latter are given in Table 8.2.

b) Dwelling time and service frequency

A vehicle's dwelling time at the platform depends on the number of passengers at alighting and boarding, the type of vehicle, and the interface between the platform and the vehicle. In the CapTA model, the dwelling time is calculated from the number of passengers boarding and alighting, according to the number of passage units (individual passenger channels along the door). The basic time for a change is 1.55 seconds/passenger, and is taken independent to vehicle and platform congestion. In contrast, we simulate a sub-optimal use of passage units (due to using foldaway seats and other operating features) which are reduced by one third.

The line model provides results disaggregated by service. Figure 8.7 shows the dwelling time simulated versus planned at Etoile station (located before La Défense)

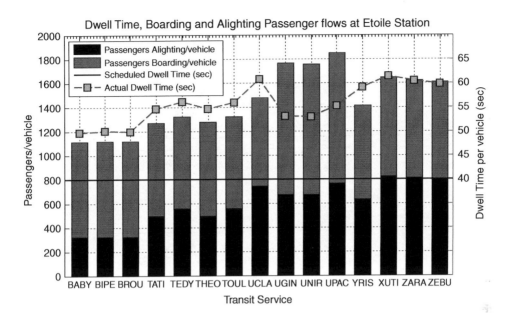

Figure 8.7 Volume of boarding and alighting passengers and dwell time of westbound vehicles at Etoile station.

for westbound vehicles. The volume of alighting passengers per vehicle is shown in blue, with the boarding passengers in brown. We can see that the dwelling time of 40 seconds (continuous line) is not respected, and that vehicles dwell for between 50 and 61 seconds, depending in the changing volume and the type of vehicle. The longest dwelling time (UCLA service) does not correspond to the services with the biggest changing volume (i.e. UGIN, UNIR and UPAC). This is because these three services use MI2N vehicles with a larger changing capacity than the UCLA service's M184. This example shows that the line operations need to be modelled at a sufficiently disaggregated level, and that in practice the assignment of vehicles to the line servicesshould depend on the vehicle capacity to match the demanded flow.

c) Stock of passengers

The platform storage and waiting model evaluates the impact of total capacity constraints on the average passenger waiting time and local choice of service (intra-line). This is done by explicitly setting out the stock of passengers on the platform. From a station i of entry, the stock of passengers σ_{is} per exit station corresponds to the number of passengers who want to get into a vehicle on a service that directly stops at the exit station. The stock depends on the exogenous flow x_{is} per exit station and the available capacity per vehicle serving z. The stock of passengers who are candidates for a service, n_{zi}, is the sum of the stock of passengers for all of the exit stations stopped at on this service. Consequently, two services with identical downstream stopping policies will

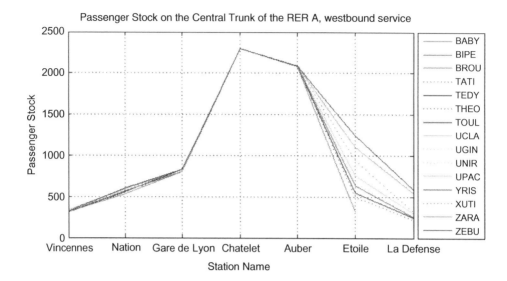

Figure 8.8 Stock of passengers for services in the central trunk (westbound).

have the same stock of candidate passengers, whatever the supply in terms of available capacity.

Figure 8.8 shows the stock of candidate passengers for all of the services on the central trunk travelling towards the La Défense station. Two particular observations reveal the behaviour of the platform model and its bottleneck submodel:

• On the profile of the stock of passengers per service: in the eastern part of the trunk, from Vincennes to Auber, the stock of passengers per service is similar, since passengers are travelling to stations in the central trunk (of whom a significant proportion alight at La Défense) and all services stop at these stations. However, at the Etoile and La Défense stations, the stock varies depending on the service because of their downstream stopping policies.

• On the evolution of stock along the line: when the total available capacity of the line is sufficient, but one or several services are saturated, the stock increases slightly compared to the flow/frequency ratio. This leads to an increase in the average waiting time (at Nation, it is 3 minutes instead of the reference 2 minutes) since some of the passengers do not succeed in getting on to the first vehicle that arrives. However, when the total available capacity is insufficient, the stock accumulates because the flow cannot be evacuated during the reference period. At Châtelet-Les-Halles, the stock reaches approximately 2,300 passengers on platform and the average waiting time is 12 minutes, which is 6 times more than the reference situation.

d) Average generalized cost

The line model, included in the CapTA model, yields average passenger cost by trip leg along a line integrating capacity constraints throughout the leg. Figure 8.9 shows the actual time (continuous line) and generalized time (dotted line) of legs to La Défense from stations on the Marne-la-Vallée and Boissy-St-Leger branches and stations on the central trunk. The time spent onboard a vehicle is show in green, the waiting time in red and total time in blue.

We observe that the time onboard a vehicle is reduced from upstream to downstream, in line with the distance between the boarding and alighting (La Défense) stations. In addition, the gap between the actual and perceived time spent on board is related to the difficulty in finding a seated place and the density of standing passengers. Depending on the discomfort coefficients, this gap ranges from 40% to 70%. A few remarks follow:

- Although leg distances differ significantly, we observe that the leg generalized time from a station on the branches is around 60 minutes. On legs beginning at the start of a branch, access to a seated place makes up for the low frequency and long distance. Conversely, closer to Paris, legs are shorter but passengers are subject to reduced comfort.
- If we compare a leg starting at Noisiel with another leg further up the line (Lognes or Torcy), the second leg emerges as more advantageous. This is due to comfort in the vehicle, because the likelihood of getting a seated place at Torcy avoids having to stand for the rest of the leg, unlike at stations down the line.
- In addition, the actual leg time simulated from Châtelet-les-Halles (TT of 21 minutes) is longer than the time from the upstream stations Gare de Lyon and Nation (TT of 18 minutes). The main reason is the important stock of passengers changing at Châtelet-les-Halles.

e) Discussion

The simulation results pertaining to RER A, the busiest line on the greater Paris network, illustrate the roles played on the field by the capacity phenomena that are captured in the CapTA model, and the need to treat them explicitly. We observe that:

- Taking dwelling time and platform occupation constraints into account reduces the frequency of all services down the line. This drop occurs at Etoile station.
- The capacity constraint of vehicles determines the waiting time and stock of passengers. The value of wait time can shoot up when the candidate passenger flow is faced with insufficient residual capacity, like at Châtelet-les-Halles, where the total stock reaches 2,300 passengers (the equivalent of a double-decker train).
- In-vehicle comfort plays an important role and has a significant impact on legs along the radial lines. On RER A, the perceived time of a leg to La Défense is around 60 minutes, whatever the starting point on a branch.

Lastly, the functional capacity of the CapTA model can be used to evaluate a project's socio-economic impact, at least in terms of demand surplus. We simulated a capacity investment project for line A of the RER, replacing current vehicles with new, higher-capacity trains. Replacing single-decker trains (MS61 and M184) with double-decker trains (M109, similar to M12N) adds 30% to 40% of passenger capacity per train, as well as changing capacity on the platform, thanks to the distribution of

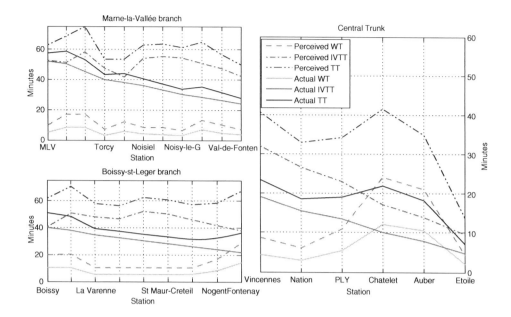

Figure 8.9 Average Actual and Perceived Time to reach La Defense station from east.

doors along the length of the train. This significantly improves the line's performance by limiting the decrease in service frequency and in-vehicle comfort. We simulated the investment project and compared the results of its assignment with those of the reference situation. The project would reduce the total generalized time by 24,150 hours during one rush hour, i.e. almost 30 million hours per year. Giving a value to this user benefit of €10/hour (conventional value for greater Paris users), the benefit would be €300 M per year, which at an annual discount rate of 4%, would justify an investment of €6G by the community: thus the investment cost, at €2G, would be repaid threefold, without counting the indirect benefits (i.e. production of value flows in the economic and social circuit, cf. chapter on territorial facilities).

CONCLUSION

Platform waiting time for passengers, in-vehicle comfort, and adapted frequency of transit services are three capacity-related phenomena among the various interactions in a PT system. We have made these explicit in the CapTA model for assigning passenger flows on a network. Its application to the greater Paris multimodal network reveals the difficult areas of passenger and vehicle traffic. Since these difficult areas concern sections that are already well known for their heavy traffic, our simulation shows the model's sensitivity to concrete effects, and the technical feasibility of simulating these effects in a study to aid planning decisions.

The CapTA model's systemic approach can be used to develop specific models targeted to the PT system's sub-systems. The simulation of the Ile-de-France network uses two line models according to sub-mode: guided modes and road modes. In addition, we can include specific models to take into account capacity features in a station linked to the feeding of passengers into the network and their transfers between lines. The CapTA model has a modular structure, so that existing models can be adapted and others included: e.g. the influence of the density of passengers on the platform and in vehicles on the basic marginal times at boarding and alighting; passengers' acceptance of greater congestion in vehicles when the residual capacity is too far below the stock on the platform.

ACKNOWLEDGEMENTS

In 2009, the EEBI Chair asked the LVMT to carry out a research operation to model capacity constraints on a PT network. Starting in 2010, in partnership with the STIF, the LVMT set up a Chair specifically devoted to the Socio-economics of Urban Public Transport for Passengers, which became the framework accommodating the development of the CapTA model. We sincerely thank the STIF for its scientific and financial support: in particular, Ektoras Chandakas's PhD and the posts held by François Combes and Alexis Poulhès have been funded by this Chair. We also thank Mr Jean Delons from VINCI Autoroutes, for his stimulating involvement in the Chair's Steering and Evaluation Committee. In addition, our thanks go to DRIEA for making available the data of the MODUS model: the part relating to PT, with our own additions to describe capacity constraints, was used for all of the applications reported here.

REFERENCES

Certu (2003), *Modélisation de déplacements urbains de voyageurs: guide de pratiques*, Certu Report, Editions de Certu, Lyon.

Chandakas E. (2012), *Note sur la Capacité du Matériel Roulant et son Affectation sur le Réseau Francilien*, Working Document, Ecole Nationale des Ponts et Chaussées, Paris Est University.

Chandakas E. (2014), *Modelling Congestion in Passenger Transit Networks*, PhD Thesis, Ecole Doctorale "Ville, Transports et Territoires", Paris-East University.

Leurent F. (2012), *On Seat Capacity in Traffic Assignment to a Transit Network*, Journal of Advanced Transportation, April 2012, Vol. 46, Issue 2, pp. 112–138.

Leurent F., Chandakas E. (2012), *The Transit Bottleneck Model*, Elsevier Procedia – Social and Behavioural Sciences, Vol. 54, pp. 822–833.

Leurent F., Chandakas E., Poulhès A. (2011), *User and Service Equilibrium in a Structural Model of Traffic Assignment to a Transit Network*, in Zak J. (ed) The State of the Art in the European Quantitative Oriented Transportation and Logistics Research – 14th Euro Working Group on Transportation & 26th Mini Euro Conference & 1st European Scientific Conference on Air Transport. Elsevier Procedia – Social and Behavioural Sciences, Vol. 20, pp. 495–505.

Leurent F., Chandakas E., Poulhès A. (2012), *A Passenger Traffic Assignment Model with Capacity Constraints for Transit Networks*, Elsevier Procedia – Social and Behavioural Sciences, Vol. 54, pp. 772–784.

Leurent F., Chandakas E., Poulhès A. (2014), *A Traffic Assignment Model for Passenger Transit on a Capacitated Network: Bi-layer Framework, Line Sub-Models and Large-scale Application*, Forthcoming in Transportation Research Part C.

Observatoire de la Mobilité en Ile-de-France, OMNIL (2012) *Synthèse de Principaux Résultats de l'EGT 2010*, Edition of the Observatoire de la mobilité en Ile-de-France, accessed in November 2012.

Spiess H., Florian M. (1989), *Optimal Strategies: A New Assignment Model For Transit Networks*, Transportation Research Part B, pp. 83–121.

TRB (2003), *Transit Capacity and Quality of Service Manual*, Online report prepared for the Transit Cooperative Research Program, available online at http://gulliver.trb.org/publications/tcrp/tcrp_webdoc_6-a.pdf. First edition 1999.

Vuchic V.R. (2006), *Urban Transit: Operations, Planning and Economics*, Wiley.

Chapter 9

Environmental information module of construction materials

Adélaïde Feraille[1], Fernanda Gomes Rivallain[1] & Yannick Tardivel[2]
[1]*Navier, Ecole des Ponts ParisTech*
[2]*SETRA*

CONTEXT

A life cycle assessment (LCA) for an urban project requires collecting data of construction materials used in buildings and infrastructure, as well as of components and equipment and numerous processes that involve energy, water, waste, transport, etc. The European base, Ecoinvent, groups these different data based on a standardized methodology and in a single format. This base mostly comprises generic European, Swiss and German data, and several people are currently working on compiling specific data. This is because some industrial processes are specific to the country in which they are carried out. In France, the INIES database contains the environmental and health characteristics of building products as well as the relevant environmental and health declarations. These declarations are produced in line with the French standard NFP01-010. However, when drawing up a LCA, it may be necessary to handle environmental data by material rather than by product. This issue, which is not covered in the INIES base, is the subject of this chapter.

In light of this observation of a lack of environmental data relating to civil engineering installations in France, the DIOGEN working group (database of environmental impacts of materials for civil engineering works), initiated by the French Civil Engineering Association, was created in 2010. The group's objective is to build a database to make available environmental information modules on materials and items used to produce French civil engineering installations. This working group is jointly led by SETRA (head: Yannick Tardivel) and IFSTTAR (head: Christian Tessier). The ParisTech-VINCI chair, "Eco-design of buildings and infrastructure" financed an 18-month post-doctorate study, carried out by Fernanda Gomes Rivallain. The research was done at the Navier laboratory (Ecole des Ponts ParisTech, Ifsttar, CNRS) and supervised by Adélaïde Féraille. The object of this chapter is therefore to present some of the results obtained by this post-doctoral study.

DIOGEN DATABASE AND ASSOCIATED METHODOLOGY

The approach taken to build the DIOGEN database (www.diogen.fr), used for all materials that we want to study, is set out below:

- Identification of environmental data needed to produce a LCA for a civil engineering installation;
- Identification of existing data in available databases (Ecoinvent, ELCD, etc.);

- Evaluation of these existing data with respect to their use (work of art): boundaries, completeness, representativeness, technological, temporal and geographical correlations;
- When an existing datum is unrepresentative or ill-suited to the French context, we have to build a new datum from existing bases, when possible using specific data supplied by companies, associations, trade unions, etc. To better understand the manufacturing processes of the different materials, we rely on experts from the various domains.

Methodology

This database gathers data adapted to the domain, in terms of environmental impacts, of a "cradle-to-gate" nature, drawing from existing data, and combining them as much as possible to end up with standard civil engineering objects.

This group comprises representatives of material producers, companies, engineering firms, and institutions. It is made up of five working sub-groups: one on cement materials, one on steel, one on wood, one on coverings and composite materials, and one on methodology.

Up to this point, the DIOGEN database conformed to standard NF P 01010 and was therefore only concerned with the 10 environmental impacts recommended by this standard: Consumption of Energy Resources (MJ), Depletion of Resources (ADP) (kg eq. Antimony), Consumption of Water (litres), Solid Waste (kg), Climate Change (kg eq. CO_2), Atmospheric Acidification (kg eq. SO_2), Air Pollution (m^3), Water Pollution (m^3), Destruction of Stratospheric Ozone Layer (kg CFC eq. R11), and Formation of Photochemical Ozone (kg eq. Ethylene).

The base is currently being adapted in line with standard EN 15804. This will mean developing certain indicators, e.g. the depletion of resources indicator will be split into depletion of abiotic component resources and depletion of abiotic fossil fuel resources.

Reliability and representativeness of the data

Given that the aim is to produce LCAs pertinent to civil engineering, it seemed important within the DIOGEN group to be capable of ascertaining the reliability of the data in the base.

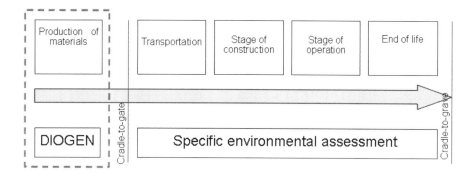

Figure 9.1 Position of the DIOGEN group.

Firstly, these data should be representative of the material studied. Their technological, geographical and temporal pertinence is essential, because it is related to supplies of natural resources, specific energy sources, etc. Similarly, Life Cycle Inventory (LCI) flow measurements should preferably be representative of the producing sites' average production rather than relating to a short period or restricted geographical selection. To limit the area for consideration, it is preferable to know the destination of the material and the certification criteria authorizing its use.

The data should be of a reliable character. Their traceability should be guaranteed and the sources used mentioned and if possible available. In addition, measures for calculating impacts should be carried out in line with procedures and a sample of the production adapted.

In the method anticipated by the DIOGEN working group, the first step is to characterize an environmental datum, and then decide on its acceptability. Since the quality requirements of a datum used in an LCA calculation can be related to the weight of this datum in the result (through critical analysis), with this approach, data of a lower quality can be used for materials with little influence, and greater focus put on the key entries.

To do this, we propose employing a confidence index by which users can determine whether an available datum is usable or not. This index will cover the different judgement criteria mentioned above. This method, set out in (Habert *et al.*) is based on the data qualification method used by Ecoinvent.

This method considers that the law of distribution for measurements and the resulting data can be modelled using normal log distribution. This type of distribution is adapted to a context where the effects of several independent factors multiply amongst themselves, which is the case for natural and industrial processes. It is then the logarithm of the variable that follows a normal distribution and not the variable itself.

The variation in the results is characterized by a Standard Deviation (SD) value. The normal log distribution assures that 95% of data are included in the $SD95^2$ interval around the value with the highest probability (often used as the estimated value). The higher the $SD95^2$ is, the greater the dispersion of the variable values is. The Standard Deviation therefore covers all of the uncertainties resulting from the different influential factors identified. We can thus characterize SD with the following formula:

$$SD95 = \sigma^2 = \exp \sqrt{([\ln(E1)]^2 + [\ln(E2)]^2 + [\ln(E3)]^2 + [\ln(E4)]^2 + [\ln(E5)]^2 + [\ln(E6)]^2 + [\ln(E7)]^2)}$$

The different Ei represent the values of requirements used to define the quality of the datum. The requirements chosen are defined based on the Ecoinvent matrix (Pedigree) and the NF EN ISO 14044 standard.

Each requirement gives an individual score based on precise criteria, reflecting whether the response brought by the datum matches the expected requirement. Five score levels are planned to take into account the different levels of response possible.

From an SD value that shows the quality of the datum, we obtain a confidence index that does or does not match the required requirement level: $Ic = f(SD95)$, Figure 9.3.

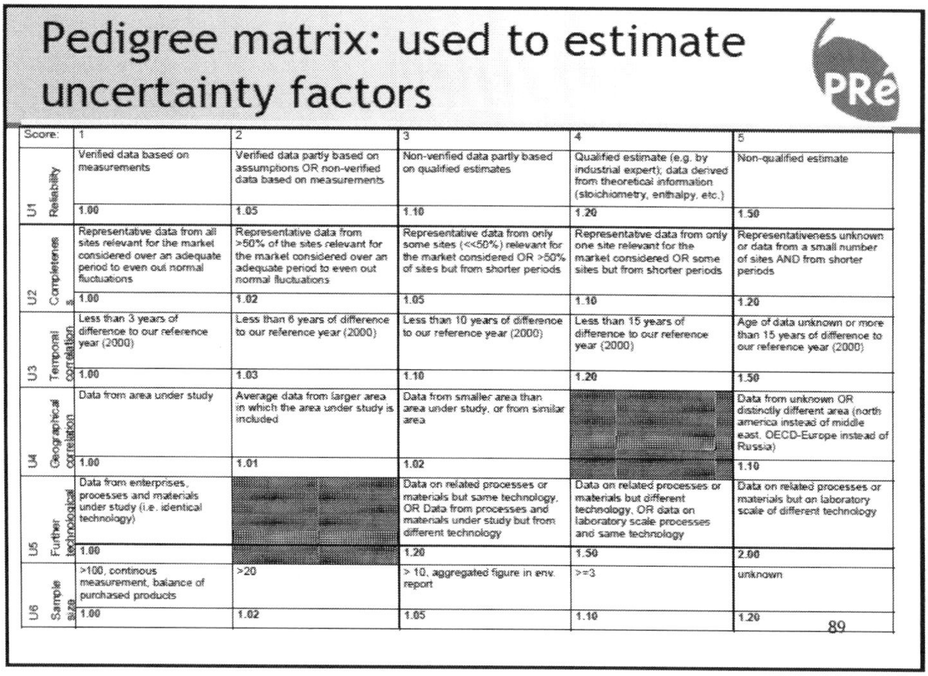

Score:	1	2	3	4	5
U1 Reliability	Verified data based on measurements	Verified data partly based on assumptions OR non-verified data based on measurements	Non-verified data partly based on qualified estimates	Qualified estimate (e.g. by industrial expert); data derived from theoretical information (stoichiometry, enthalpy, etc.)	Non-qualified estimate
	1.00	1.05	1.10	1.20	1.50
U2 Completeness	Representative data from all sites relevant for the market considered over an adequate period to even out normal fluctuations	Representative data from >50% of the sites relevant for the market considered over an adequate period to even out normal fluctuations	Representative data from only some sites (<<50%) relevant for the market considered OR >50% of sites but from shorter periods	Representative data from only one site relevant for the market considered OR some sites but from shorter periods	Representativeness unknown or data from a small number of sites AND from shorter periods
	1.00	1.02	1.05	1.10	1.20
U3 Temporal correlation	Less than 3 years of difference to our reference year (2000)	Less than 6 years of difference to our reference year (2000)	Less than 10 years of difference to our reference year (2000)	Less than 15 years of difference to our reference year (2000)	Age of data unknown or more than 15 years of difference to our reference year (2000)
	1.00	1.03	1.10	1.20	1.50
U4 Geographical correlation	Data from area under study	Average data from larger area in which the area under study is included	Data from smaller area than area under study, or from similar area		Data from unknown OR distinctly different area (north america instead of middle east, OECD-Europe instead of Russia)
	1.00	1.01	1.02		1.10
U5 Further technological correlation	Data from enterprises, processes and materials under study (i.e. identical technology)		Data on related processes or materials but same technology, OR Data from processes and materials under study but from different technology	Data on related processes or materials but different technology, OR data on laboratory scale processes and same technology	Data on related processes or materials but on laboratory scale of different technology
	1.00		1.20	1.50	2.00
U6 Sample size	>100, continous measurement, balance of purchased products	>20	> 10, aggregated figure in env. report	>=3	unknown
	1.00	1.02	1.05	1.10	1.20

Figure 9.2 Pedigree Ecoinvent matrix.

Lastly, we link the confidence index to the class of data, Figure 9.4, which shows the acceptability of the datum in a specific context.

After presenting the approach taken by the DIOGEN working group and the associated methodology, we will present the principal results on materials to which our team made a particular contribution.

STEEL MATERIALS

The research focused on reinforcing steel and was the subject of two publications (Gomes *et al.*, 2012; Gomes *et al.*, 2013). The aim was to obtain environmental data on this type of steel sold on the French market in 2011, and then compare the results with Ecoinvent and Wordsteel data. A distinction is made between two types of steel bars, because they are associated with different production processes: B500 A and B B500 (NF A35-080-1, 2010).

Methodology

The system studied was one kilogram of reinforcing steel of B500 A and B500 B. The boundaries of the study are shown in Figure 9.5. It is limited to the production and on-site delivery of one kilogramme of steel bars. The life cycle considered is therefore

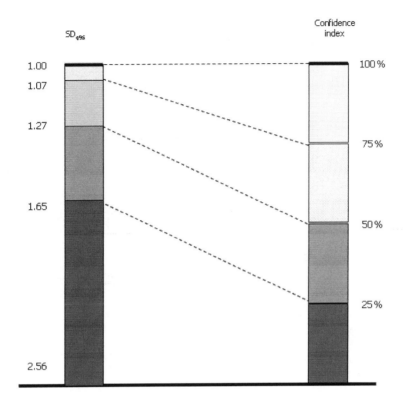

Figure 9.3 Relationship between Standard Deviation and confidence index.

from cradle to construction site gate. The main difference between the two products studied (B500A and B500B) is that producing B500A steel involves cold rolling, which can be carried out in different factories from those doing hot rolling. A supplementary transport phase must be included.

The reinforced steel bars B500 B can also be produced by cold rolling. In this study, we only compare 500A and 500B steel produced by hot rolling because we consider that this hypothesis represents extreme production cases.

To build a precise inventory of the reinforcing steel sold in France, the Ecoinvent database was adapted to the French context. This was done with the help of steel experts, adding or deleting some processes and modifying the contribution of other processes. These experts know all about quality and thus the processes involved in producing steel because they are responsible for certifying the reinforcing steel sold in France (AFCAB).

Lastly, to evaluate the robustness of the hypotheses, values for which we cannot be certain of their exactness were modelled using a variable with an average, a maximum value, and a minimum value. Using these intervals, we can make a Monte Carlo analysis to obtain an average value of the environmental load of the reinforced steel product and a standard deviation for this value.

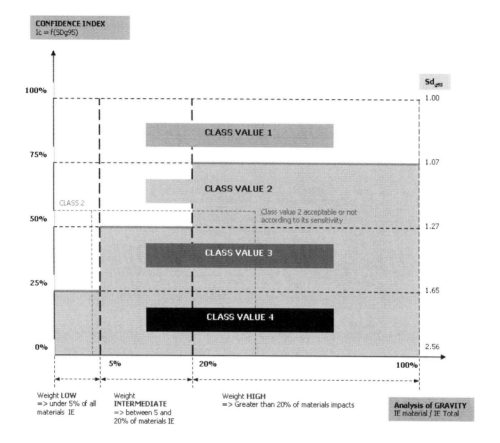

Figure 9.4 Domain of data usage.

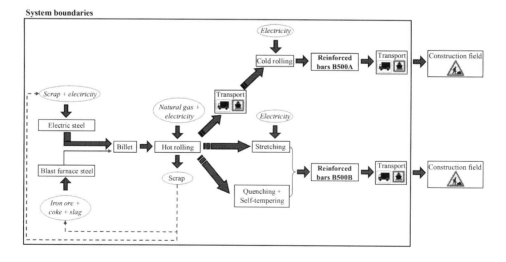

Figure 9.5 Limits of the reinforcing steel product system.

We chose to apply the impact analysis method CML 2011 (Guinee, 2002) with a particular focus on the following indicators: abiotic resources depletion (kg eq. antimony), atmospheric acidification (kg eq. SO_2), eutrophication (kg eq. PO_4^{-3}), climate change (GWP100) (kg eq. CO_2), ozone layer depletion (ODP) (kg eq. CFC-11), human toxicity (kg eq. 1,4-DCB (1,4-Dichlorobenzene)), aquatic ecotoxicity (kg eq. 1,4-DCB), marine toxicity (kg eq. 1,4-DCB), terrestrial toxicity (kg eq. 1,4-DCB) and photochemical oxidation (kg eq. C_2H_4). The simulations were carried out using SimaPro software.

Data used

The hypotheses used to build the inventories of B500 A and B500 B reinforcing steel are shown in Tables 9.1 and 9.2.

Two steelmaking processes exist: blast furnace (BF) and electric arc furnace (EAF). Almost 100% of reinforcing steel consumed in France is made using the electric arc process [AFCAB, personal paper]. It is worth noting that the EAF process represents recycled steel because the main raw material is scrap metal, whereas the BF process uses iron ore as a raw material (Figure 9.5).

Reinforcing steel is produced from non-alloyed steel billets (NF EN 10 020, 2000).

Because steel is not always produced in the same country, a reference energy mix was modelled according to the origin of the different steels used in France (relative to their mass). Tables 9.3 and 9.4 show the hypotheses considered.

For steel "processing", the hot rolling inventory was built taking from the Ecoinvent hot rolling processes (Ecoinvent Report part XV, 2009) only those sub-processes that were pertinent for making B500 A and B500 B reinforced bars. These are given in detail in Table 9.5. Similarly, the Ecoinvent modelling of the rolling process – using cold rolling – was studied to identify the pertinent sub-processes for reinforced bars.

Table 9.1 Inventory for type B500 A reinforced bars.

Reinforced bars type B500 A

Parameters	Origin of data	Mean	Minimum	Maximum
Steel process nature				
electric steel (%)	Ecoinvent process: *"steel, electric, un-and low-alloyed, at plant/RER"*	98	95	100
blast furnace steel (%)	Ecoinvent process: *"steel, converter, unalloyed, at plant/RER"*	2	5	0
Electricity mix (%)	Ecoinvent process: *"electricity, medium voltage, at grid"*	Reference mix (table 9.5)	French electricity	German electricity
Treatment type				
hot rolling (1 kg)	Ecoinvent process changed (table 9.3): *"hot rolling, steel, RER"*	–	–	–
cold rolling energy (kWh/t)	Industry data (table 9.4)	–	–	–
Transport distance (km)				
to the cold rolling plant	Ecoinvent process – *"transport, lorry >32t, EURO 4/RER"*	300	0	1000
to the construction field (Table 9.7)	Ecoinvent process – *"transport, lorry >32t, EURO 4/RER"*	686	0	1445

Table 9.2 Inventory of B500 B reinforced bars.

Reinforced bars type B500 B

Parameters	Origin of data	Mean	Minimum	Maximum
Steel process nature				
electric steel (%)	Ecoinvent process: "*steel, electric, un-and low-alloyed, at plant/RER*"	98	95	100
blast furnace steel (%)	Ecoinvent process: "*steel, converter, unalloyed, at plant/RER*"	2	5	0
Electricity mix (%)	Ecoinvent process: "*electricity, medium voltage, at grid*"	Reference mix (table 9.6)	French electricity	German electricity
Treatment type				
hot rolling (1 kg)	Ecoinvent process changed (table 9.3): "*hot rolling, steel, RER*"	–	–	–
stretching (%)	Industry data: (1/2) of "*cold rolling, steel, RER*" energy	50	0	100
quenching + self tempering (%)	Industry data	50	100	0
Transport distance (km)				
to the construction field (table 9.7)	Ecoinvent process – "*transport, lorry >32t, EURO 4/RER*"	686	0	1445

Table 9.3 Electricity mix hypotheses to produce type B500 A reinforced bars.

France	Germany	Belgium	Other countries
3/4	1/10	1/8	1/100

Table 9.4 Electricity mix hypotheses to produce type B500 B reinforced bars.

France	Germany	Spain	Italy	Other countries
2/3	1/8	1/10	1/16	1/20

Cold rolling was then modelled by electricity consumption, as shown in Table 9.6 with the production mix of each country considered in Ecoinvent; the proportion for each country is shown in Table 9.3. These values are taken from specific industrial data.

The specific production process of B500 B reinforced bars either involves stretching or quenching plus self-tempering. Industrial data are not very precise concerning the relative proportion of each of these processes. This proportion was therefore introduced as a parameter in the LCA model. The stretching process is modelled by an electricity consumption that represents half of the energy consumption of cold rolling. The impacts of the "quench and self-temper" process are not significant.

Transportation distances were estimated according to the distances between the capitals of steel-making countries and Paris. The distance considered for French

Table 9.5 Process associated with hot rolling.

Process involved
Ecoinvent process steps
Hot rolling, steel, furnace
Hot rolling, steel, descaling
Hot rolling, steel, hot rolling
Hot rolling, steel, wastewater treatment plant
Hot rolling, steel, overall
Hot rolling, steel, packaging

Table 9.6 Process used for cold rolling.

Process involved	Amount (kWh)
Ecoinvent process	
Electricity, medium voltage, at grid	0.035

Table 9.7 Transportation distance (km) from factory gate to the construction site gate.

France	Germany	Spain	Italy	Luxembourg
500	1053	1272	1445	400
Switzerland	Belgium	United Kingdom	Turkey	Netherlands
585	309	451	3870*	517

*2800 km by boat and 1070 km by truck.

steelmaking is 500 km. This approach overestimates transportation distances, given that the main foreign producers (Germany, Spain and Italy) are very close to the French border. All transportation is assumed to be by truck except for Turkey, where part of the distance is assumed to be covered by boat (Table 9.7). Distances range from a minimum value of 0 km to a maximum value of 1445 km. This corresponds to the distance between Paris and Rome, which is the maximum distance considered for transport by truck.

Additional transportation distance, which must be included for B500A because the factories in which cold processing and hot processing are carried out are not always the same, was modelled with truck transportation for a distance ranging from 0 to 1000 km with an average value of 300 km.

Results and interpretation

The results obtained are shown in Table 9.8. They can be compared to the values obtained by WorldSteel Association and those of the reinforced steel inventory available in Ecoinvent. To make this comparison possible, transportation from factory door to building site is not taken into account. Lastly, because no distinction is made between B500A and B500B in these databases, a single reference value is calculated for the existing bases.

Impact values are presented in Figure 9.6.

The results obtained with the modelling developed in this study are much lower or equal to the Ecoinvent values for all indicators except terrestrial ecotoxicity. This exception is due to the high proportion of electric arc furnace (EAF) steel included in this study. Currently, in Ecoinvent steel bars are modelled with a ratio of EAF steel and BF steel of respectively 37% and 63%. However, we know that EAF steel involves high mercury emissions (Pirrone and Mahaffey, 2005), which leads to more significant impacts on terrestrial ecotoxicity.

For all other impact categories, and in particular for acidification and global warming potential, blast furnace (BF) steel has much greater impacts than electric arc furnace

Table 9.8 Environmental impact indicators for steel bars.

Impacts	Unit	Worldsteel	Ecoinvent	B500 A	B500 B
Abiotic depletion	kg Sb eq.	0.00540	0.01264	0.00586	0.00567
Acidification	kg SO_2 eq.	0.00336	0.00504	0.00231	0.00230
Eutrophication	kg PO_4^{-3} eq.	0.00025	0.00309	0.00134	0.00135
Global warming (GWP100)	kg CO_2 eq	110.199	148.607	0.69700	0.67200
Ozone layer depletion (ODP)	kg CFC-11 eq.	8.4E–09	6.1E–08	6.9E–08	6.2E–08
Human toxicity	kg 1.4-DB eq.	0.01603	0.78167	0.87500	0.86300
Fresh water aquatic ecotoxicity	kg 1.4-DB eq.	0.00110	0.95638	106.000	106.000
Marine aquatic ecotoxicity	kg 1.4-DB eq.	313.019	1292.80	1170.00	1180.00
Terrestrial ecotoxicity	kg 1.4-DB eq.	0.00348	0.02631	0.06440	0.06430
Photochemical oxidation	kg C_2H_4 eq.	0.00044	0.00080	0.00022	0.00023

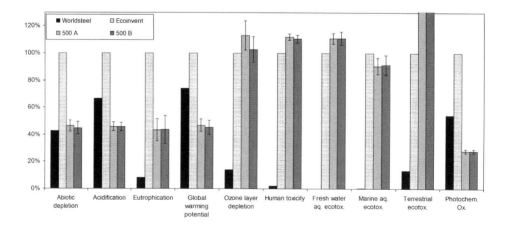

Figure 9.6 Comparison of environmental data of a cradle-to-factory door type. The intervals shown for 500 A and 500 B represent sensitivity to various production parameters (see tables 9.1 & 9.2 for more details).

(EAF) steel. This is why the impact values for French reinforced bars, mostly made using EAF, have around 50% lower values than in Ecoinvent.

A comparison with the WorldSteel Association data shows the same trend. Although the split between EAF and BA is not stipulated in the ELCD, it is probably lower than that of French reinforced bars.

In addition, the fact that values are lower by one or two orders of magnitude – even for eutrophication which is known to be a more robust indictor than toxicity indicators – can be due to an incomplete inventory (notably for indirect emissions or infrastructure considerations).

Consequently, based on this comparison, it appears clear that the impact values proposed in this study are lower than those of Ecoinvent and Wordsteel, apart from terrestrial ecotoxicity, which tends to be the main problem caused by intensive use of recycled steel in electric steelmaking. The problem of radioactive waste has not, however, been assessed.

Lastly, there is no significant difference between B 500A and B 500B, even with an alternative distance of 0 to 1445 km.

CONCLUSIONS

In conclusion, this study shows that most specific impacts for reinforced bars sold in France are lower than in the generic databases, probably because the French context is different from that of other European countries. The human toxicity and ecotoxicity aspects appear to be equivalent, and it would be worth examining damage-focused indicators (e.g. damage to human health, damage to ecosystem quality). A standardization process could facilitate a more global comparison of all indicators: the steelmaking contribution may be more significant for some indicators than for others.

On the other hand, there is no difference between 500A and 500B and no major variation with a transport modification. Consequently, we can use the data proposed in this study for all civil engineering structures built in France, whatever the category of steel and the distance from the factory.

This study therefore means we can considerably simplify the environmental assessment of reinforced bars used in France.

A study of this type is being done on cable steel.

Along with comparing data from WorldSteel and Ecoinvent with constructed data, it would be interesting to incorporate additional indicators. Indicators on resources and waste (including radioactive) and damage-focused indicators (health, biodiversity) will be studied in the near future so that this data can be integrated into the DIOGEN database.

BRIDGE BEARING

The preceding paragraph presents the result of a study of a specific material that gave rise to a DIOGEN report. When undertaking life cycle assessments for a building, information is required on the different types of material or objects that make up the building. Thus, a study of a bridge bearing was done, which resulted in DIOGEN reports, but the investigation continues on several other materials or objects, such as coverings, proofing materials, and expansion joints.

This research was carried out by Hélène Hillion and supervised by Fernanda Gomes; it resulted in a presentation at the GC meetings in March 2013 (Hillion *et al.*).

Following advice by experts, for the AFGC database a type B reinforced elastomeric bridge bearing was chosen, measuring $400*500$ mm^2 which is a standard size. The study focused on two particular cases: one with 4 internal elastomeric sheets, the other with 8 internal sheets.

The thickness of the external sheets (e, Figure 9.7) corresponds to the minimum thickness imposed by standard NF EN 1337-3: 2.5 mm. The thickness of the internal sheets (ti, Figure 9.7) is 12 mm. The thickness of the steel reinforcements (ts, Figure 9.7) is 4 mm. The thickness of the coating (e', Figure 9.7) corresponds to the minimum thickness required by standard NF EN 1337-3: 4 mm. The study therefore focused on bridge bearings with a total height (Tb, Figure 9.7) of 73 mm and 137 mm respectively.

The functional unit chosen for this environmental information module is a type B reinforced elastomeric bridge bearing measuring $400*500$ mm^2 with the aim of transmitting the vertical loads (own weight and traffic weight) without excessive pressure on the bearing support and allowing horizontal movements in all directions.

Companies manufacturing bridge bearings keep their data confidential. Manufacturing data was however obtained from a company quality manual dating from the same period as the present study. Some data are still unknown and are indicated below. Figure 9.8 shows the manufacturing process of a reinforced elastomeric bridge bearing with details of the different manufacturing stages.

The stages detailed above are summed up in Figure 9.9.

We study two different cases according to the number of elastomeric sheets n. In all cases, the bridge bearing considered will be reinforced elastomeric and measure $400*500$ mm^2: 4 internal sheets; 8 internal sheets.

Figure 9.10 shows for the 4-sheet case how the manufacturing stages are distributed for the 13 impact categories.

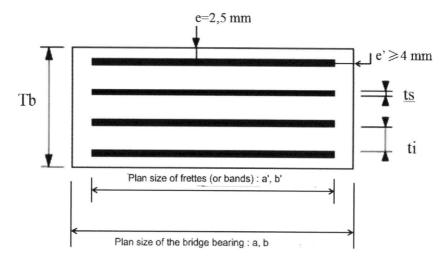

Figure 9.7 Diagram of reinforced elastomeric bridge bearings (Pirrone and Mahaffey, 2005).

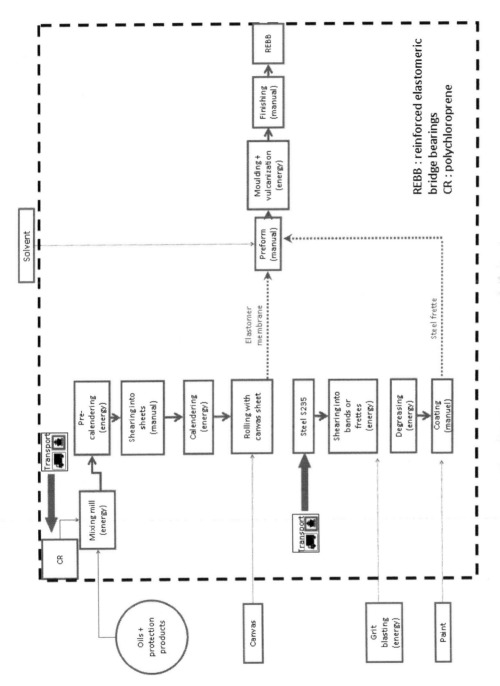

Figure 9.8 Diagram of processes taken into account.

Process	Fabrication of a bridge bearing (type B, dimension 400*500)			
Outgoing product	Bridge bearing (type B, dimension 400*500)			
		Quantity	Unit	Comments
Bridge bearing (type B, dimension 400*500)				
	Preparation of elastomer sheets (for 1 kg)			
	synthetic rubber, at plant	1	kg	The process takes into account the energy of the open roll mixer and the energy of the press (precalender and calendering)
	Preparation of frettes (pour 1 kg)			
	steel, converter, unalloyed, at plant	0.98	kg	Steel to make frettes
	steel, electric, un- and low-alloyed, at plant	0.02	kg	Steel to make frettes
	hot rolling	1	kg	Fabrication process of steel plates
	sheet rolling (prise en compte uniquement du sous processus annealing)	1	kg	Fabrication process of steel plates
	electricity, medium voltage, at grid	7.86E-3	kWh	Shearing of plates before transport
	electricity, medium voltage, at grid	7.86E-3	kWh	Shearing into frettes of steel plates
	Transport			
	transport, lorry 16-32t, EURO 4	0.883	t*km	Transport ok 1 kg of elastomer
	transport, lorry >32t, EURO 4	0.378	t*km	Transport ok 1 kg of steel
	Molding: valcanisation			
	electricity, medium voltage, at grid	3.582	MJ	Energy of the valcanisation press

Figure 9.9 Table summarizing processes.

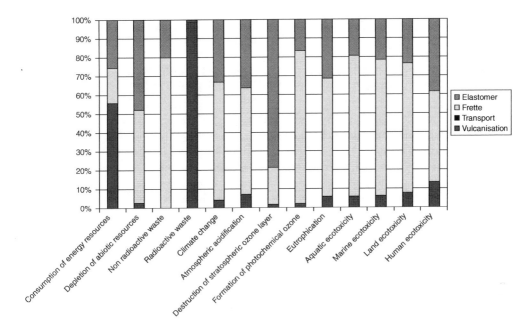

Figure 9.10 Case I – distribution by manufacturing phase of bridge bearing.

We can see on this figure that for most impact categories, the reinforcement manufacturing phase has the greatest impact. For radioactive waste, the biggest impact is during the vulcanisation stage. For the destruction of the stratospheric ozone layer category, the elastomeric manufacture impacts the most.

Transporting raw materials to the manufacturing factory generates a negligible impact compared to that of transforming raw materials into a finished product.

Most of the work carried out on the two studies presented above was done by the team at the Navier laboratory as part of the AFGC working group. The Navier team has also done significant research on galvinisation (Gomes *et al.*, June 2012) and steel in the form of sheets or plates (Gomes *et al.*, October 2012), always taking the scientific approach described in the first paragraph of this chapter.

However, the team's contribution, part of which was directly financed by the ParisTech chair "Eco-design of buildings and infrastructure", goes much further. They obviously participated in drawing up the methodology presented in the first part of this chapter, and also contributed to other working sub-groups and thus helped establish a number of DIOGEN reports on cement for ATHIL and on wood in collaboration with FCBA.

CONCLUSION

Environmental assessment is now used to define the principles of eco-design and in building work markets. Excluding the impacts of the circulation of vehicles, which are by far the largest, environmental reports of civil engineering installations clearly show that materials production weighs most heavily in LCA study results, and it is thus necessary to obtain representative, reliable data. The use of unsuitable data brings the risk of making counterproductive choices.

The observation that environmental data available to the civil engineering sector is inadequate led us to build a method to evaluate the quality of data and in particular highlight the significant differences between data from generic bases and data required in this specific sector of activity, orientated towards creating prototypes rather than standardizing industrial activities.

The research presented here illustrates our response to this observation. In opting to take a scientific approach to calculate data combined with precise knowledge of the field studied, including regulations, we identify and assess all of the processes associated with putting together materials, including the specific features of producing countries (e.g. type of energy used) and the technologies used, on a perimeter adapted to the material considered.

This approach to producing suitable, robust results is extended to all of the work carried out on the other families of materials covered by the DIOGEN database. The ultimate objective is therefore to obtain a representative, reliable environmental assessment using data whose quality is proportional to their weight in the assessment. It would be interesting to extend this approach to other indicators, in particular energy and water consumption, waste production, and damage-focused data (concerning health and biodiversity). This may require greater precision in drawing up inventories: the INIES database only imposes 168 flows, and e.g. dioxins can be grouped with other volatile organic compounds despite the fact that their toxicity is much higher. The neighbourhood study requires integrating both building products and materials used in civil engineering. It could therefore be useful to work on standardizing the different databases.

REFERENCES

AFCAB. Certification Acier pour Béton Armé. RCC03, AFNOR 58pp.

Ecoinvent Report part XV, 'Metal processing', Data v2.1, Swiss Centre for Life Cycle Inventories, Dübendorf, 2009.

Gomes, F. and Brière, R. A Feraille, G Habert, S Lasvaux, C Tessier, *Adaptation of environmental data to national and sectorial context: application for reinforcing steel sold on the French market*, International journal of life cycle assessment, June 2013, Volume 18, Issue 5, pp. 926–938.

Gomes, F., Brière, R. and Habert, G. A Feraille and C Tessier, *Environmental evaluation of reinforced steel sold on the French market*, Nantes juillet 2012.

Gomes, F., Feraille, A., Tardivel, Y., Tessier, C. and Neel, L. *DIOGEN: database of environmental impacts of materials for civil engineering constructions – Application for galvanization*, Intergalva 2012, Paris, 10–15 June 2012.

Guinée, J.B., Gorrée, M., Heijungs, R., Huppes, G., Kleijn, R., van Oers, L., Wegener Sleeswijk, A., Suh, S., Udo de Haes, H.A., de Bruijn, H., vanDuin, R. and Huijbregts, M.A.J. 2002. *Life Cycle Assessment: An Operational Guide to the ISO Standards*. Kluwer Academic Publishers, Dordrecht (NL).

Habert, G., Tardivel, Y. and Tessier, C. *DIOGEN: base de Données d'Impacts environnementaux des matériaux pour les Ouvrages de GENie civil*, GC'11.

Hillion, H., Gomes, F., Rizard, F. and Feraille, A. *Module d'information environnemental des appareils d'appui de pont en élastomère fretté*, GC'13.

NF A35-080-1: Reinforcing steel – Weldable steel – Part 1: bars and coils, October, 2010.

NF EN 10 020: Definition and classification of grades of steel, September, 2000.

Pirrone, N. and Mahaffey, K.R. *Dynamics of mecury pollution on regional and global scales – Atmospheric processes and human exposures around the world*, Eds Springer (2005).

Chapter 10

Retrofitting buildings

Mathieu Rivallain[1], Bruno Peuportier[2] & Olivier Baverel[1]
[1]*Ecole des Ponts ParisTech*
[2]*MINES ParisTech*

This chapter is the result of research for a PhD thesis by Mathieu Rivallain, jointly supervised by Bruno Peuportier (MINES ParisTech) and Olivier Baverel (Ecole des Ponts ParisTech). This piece of work received support from the VINCI ParisTech Chair and was funded by the French Ministry of Ecology, Sustainable Development and Energy.

INTRODUCTION

a) A major challenge in the energy transition

As we face the threat of wide-scale climate change, with some natural and energy resources on the point of depletion and fuel poverty affecting 7 million people in France, the use of buildings – in heating, air conditioning, ventilation, hot water production, lighting and equipment – is responsible for over 40% of global consumption of final energy (UNEP, 2003) and 44% in France.

In addition, over 60% of the buildings of 2050 have already been built! The average renewal rate of existing houses is less than 1% per year in most developed countries (Meijer *et al.*, 2009). Thus, along with significant efforts to encourage new buildings, led by changes in regulations (RT2012 then RT2020) and the extensive development of labels and benchmark systems, it remains essential to retrofit existing buildings if we intend to reach environmental targets like quartering greenhouse gas emissions from 1990 to 2050.

Experience has shown that heating needs can be significantly cut down, by as much as a factor of 10, with a systemic retrofit programme (Sidler, 2009) and at a reasonable economic cost.

Retrofitting existing buildings is thus a major issue for sustainable development and the energy transition in the building sector.

b) A complex and systemic issue

Retrofitting is however a complex issue that requires a systemic approach.

Retrofit strategies should attempt to integrate, if possible in a time sequence, combinations of retrofit interventions that target lower energy needs, efficient equipment, and the use of renewable energy. These measures must match projects' local specificities.

In addition, although reducing energy consumption is the main objective, energy-saving improvements should cost-effectively combine this performance with environmental issues, equal or better comfort conditions for occupants, architectural quality or usage of space, and their heritage value, even beyond the regulatory framework. As N. Kohler and U. Hassler (2002) point out, buildings represent a society's greatest physical, economic, social and cultural value. Transformation should not translate into a breakdown of this capital.

What is more, energy-saving upgrades interact with buildings' maintenance, renovation and transformation cycles in line with the evolving aspirations of their occupants.

Thus, the preservation of living areas, energy consumption levels, the degree of deterioration, occupancy rate, ease of transformation, cultural value, rent and financial reserves, potential savings, capacity for organizing decisions, industrial and sectorial capacity, etc. are all parameters that determine what is possible in terms of energy improvements.

In the face of this complexity, the extent of the challenge and the size of the building stock, building professionals need support to define efficient retrofit strategies on the life cycle of existing buildings. Most operational approaches are currently based on iterative simulations, mainly guided by experience (Alanne, 2004).

c) Positioning and objectives of the research

In terms of positioning, the studies presented here are not concerned with the quality of building operations, or controlling energy performance achieved in situ (problem of performance guarantee), which are also fundamental retrofit issues. Instead, they focus on aiding decisions to improve the energy performance of existing buildings.

The problem usually encompasses two complementary situations, i.e. managing buildings in use in real time, and aiding design decisions in the project's early stages.

Real-time approaches employ direct monitoring or anticipation of the building's performances and external conditions (including climate) to adapt the management strategy and optimize system controls and settings. Chapter 11 of this publication focuses on this aspect.

Early decision aid targets a building's performance through designing, identifying and specifying components or systems to integrate into the building project to improve performance. The present contribution looks at aiding design decisions by identifying efficient retrofit strategies on the scale of a building, and as a prerequisite to optimizing the building's dynamic management.

In addition, a retrofit process assumes different technical stages, and we assume here that an audit has been previously carried out of the existing building in its initial state, and that basic retrofit measures have been pre-selected in light of the project's local specificities.

This chapter contributes to the development of methodologies and robust decision-making tools for identifying efficient energy upgrade strategies based on multi-criteria and the life cycles of existing buildings. Readers who would like to know more should refer to (Rivallain, 2013) for a more detailed description of the modelling, hypotheses and optimization algorithms.

DECISION-AIDING FOR RETROFITS

The strategic planning of retrofit interventions constitutes a decision-making problem whose nature we need to understand to in order to choose a methodological approach. We therefore set out the criteria and variables considered (objective functions and search space), and the models used to evaluate the performances of the potential strategies, in order to motivate the choice of methodological approach.

a) Decision-making criteria

Although the key objective of energy retrofits is to reduce energy consumption, they must also fulfil numerous objectives – just like new buildings – related to respecting regulatory constraints and clients' new expectations: structural stability, fire safety, durability, the quality of interior spaces, comfort in terms of hygrothermics, light, visuals, acoustics, indoor air quality, energy efficiency, environmental balance over the life cycle, economic and financial profitability, etc. (Kolokotsa *et al.*, 2009).

To avoid impact transfers, a multi-criteria approach is not however enough. The different stages of the life cycle should be integrated into the analysis so that, for example, a reduction in operating energy consumption should not come at the price of considerable energy consumption to produce the retrofit's components.

The performance of potential retrofit strategies should therefore be evaluated on a multi-criteria basis and over a building's life cycle. In a sustainable development perspective, the objective functions considered here target the following over the life cycle:

- Cumulated primary energy consumption and environmental impact indicators, including climate change potential at 100 years, the potential for abiotic resources depletion, potential for atmospheric acidification, etc. All of the indicators used in the EQUER building life cycle assessment software (Popovici, 2006) are also proposed.
- Economic and financial indicators: investment cost (cumulated over the retrofit phases) and overall cost over the life cycle. The overall life cycle cost is the sum of the cumulated costs of the investment, the building's operation and end-of-life treatment of components deposited or used in the retrofit.
- Occupants' comfort. The thermal comfort of occupants in the summertime is assessed using an adaptive approach, as a function of indoor temperatures and the average outside temperature over the previous seven days (NF EN 15251, 2007).

It is vital that retrofits respect regulations (especially on security and human health), focus on how spaces will operate and adapt them to modern usage, and take into account architectural quality and heritage value. However, these criteria tend to be considered as constraints on the search space rather than objectives to be optimized. Thus, we consider here that they should be employed when defining alternatives or pertinent retrofit measures. A downstream analysis of the strategies proposed can supplement the discrimination regarding these criteria.

The problem involves multiple criteria and some targets appear contradictory; defining efficient strategies for sequential energy retrofits necessarily involves a degree of compromise.

b) Decision alternatives and energy retrofit measures

Identifying a decision aiding approach also involves stipulating the decision variables and defining solutions, being programmes for sequential energy rehabilitation. These are characterized by their composition and their temporal phasing.

The composition refers to the combination of energy retrofit measures used, which systemically targets the following:

- The building envelope (thermal insulation of facades, bottom floor and roof; air tightness; replacement of windows and doors; glazed surface according to orientation);
- Replacement of ventilation, heating and water-heating equipment (including a share of solar production).

For each of these retrofit measures, alternatives are considered. These are modelled by discrete variables, in line with obvious industrial constraints.

The measures and alternatives, considered when putting together a retrofit programme, are assumed to have been previously identified based on the local features of the project and thus on an energy audit that will have been conducted on the existing building (geometric and physical characteristics, urban environment, local climate, usage scenarios of the premises, etc.).

Besides, controls are not the object of decision alternatives here. They are more closely connected to the building's real-time management than to design decision aids. Chapter 11 of this book looks specifically at dynamic energy management.

Phasing corresponds to the temporal sequence of implementing these measures. In mathematical terms, solutions are permutations of discrete variables, making the problem combinatorial discrete by nature.

c) Life cycle model

The performances of retrofit strategies are assessed according to the decision criteria mentioned above and following a life cycle model, whose main stages are shown below in Figure 10.1.

The system studied is limited to the elements discriminating alternative retrofit strategies.

Each retrofit phase necessarily involves producing retrofit components (1), end-of-life treatment of the components deposited (2) and energy consumed in using the building during the phase concerned (3). At the end of the implementation of the whole programme, the energy consumption related to the building use over the extended life cycle (4) and the end-of-life treatment of retrofit components (5) must be evaluated. At this stage, the life cycle assessment model does not include the transport of materials (from factory to building site), the environmental impacts of implementation, or maintenance over the life cycle.

The operating phase is modelled by energy consumption relative to heating, ventilation and domestic hot water production. Heating and thermal comfort requirements are evaluated using dynamic thermal simulations (with the PLEIADES+COMFIE software) (Peuportier & Blanc-Sommereux, 1990).

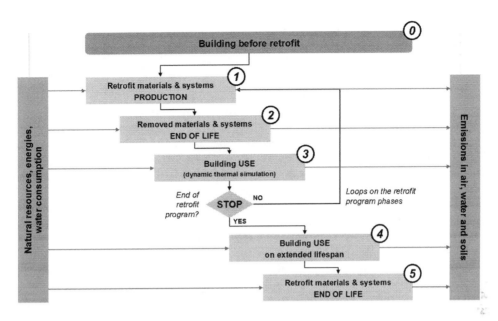

Figure 10.1   Life cycle model for assessing retrofit strategy performance.

At each of these life cycle phases, matter and energy are consumed, generating emissions in different environmental compartments. Consumption of natural and energy resources and the emissions generated are then cumulated and translated into environmental and economic impacts using life cycle assessment (LCA) and life cycle cost analysis (LCCA) databases. Environmental indicators are calculated using the Ecoinvent database (Version 2.0) (Frischknecht *et al.*, 2004). Databases and price estimates are used for the financial data.

Using dynamic thermal simulation involves the implicit character of objective functions; we do not have any analytical expressions for the decision criteria studied.

d) Multi-criteria approaches to decision aiding on sequential energy retrofits

The decision aid problem, here modelling the strategic planning of an energy retrofit, depends on multiple criteria and combinations and is characterized by a discrete search space and implicit objective functions. In terms of decision aid, the nature of the problem is not without consequences. Its multi-criteria character imposes compromises in the case of non-concurrent objectives. In addition, although the solutions are explicitly indentified by their composition and phasing, their performances are not known a priori. This does not therefore simply involve expressing an inter-criteria preference with a view to making a choice, it means identifying efficient solutions.

Faced with this type of problem, different approaches can support decision aiding. They can be classed into two types of methodology: approaches based on preference and generative approaches (Deb, 2001).

Preference-based approaches include standard transformations of multi-criteria decision aiding problems into mono-criterion optimization problems, i.e. weighting approaches, programming by objective, ε-constraints, etc. Despite being relatively intuitive and simple to implement, these transformations are not neutral. They require some knowledge of the problem and its solutions to establish the weighting coefficients, constraints or goal, for example. This information constitutes the expression of a type of preference, introduced early on in the research process. In addition, these procedures result in identifying a single solution for each simulation. Lastly, weighting or ε-constraints are techniques that are sensitive to the mathematical properties of the problem's convexity. If the problem is non-convex, this type of technique cannot identify some solutions. They are then not accessible to decision makers despite their potential to represent promising compromises (Deb, 2001).

Generative approaches aim to provide decision-makers with a set of efficient solutions, describing the compromise solutions that can be considered. The latter are generally represented in the form of a Pareto frontier. The Pareto frontier designates all of the non-dominated solutions within a given set of solutions. By definition, a solution is said to be non-dominated if no alternative solution exists – within the search space – that is at least as good for all of the objectives and strictly better for at least one of them (Deb, 2001). Figure 10.2 shows the Pareto frontier and the dominated solutions of a minimization problem with two objectives.

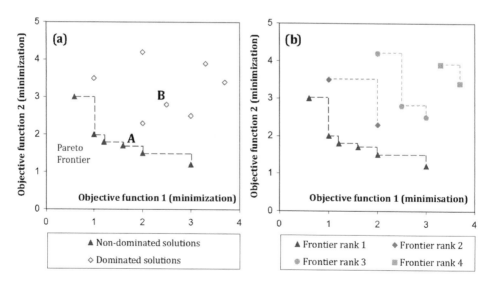

Figure 10.2 Pareto frontier and dominated solutions for a minimization problem with two objectives.

In graph a), no better solution exists within the search space than the Pareto frontier solutions identified for the two objectives. For example, solution A, on the Pareto frontier, minimizes performances for the two objectives considered in relation to solution B; B is therefore a dominated solution. In graph b), the solutions are classified by Pareto frontiers in increasing ranks. The non-dominated solutions constitute the frontier of rank 1. The successive frontiers are identified by seeking non-dominated solutions within the sub-set formed by solutions that do not belong to any of the frontiers described above.

The present contribution focuses on aiding decisions through a generative approach. The search for a Pareto frontier can thus benefit from a multi-criteria optimization. Given the nature of the problem, a classification of optimization methods proposed by Colette *et al.* (2002) suggests using metaheuristics, for example. These approximate methods of stochastic optimization are suited to seeking out efficient solutions in large search spaces. Faced with a given problem, the pertinence of one particular metaheuristic rather than the others remains an open problem (Dréo *et al.*, 2003). We chose to develop a genetic algorithm on the basis of previous conclusive experiments in the building design domain (Pernodet Chanterelle, 2010).

One of the original aspects of this study lies in the identification of compromise surfaces involving energy, environmental, economic and societal criteria to identify efficient strategies for sequential energy retrofits. Based on this rich information, decision-makers can make choices while asserting their preferences. The expression and modelling of preferences can still make use of complementary methods such as multi-attribute value theory (MAVT) or outranking (Roy, 1985), to discriminate between the non-dominated strategies identified.

PRESENTATION OF THE CASE STUDY

The decision aiding approaches presented below have been implemented to study sequential energy retrofit strategies for different buildings. The example considered for the case study in this chapter is a residential apartment building, referred to as the so-called "Barre Grimaud".

For information, apartment buildings in France provide 14 million homes, 63% of which were built before 1974. The average energy consumption for heating is 330 kWh/m².yr (ADEME, 2011).

The Barre Grimaud is a five-storey apartment block in the Parisian suburbs. The ten apartments are divided over a living floor area of 792 m². This building was delivered in 1974, i.e. before the introduction of the first thermal regulation in France (1975). The construction period explains the composition of the block's envelope, the heating needs and consumption, evaluated as respectively 178 kWh/m².yr and 261 kWh/m².yr (dynamic thermal simulation using PLEIADES+COMFIE for a temperature setting of 19°C). The energy bills reveal higher consumption than this, but it was not possible to ascertain the actual temperature of the apartments, probably above 19°C.

a) Description of the Barre Grimaud before the retrofit

Table 10.1 describes the composition of the envelope and the systems used for heating, ventilation and domestic hot water production in its initial state. Before the retrofit, the building's envelope had very little insulation.

In this study, heating was set at a temperature of 19°C from early October to end April, in all of the rooms day and night. During the summer, solar protections (shutters) and night free cooling were used for thermal comfort. Occupancy scenarios were established independently from the retrofit programme evaluated.

Table 10.1 Barre Grimaud, characteristics of the envelope and systems before energy retrofit (thickness is given in [mm]; compositions details are given from outside to inside).

Components	State before energy retrofit
Outside walls	Plaster coating (20) + solid concrete blocks (150) + air layer (10) + plaster (50)
Bottom floor	Solid concrete slab (150) on cellars + mortar (50) + tiling (10)
Intermediate floors	Solid concrete slab (150) + mortar (50) + tiling (10)
Roof terrace	Gravel (30) + bitumen (4) + solid concrete slab (150)
Openings	Single glazing, PVC frames
Ventilation	Non-modulated mechanical ventilation
Heating	Collective gas boiler, installed prior to 1988 (HHV yields considered: 80% at nominal load, 65% at partial load of 30%)
Hot water production	Individual gas boiler

The Barre Grimaud was modelled in the form of three homogeneous thermal zones, represented on Figure 10.3 and corresponding respectively to the ground floor, intermediate floors and the top floor, under the roof.

Figure 10.3 Barre Grimaud, view of the south face and 3D representation of homogeneous thermal zones.

b) Composition and phasing of retrofit programmes

The retrofit programmes studied for the Barre Grimaud are defined as ordered combinations of elementary measures chosen from the 8 classes of measures shown in Table 10.2.

For a given retrofit programme:

- The variation ratios of glazed surfaces can be differentiated depending on the façade;
- The size of the condensing gas boiler is adapted to the building's current heating needs, during the retrofit phase corresponding to the replacement of the heat generator (thus depending on the action carried out during the preceding phases).

In terms of phasing, each of the measurement classes is considered at a different phase of the retrofit, except for the modification of opening surfaces. The openings are

Table 10.2 Elementary retrofit measures considered for the Barre Grimaud (thicknesses of external thermal insulation (ETI) are given in [mm]).

Class of measurement	Retrofit alternatives
Outside walls	ETI mineral wool: 100, 120, 150, 180, 200, 250 or 300 mm
Roofing	ETI polyurethane: 100, 150, 200, 250, 300, 350 or 400 mm
Lower flooring	ETI extruded polystyrene: 100, 120, 150, 180, 200 or 250 mm
Type of openings	Low-emission double or triple glazing, wooden frames
Surface of openings	North: increase factor: 0.8, 1 or 1.5
	West, South, East: increase factor: 0.8, 1, 1.25 or 1.5
Ventilation	Dual flow or Hygro-B-type ventilation
Heating	Condensing gas boiler (HHV yields considered: 90% at nominal load, 95% at partial load of 30%)
Hot water production	Solar fraction of production: 35%, 55% or 75%

resized here when the type of glazing is replaced, in line with obvious economic constraints. Like the windows, the outside walls on all façades are respectively retrofitted during the same phase. In this case study, the different retrofit phases are considered in sequence and separated by one year.

Based only on the above hypotheses, we could generate over 27.3 billion different energy retrofit programmes for the Barre Grimaud over a period of 50 years. This exploratory study thus illustrates how important it is to implement efficient methods.

c) Objective functions

In this case study, 7 objective functions are simultaneously considered to evaluate the performance of retrofit programmes in the Barre Grimaud extended life cycle (arbitrarily set at 50 years):

- Cumulated consumption of primary energy [MJ]   (Frischknecht *et al.*, 1996)
- Climate change potential at 100 years [kg CO_2 eq.]   (GIEC, 2007)
- Abiotic resources depletion potential [kg Sb eq.]   (Guinee *et al.*, 2001)
- Acidification potential [kg SO_2 eq.]   (Guinee *et al.*, 2001)
- Investment cost [k€]
- Overall (or global) cost over the life cycle [k€]
- Adaptive thermal discomfort in summer [degrees.hours]   (NF EN 15251, 2007)

At a earlier phase of the building work, additional objectives could be considered from among the decision criteria mentioned in the previous section.

In this chapter, the Barre Grimaud is the building support for implementing the methods and tools developed. The decision aid approach proposed in the following section is based on multi-criteria optimization by genetic algorithm.

IDENTIFYING COMPROMISE SURFACES

We set out an approach to decision aiding based on multi-criteria optimization by genetic algorithm to identify a Pareto frontier, i.e. a set of efficient strategies for a sequential energy retrofit. This set describes the accessible compromises in terms of energy consumption, impacts on the environment, economic and financial performance, and thermal comfort of the atmosphere.

a) An optimization approach by genetic algorithm

Genetic algorithms are stochastic optimization methods directly inspired by the mechanisms of the theory of evolution. We give a brief introduction below.

Representation of solutions and their application to sequential energy retrofits

By analogy, the solutions of a genetic algorithm are called individuals and represented by chromosomes. Chromosomes are made up of genes that model the parameters of the system to be optimized. Alleles, coded to the genes, transcribe the potential values that these descriptive parameters can take (Goldberg, 1989). Representing solutions by a given genotype should therefore allow us to fully characterize them.

In our case, the solutions are sequential energy retrofit programmes. Each solution is represented by a pair of chromosomes: the first codes the composition of the programme and the second codes its temporal phasing. Each gene in the chromosome composition represents a specific retrofit measure to be implemented. The alleles are thus alternatives for a given measure (e.g. type of window on the north façade). Each gene of the phasing chromosome corresponds to the position of a given retrofit measure in the temporal implementation sequence.

Algorithm principle and its application to sequential energy retrofits

Genetic algorithms are based on a simple principle. They establish the exploration of the search space for the evolution of a population of solutions, over several generations, by selecting the most suitable individuals.

For each generation, the performances of individuals in the population are evaluated. The best ones are selected for reproduction. A population of individual children is therefore generated by crossover and mutating the genetic material of the parents. Lastly, a selection process is applied to constitute the population of the next generation, from the current population and the children generated. Starting with a random initial population, the generational evolution thus gradually improves the performances of the solutions and the description of the accessible compromise surface.

Multi-criteria optimization by genetic algorithm covers a broad range of techniques. The NSGA-II (Non Dominated Sorted Algorithm), used in this research has performed well on different test problems (Deb *et al.*, 2000) (Zitzler *et al.*, 1999).

This algorithm uses an original strategy for the replacement selection. First, solutions are classified into increasing rank fronts. The non-dominated solutions belong to rank front 1. Then, from the dominated solutions the increasing ranks are iteratively identified on the basis of the relation of dominance. Then, for each solution, a "crowding distance" (Deb, 2001) is calculated to characterize its relative distance from its closest neighbours on the front it belongs to. Lastly, these solutions are selected according to their Pareto ranking, then their crowding distance. This strategy responds to the dual challenge of multi-criteria optimization, i.e. identifying efficient solutions and describing the overall compromise surface throughout.

In line with the specific representation of the solutions – sequential energy retrofit programmes – on two chromosomes, the crossover operators are differentiated for the composition and phasing chromosomes. A simple crossover at two cutting points is used for the chromosome composition. The Davis crossover (Murata *et al.*, 1995) is implemented for the phasing chromosome in order to generate viable crossovers and partially preserve the parents' temporal sequences.

b) Barre Grimaud case study

A decision-making approach through multi-criteria genetic optimization has been used to study the sequential energy retrofit of different buildings. The case study presented here concerns the Barre Grimaud, described in the preceding section.

Parameterization of the genetic algorithm

In this case study, multi-criteria genetic optimization has been carried out with the following set of parameters:

- Number of individuals in the current population: 200;
- Number of individuals of the offspring, generated at each generation: 200;
- Probability of genetic crossover on the chromosome composition: 80%;
- Probability of crossover on the phasing chromosome: 80%;
- Probability of genetic mutation on the chromosome composition: 1%;
- Probability of mutation on the phasing chromosome order: 1%.

The values were fixed in line with recommendations in published works (Vincenti *et al.*, 2010).

Results

Efficient retrofit strategies are identified by the algorithm NSGA-II developed, from an initial random population. The solutions obtained then form a compromise hypersurface in dimension 7, in line with the 7 objective functions considered.

In Figure 10.4, this hypersurface is projected onto the 21 planes formed by the pairs of criteria to make it easier to visualize compromises accessible to the decision. We show here the sets of non-dominated solutions identified at generations 1, 10, 20, 40, 60, 80 and 100. These different generational fronts can be used to graphically observe the search's progression in terms of both the performance of the solutions, and the description of the problem's compromise surface.

Different observations can thus be established:

- All of the concerned graphs highlight the necessary compromise between the investment cost and the reduction of impacts on the Environment. Overall, in the search area studied here, the more an energy retrofit programme contributes to mitigate environmental impacts, the more it costs in investments. However, the relationship is not affine and the marginal reduction of impacts starts decreasing rapidly starting at a certain level of investment.

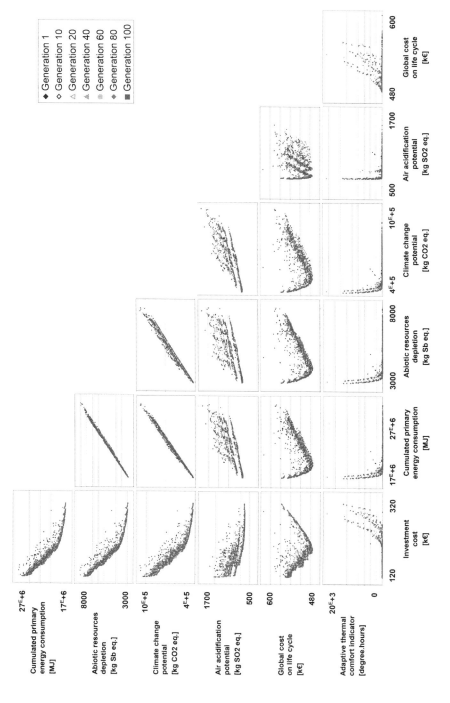

Figure 10.4 Evolution of the Pareto frontier over different generations.

- In addition, there is a global minimum in terms of overall cost over the extended life cycle (sum of investments and usage cost of the building, over 50 years in this case study). Given the hypotheses considered, the retrofit strategies with the most costly investments or, on the contrary, those with the lowest energy performance, do not prove to be optimal in terms of overall cost.

- What is more, significant correlations are observed between the different energy and environment indicators: cumulated primary energy consumption, climate change potential at 100 years, abiotic resources depletion potential, and to a lesser extent, atmospheric acidification potential. These indicators are largely correlated to the energy performance over the operating phase of the retrofitted building life cycle.

- All of the non-dominated sequential energy retrofit strategies guarantee a high level of thermal comfort in summer. The number of hours during which the inside temperature of at least one thermal zone exceeds 28°C, comes to less than 600 hours over 50 years, or 12 hours per year, for all of the solutions. This remark should be considered with caution. Indeed, it relates to Parisian climate conditions, except for heat wave periods, and in a context of suitable occupant behaviour (i.e. use of solar protections, shutters, and night cooling ventilation in summer).

On these Pareto frontier projections, three specific retrofit strategies can be identified: solutions "A, B and C", systematically shown in Figures 10.5 to 10.7. These three solutions illustrate the above remarks. They respectively correspond to the following sequential energy retrofit programmes:

Solutions **A, B** and **C** are characterized here by one single type of glazing on all of the façades.

Figure 10.5 illustrates the compromise between the investment cost and the cumulated primary energy consumption over the extended life cycle of the building to be retrofitted. The most efficient solutions in terms of reducing primary energy consumption also turn out to be the most expensive in terms of investment (solution **A**). As the above table shows, strategy A sequentially involves:

1   Replacing the existing heating system with a condensing gas boiler;
2   ETI on façades (mineral wool, $R = 7.3\ m^2.K/W$);
3   ETI on roofing (polyurethane, $R = 10\ m^2.K/W$);
4   Integration of solar thermal production for domestic hot water (solar fraction: 75%)
5   ETI on bottom floor (extruded polystyrene, $R = 8.2\ m^2.K/W$);
6   Replacing existing windows with triple-glazing;
7   Heat recovery ventilation.

On the same graph, solution **B** shows a compromise solution, characterized by a significant reduction in the investment costs for a limited increase in the cumulated primary energy consumption. Solution **B** differs from solution **A** by:

- Different phasing: the existing heating system is replaced during the final retrofit phase;

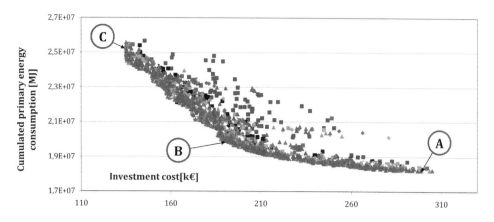

Figure 10.5 Pareto frontiers between investment cost and cumulated primary energy consumption.

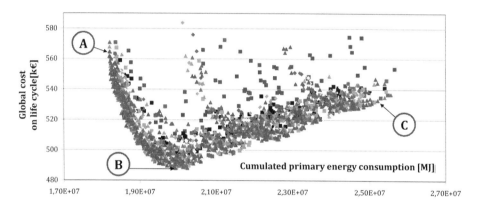

Figure 10.6 Pareto frontiers between overall cost and cumulated primary energy consumption.

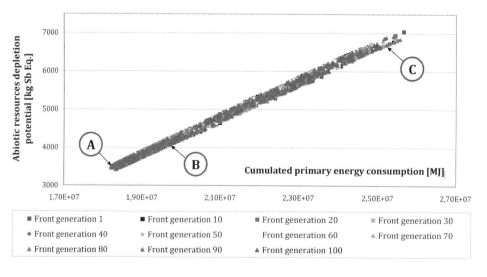

Figure 10.7 Pareto frontiers between cumulated primary energy consumption and abiotic resources depletion potential.

- Much lower levels of insulation: the thermal resistance of the envelope insulated from the outside reaches 2.9 m^2.K/W, 3.3 m^2.K/W and 3.4 m^2.K/W for the façades, roof and bottom floor respectively;
- Replacement of existing windows with double glazing.

Solution C represents the overall minimum in terms of investment cost, but is characterized by one of the lowest energy performance levels. Logically, and as detailed in Table 10.3, retrofit programme C involves the lowest level of insulation, the least efficient equipment, and the smallest solar fraction of domestic hot water (DHW) production, among the retrofit alternatives for the considered research area.

Table 10.3 Detail of retrofit programmes A, B & C.

	Class of measurement	*Retrofit alternatives*	*Phase No.*
A	Façade walls	ETI mineral wool: 300 mm	2
	Roofing	ETI polyurethane: 300 mm	3
	Bottom floor	ETI extruded polystyrene: 250 mm	5
	Openings	Low-emission triple glazing (argon filled)	6
	Ventilation	Heat recovery	7
	Heating	Condensing gas boiler	1
	DHW production	DHW Solar fraction of production: 75%	4
B	Façade walls	ETI mineral wool: 120 mm	1
	Roofing	ETI polyurethane: 100 mm	2
	Bottom floor	ETI extruded polystyrene: 100 mm	3
	Openings	Low-emission double glazing (argon filled)	5
	Ventilation	Heat recovery	6
	Heating	Condensing gas boiler	7
	DHW production	DHW Solar fraction of production: 75%	4
C	Façade walls	ETI mineral wool: 100 mm	2
	Roofing	ETI polyurethane: 100 mm	1
	Bottom floor	ETI extruded polystyrene: 100 mm	5
	Openings	Low-emission double glazing (argon filled)	3
	Ventilation	Hygro-B (humidity controlled)	6
	Heating	Condensing gas boiler	7
	DHW production	DHW Solar fraction of production: 35%	4

The performance analysis of these three solutions on the following graphs reveals interesting insights.

The most energy-efficient solutions (**A**) are not however optimal in terms of overall cost over the 50-year life cycle considered starting from the beginning of the retrofit work (Figure 10.6). On the contrary, the intermediate solution **B** represents the overall minimum in terms of overall cost over the extended life cycle; the lower energy performance for usage is compensated by the reduction in retrofit investments. Concerning the compromise between cumulated primary energy consumption and overall cost, the least energy efficient solutions, such as retrofit programme **C**, are not efficient from a Pareto point of view.

In addition, we can see on Figure 10.7 that the difference in performance between solutions **A** and **B** in terms of environmental indicators (such as cumulated primary energy consumption , abiotic resource depletion potential and climate change potential at 100 years) remains relatively low compared to the difference between **A** and **C** or **B** and **C**.

Lastly, with respect to the hypotheses chosen in this simulation, solutions **B** and **C** represent 0 degree-hours of discomfort over the 50-year life cycle, guaranteeing a higher summertime thermal comfort level than solution **A**, even though **A**'s level is still reasonable.

With their different combinations of retrofit measures and building work phasing, programmes **A**, **B** and **C** represent three different local optima (on at least one of the decision criteria) and thus three distinct types of preference on the decision criteria.

These analyses exhibit trade-off relations between the different decision criteria considered in this case study. Retrofit programmes **A**, **B** and **C** illustrate this compromise and allow us to identify the nature of the associated strategies (combinations of retrofit measures and phasing). Identifying these solutions necessarily involved constructing a (life cycle) model to evaluate them, based on multiple hypotheses. In the next section, we propose a study of how sensitive efficient retrofit strategies are to some of these hypotheses.

SENSITIVITY OF EFFICIENT STRATEGIES

In essence, the strategic planning of retrofit interventions on a building or stock of buildings is a prospective exercise, based on uncertain hypotheses, whose influence is measured over long periods, over the building's extended life cycle. Thus, decision-makers must be able to rely on elements that qualify how sensitive solutions are to the chosen hypotheses so that they can privilege the most robust strategies.

The sensitivity study must make it possible to apprehend the modelling parameters' influence on the efficient solutions, both in terms of nature and performance. This assumes identifying exact solutions that are independent from the research or optimization process used.

Although genetic algorithms have proved their efficiency for numerous industrial problems, they are still an approximate optimization method. Dynamic programming is therefore implemented for a more precise study. Alternative techniques may be considered, but we have chosen this algorithm because it can be directly applied to the discrete finite processes of sequential decisions, which include sequential energy retrofits of buildings.

The sensitivity study here focuses on the influence of three parameters: the heating temperature setting (rebound effect on occupants' thermal comfort); evolution scenarios for energies cost; and the lifespan of the retrofit components.

a) Dynamic programming and application to sequential energy retrofit

Dynamic programming first came to the forefront thanks to the work of Bellman (1957). It is particularly suited to the optimization of sequential decision-making processes. The technique is based on Bellman's recursive equation to find efficient pathways from non-dominated sub-pathways: "An efficient global pathway can only be formed from partial non-dominated pathways" (Vanderpooten, 2011). The advantage of using such algorithm is obvious for combinatorial problems that require comparing a high number of solutions or potential strategies (Duharcourt, 1969).

The constitution of a retrofit programme can be interpreted as a finite discrete sequential decision process. One or several basic measures are implemented at each stage of the retrofit. During each phase of the retrofit, the performances associated with a given decision thus depend exclusively on the current state of the building, which is entirely determined by its initial state and the sequence of measures previously implemented. Dynamic programming therefore seems particularly suited to sequential energy retrofits of existing buildings.

To apply Bellman's theory, the decision-making process is represented in the form of a sequential graph (see example in Figure 10.8) in which:

- The peaks, or the system's sequential states, represent the combinations of measures implemented during the previous decision phases;
- The arcs, or decisions, are formed of the retrofit measures that can be selected.

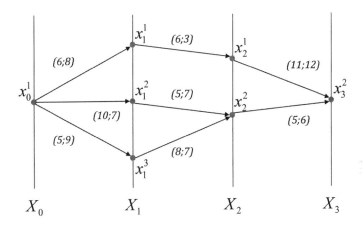

Figure 10.8 Example of sequential graph with multiple criteria. At each stage of decision X_i, different decisions (arcs of the graph) can be taken, bringing the system to a new state x^j_{i+1}. With multiple criteria, the value or performance of a decision is not a single real number but a vector of real numbers.

Bellman's equation, associated with the dominance relationship, can therefore be used to identify the strategies of retrofit solutions, like efficient pathways on this sequential graph.

An algorithm based on dynamic programming was also implemented. In addition to sensitivity problems, the knowledge of exact solutions allowed us to qualify the precision of the approximate methods, both in terms of nature and performance of solutions and description of the compromise surface on its extent. The approximate solutions identified by the genetic algorithm could therefore be validated on several study cases (Rivallain, 2013).

b) Three case studies on the sensitivity of solutions

Next, the identification of exact solutions allowed us to carry out sensitivity studies. These focused on 3 modelling parameters: the heating temperature settings; the evolution of prices of energy consumed for building usage; and the lifespan of components used in the retrofit.

Efficient retrofit strategies and the rebound effect on temperature settings

Occupants' behaviour is a complex issue influenced by activity, age, cultural habits and individual aspirations. However, it is a fundamental to energy retrofits. This is because the financing of operations often relies on a forecast of the savings that will be made during the building's usage. Then, experience has frequently revealed significant differences between the energy performance simulated in design, and the actual consumption revealed in operations. Usage conditions are partly responsible for these differences.

The heating savings generated by housing's energy performance are thus often reinvested into raising the thermal comfort and heating settings: this is the rebound effect. Hass and Biermayr (2000) estimate that the rebound effect leads to a 20% to 30% increase in heating consumption in Austria.

Since buildings are designed to serve the activities of their occupants (and not the other way round), the variability of behaviour patterns should be integrated into an analysis to study the influence of temperature settings on the type and performance of retrofit strategies. In the case study concerning the Barre Grimaud, the compromise surfaces relative to temperature settings of 19°C, 22°C and 24°C were thus identified then compared.

Figure 10.9 shows the compromise surfaces for the settings 19°C and 22°C and clearly illustrates that the higher the temperature setting, the greater the impact on the environment and the overall cost over the life cycle.

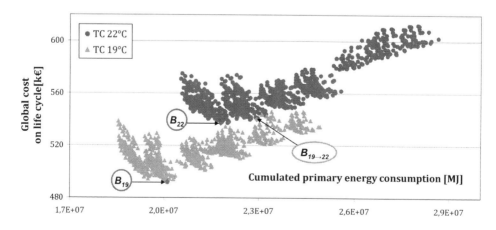

Figure 10.9 Energy/overall cost compromise and influence of an increase in the heating setting, for an optimal solution at 19°C.

If a rebound effect occurs, in order to reach a minimum in terms of overall cost over the life cycle, the nature of the retrofit solutions has to change, i.e. reinforced thermal insulation for a temperature of 22°C (B_{22}: 150 mm of EIT) compared to 19°C (B_{19}: 100 mm).

To illustrate the rebound effect, we observe that the use of a temperature of 22°C in the Barre Grimaud with a retrofit following solution B_{19} leads to a rise of 14% in cumulated primary energy consumption (including the manufacture, replacement and end-of-life of building products), 30% in climate change potential at 100 years, 18% in acidification potential, and 10% in overall cost over the life cycle (comparison with solutions B_{19} and $B_{19 \to 22}$) (Rivallain, 2013).

Intermediate strategies, implementing average levels of thermal insulation to the façades and bottom floor (from the considered decision space, i.e. ETI 150 mm), and often corresponding to solutions close to the optimum in terms of overall cost, prove highly robust to measured variations in heating settings imposed by occupants.

Influence of energy price changes

For the case study considered, existing buildings are used for 50 years from the start of energy retrofit operations. Over this period, using the premises requires energy consumption. One of the strong hypotheses of the model thus depends on the changing costs of energy consumed to satisfy this usage.

Based on DGEC forecasts (used by default), alternative scenarios have been built. A first, conservative scenario called "DGEC −50%" was built as follows:

- For gas: 50% reduction in multi-annual growth rates;
- For electricity: basic DGEC scenario.

The second, more pessimistic scenario, called "DGEC +50%", in contrast considers:

- For gas: 50% increase in multi-annual growth rates;
- For electricity: introduction of UFE (Union Française de l'Electricité) hypotheses from the study, "Electricité 2030, quels choix pour la France" (2010).

No change in the electricity mix is introduced at this stage, and the environmental balance of the kWh of electricity consumed in 2050 is the same in 2010. A complementary sensitivity study could challenge this hypothesis.

The efficient strategy groups identified for the Barre Grimaud, for each of the energy price change scenarios considered are shown on the following figure, by projection.

On Figure 10.10, performance levels in terms of overall cost appear very distinct for the different scenarios. In particular, low energy performance solutions (type C) are highly sensitive to price changes for the energy consumed for usage, which can considerably reduce their performance in overall cost over the life cycle. On the other hand, the nature of specific solutions (composition and phasing) of type A, B or C does not depend on the scenario considered for the chosen hypotheses.

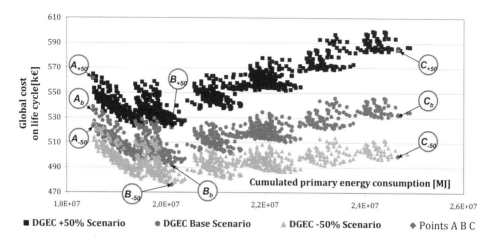

Figure 10.10 Scenarios of energy price changes and compromise surfaces between overall cost, cumulated consumption of primary energy and investment cost.

Control the life spans of components for an efficient energy retrofit

Lastly, we consider here the sensitivity of results to the life spans considered for energy retrofit components.

By default, these were considered as at least equal to 50 years, thus entailing no replacements over the entire life cycle studied. The choice was mainly motivated by the difficulty in anticipating the state of technologies in 20 to 30 years when they are replaced, and the consequential impact on the discrimination of retrofit strategies.

An alternative scenario, based on the identical replacement of the components, according to 20-to 30-year maintenance cycles, was thus introduced to challenge this hypothesis. The life spans considered are thus (Thiers and Peuportier, 2012):

- 30 years: for the insulation and coatings, openings and ventilation equipment;
- 20 years: for the heating systems and thermal solar panels.

The following figures represent two projections of compromise hypersurface identified.

Although performances are at about the same level for some environmental indicators (cumulated primary energy consumption, abiotic resource depletion potential, climate change potential at 100 years) largely correlated to energy performance in usage, we can see a clear difference between the two Pareto frontiers in terms of economic criteria: investment cost and overall cost over the life cycle. These differences can obviously be put down to the replacement of retrofit components, which leads to a multiplication of cumulated investment costs by a factor higher than 2.

The more energy efficient the strategies are, the greater the differences in performance, in line with the cost of the associated retrofit components.

It is essential to note that the optimum overall cost, when components are replaced at 20 to 30 year periods, corresponds to a much lower energy performance solution.

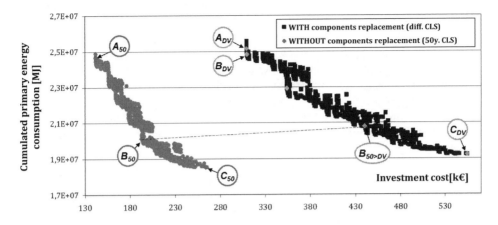

Figure 10.11 Maintenance and investment/energy compromise scenarios.

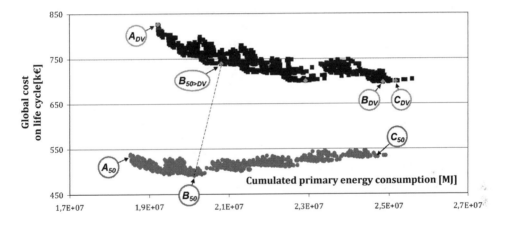

Figure 10.12 Maintenance and energy/overall cost compromise scenarios.

In conclusion, life spans and maintenance scenarios have a clear impact on the performance levels of compromise surfaces and solutions including the financial profitability of energy retrofit strategies. Looking for solutions to optimize energy performance in usage, environmental balance, and overall cost, encourages advances in terms of life span, in relation to the 20- to 30-year levels often considered.

CONCLUSION AND PERSPECTIVES

Renovating existing buildings represents a significant challenge for sustainable development and energy transition. It is an eminently complex problem that requires intervention on the built heritage and our living environment, while guaranteeing energy performance with profitable investments and limited impacts on the Environment, preserving and contributing to the comfort of occupants.

This chapter presents a contribution to decision aiding on strategically planning retrofit interventions. Efficient sequential energy retrofit programmes are identified here using different optimization techniques on a multi-criteria basis over the life cycle of existing buildings. Solutions are characterized by both their nature and their performance. One of the original features of this study is that it identifies compromise surfaces involving criteria on energy, the environment, economics and society. These Pareto frontiers constitute rich information by which deciders can make their own choices.

The approach developed makes a modest contribution to aiding decisions. The transformation of building stocks, with a view to sustainable energy retrofits, once again raises multiple questions that point to different development perspectives.

A study of the robustness of efficient strategies thus merits continued research to anticipate phenomena like resilience to climate change, the influence of maintenance cycles, and sensitivity to the dynamic evolution of techniques. The potential pathologies following retrofit interventions also need to be fully understood.

Economic appraisals should also be coupled with the relevant finance mechanisms.

Buildings, as objects with long life spans, constitute assets with considerable intergenerational value in economic, physical, social and cultural terms. As a result, energy retrofits should be part of global, holistic long-term heritage management strategies developed to ensure that their quality and multidimensional value last through time.

Reflection on strategic planning should therefore take on a new scale to introduce complementary decision variables specific to building stocks, and to seek overall policy efficiency.

In addition to identifying compromises, decision aid should result in progressively building up the choices made. An interactive process by which preferences are progressively expressed could provide support for dialogue between project managers, architects and engineers.

Lastly, although successful energy retrofits will involve informed decisions, they will also depend on cross-cutting management of performance right through to the operating guarantee, which is a topical subject if ever there was one!

REFERENCES

ADEME Agence de l'Environnement et de la Maîtrise de l'Energie (2011). Les chiffres clés du bâtiment: énergie-environnement.

AFNOR (2007). Indoor environmental input parameters for design and assessment of energy performance of buildings addressing indoor air quality, thermal environment, lighting and acoustics. NF EN 15251.

Alanne, K. (2004). Selection of renovation actions using multi-criteria "knapsack" model, Automation in Construction 13, pp. 377–391.

Bellman, R. (1957). Dynamic Programming. Princeton University Press, New Jersey.

Collette, Y., Siarry, P. (2002). Optimisation multiobjectif. Eyrolles, 2002.

Deb, K. (2000). An efficient constraint handling method for genetic algorithms. Computer methods in applied mechanics and engineering, 186, (2–4), pp. 311–338.

Deb, K. (2001). Multi objective optimization using evolutionary algorithms, Wiley.

Dréo, J., Petrowski, A., Siarry, P., Taillard, E. (2003). Métaheuristiques pour l'optimisation difficile. Recuit simulé, recherché avec tabous, algorithmes évolutionnaires et algorithmes génétiques, colonies de fourmis. Eyrolles.

Duharcourt, P. (1969). Introduction a la programmation dynamique. Revue économique, Vol. 20, No. 2, pp. 182–234. Published by: Sciences Po University Press.

Frischknecht, R., et al. (1996). Ökoinventare für Energie systeme, Eidgenössische Technische Hochschule, Zürich, 1996, 1817 p.

Frischknecht, R., Jungbluth, N., Althaus, H.J., Doka, G., Heck, T., Hellweg, S., et al. (2004).Overview and methodology, ecoinvent report No. 1. Dübendorf, Swiss: Swiss Centre for Life Cycle Inventories.

Goldgerg, D. E. (1989). Genetic Algorithms in Search, Optimization, and Machine Learning. Addison Wesley.

Guinée, J.B., Gorrée, M., Heijungs, R., Huppes, G., Kleijn, R., de Koning, A., van Oers, L., Wegener Sleeswijk, A., Suh, S., Udo de Haes, H.A., de Bruijn, H., van Duin, R., Huijbregts, M.A.J., Lindeijer, E., Roorda, A.A.H., Weidema, B.P. (2001). Life cycle assessment; an operational guide to the ISO standards; Ministry of Housing, Spatial Planning and Environment (VROM) and Centre of Environmental Science (CML), Den Haag and Leiden, The Netherlands, 704 p.

Hass, R., Biermayr, P. (2000). The rebound effect for space heating: empirical evidence from Austria. Energy Policy, 28 (6–7), pp. 403–410.

IPCC, Foster, P.M. (2007). Change in atmospheric constituents and in radiative forcing. In: Salomon, S., Qin, D., Manning, M., Chen, Z., Marquis, M., Averyt, K.B., Tigor, M., Miller, H.L., Climate Change 2007: The physical basis. Contribution of the working group I to the Fourth Assessment Report of the IPCC, Cambridge University Press.

Kohler, N., Hassler, U. (2002). The building stock as a research object. Building research and information, 30(4), pp. 226–236.

Kolokotsa, D., Diakaki, C., Grigoroudis, E., Stavrakakis, G. and Kalaitzakis, K. (2009). Decision support methodologies on the energy efficiency and energy management in buildings. Advances in building energy research 3(1), pp. 121–146.

Meijer, F., Itard, L. and Sunikka-Blank, M. (2009). Comparing European residential building stocks: performance, renovation and policy opportunities. Building Research & Information, 37(5–6), Special Issue: Research on Building Stocks.

Murata, T., Ishibuchi, H. (1995). MOGA: Multi objective genetic algorithms. In proceedings of the Second IEE Interna-tional Conference of Evolutionary Computation, pp. 289–294.

Pernodet Chanterelle, F. (2010). Méthode d'optimisation multicritère de scénarios de réhabilitation de bâtiments tertiaires. Application à l'évaluation de dispositifs de rafraîchissement hybride. Thèse de doctorat CSTB, ENTPE.

Peuportier, B., Blanc-Sommereux, I. (1990). Simulation tool with its expert interface for the thermal design of multizone buildings. International Journal of Solar Energy, 8, pp. 109–120.

Popovici, E.C. (2006). Contribution to the life cycle assessment of settlements. PhD Thesis, MINES ParisTech.

Rivallain, M. (2013). Etude de l'aide à la décision par optimisation multicritère des programmes de réhabilitation énergétique séquentielle des bâtiments existants. Thèse de doctorat. Ecole des Ponts ParisTech, MINES ParisTech.

Roy, B. (1985). Méthodologie multicritère d'aide à la décision, Ed. Economica.

Sidler, O. (2009). La rénovation à très basse consommation d'énergie des bâtiments existants. Formation Institut NegaWatt.

Thiers, S., Peuportier, B. (2012). Energy and environmental assessment of two high energy performance residential buildings. Building and Environment, 51, pp. 276–284.

UNEP: United Nations Environment Program (2003). Industry and environment.

Vanderpooten, D. (2011). Aide multicritère à la décision et optimisation multiobjectif. Journée multicritère/multiobjectif, Nancy, 21 septembre 2011.

Vincenti, A., Ahmadian, M.R., Vannucci, P. (2010). BIANCA: a genetic algorithm to solve hard combinatorial optimisation problems in engineering. Journal of global optimization, 48(3), pp. 399–421.

Zitzler, E., Deb, K., Thiele, L. (1999). Comparison of multiobjective evolutionary algorithms: Empirical results. Technique 70, Computer Engineering and Networks Laboratory (TIK), Swiss Federal Institute of technology (ETH) Zurich.

Chapter 11

Energy efficient building control strategies

Bérenger Favre & Bruno Peuportier
MINES ParisTech

Low-energy buildings, incorporating high levels of thermal insulation, are very sensitive to climate variations (solar gains) and occupants' behaviour (internal gains). They therefore require refined management to satisfy greater requirements of comfort and every savings. Buildings represent 60% of total electricity consumption and incur very high seasonal variations in demand, in particular for electric heating (peaks of over 100,000 MW in winter in France, compared to around 40,000 MW in summer). Reducing these peaks would diminish the environmental impacts corresponding to carbon-intense production.

The aim is therefore to devise management strategies that, for a given comfort level, can minimize energy requirements over a time period including a future period: e.g. deciding at each time step whether or not to store energy in order to minimize a building's consumption due to an anticipated change in the climate, usage scenario, or even electricity production. Meteorological forecasts can be used to anticipate heat waves. In this case, a night free cooling strategy can be implemented and associated with high thermal mass, to reduce temperature peaks, especially in summertime.

The first step is to define the quantity(ies) to optimize, also known as the "cost function": final and primary energy consumption, cost, degrees-hours of discomfort, CO_2 emissions, generation of radioactive waste, etc. If a stochastic meteorological forecast model is used then optimization concerns the expected value of these quantities, since the data include a stochastic element.

The models used to evaluate the cost function are as follows:

-   building model (a reduced model is more suitable because a great number of calculations need to be carried out),
-   climate model (stochastic evolution model, e.g. a Markov chain, or use of meteorological forecasts),
-   occupancy model (e.g. thermostat set points over time, possibly random actions),
-   electricity production model and corresponding CO_2 emissions.

A method for generating an "optimal" management strategy (e.g. to minimize energy consumption over the period considered) is the dynamic programming method. This involves defining a set of possible actions for each time step, restricted to a reasonable number of configurations (on/off of a piece of equipment, level of opening of a blind, free cooling, storage, etc.). The quantity to be optimized can be compared for these different configurations over a time step, then little by little over successive time steps. The Bellman equation can reduce the number of calculations. A control strategy can be defined from analyzing the results, and then comparing it to a more standard approach (i.e. without anticipation).

The present chapter describes the method and its limits, then several potential applications, considering action on heating, solar protection, and ventilation. This work corresponds to a thesis by Bérenger Favre [FAVR13].

PRESENTATION OF THE METHOD

Different models used

The energy management of a building can only be optimal if we anticipate the evolution of the various elements interacting with it. In addition to the building's thermal dynamic model, we must therefore describe the occupancy and local meteorology models feeding into it, as well as the electricity production model.

Dynamic thermal model of the building

A building is divided into thermal zones, defined as areas with homogeneous temperature. In each of these zones, each partition (wall, floor, ceiling) is also divided into small volumes constituting nodes that are sufficiently fine to be considered at a homogeneous temperature. Another node exists for the air temperature and the zone's furniture. A heat balance is then applied to each node of the building thus modelled. It takes into account:

P_{cond}: losses (or gains) by conduction in the walls, floor and ceiling, etc.
P_{sol}: gains from solar radiation through the windows
P_{conv}: losses (or gains) by convection on the wall surface
P_{in}: gains from heating, occupancy and internal gains transmitted into the air
P_{ponts}: losses (or gains) due to thermal bridges
P_{ventil}: losses (or gains) from air exchanges

By carrying out a heat balance on the air node in a thermal zone, the energy stored is:

$$C_{air}\dot{T}_{air} = P_{in} + P_{cond} + P_{ponts} + P_{ventil} + P_{sol} + P_{conv} \tag{11.1}$$

C_{air} being the heat capacity of the air node including furniture.

Proceeding analogously for each node in each zone of the building, and linearizing the radiation exchanges, we obtain the following system of continuous linear equations [PEUP 90]:

$$C\dot{T} = AT + EU \tag{11.2}$$

where T is the discretized temperature field, U is the vector of driving forces (outside temperature, solar radiation, heating or air conditioning power, etc.), C is the diagonal matrix of thermal capacities, A is the matrix containing the terms of exchange between nodes, and E is the matrix containing the terms of exchange between nodes and driving forces.

In steady state, $\dot{T} = 0$ and in this case $T = A^{-1}EU$. It is then possible to write the temperature field as follows:

$$T = T_0 - A^{-1}EU \tag{11.3}$$

where $A^{-1}EU$ is the temperature in steady state and T_0 the dynamic part of the temperature field. We thus obtain:

$$C\dot{T}_0 = C^{-1}AT_0 + A^{-1}E\dot{U} \tag{11.4}$$

We can resolve this system more easily by making it so that the derivative of a node temperature is only connected to this temperature. The transition matrix that allows this transformation is called P. Thus:

$$T_0 = PX \tag{11.5}$$

Applied to the equation (11.4), we obtain:

$$\dot{X} = FX + B\dot{U} \tag{11.6}$$

where F is a diagonal matrix whose ith term is $-1/\tau_i$, τ_i being the ith time constant of the system. Terms corresponding to time constants lower than a third of the time step in the simulation can be disregarded in the dynamic part of the temperature field, since the steady state is almost achieved. After comparisons with complete models, the number of time constants is reduced to 10. This model reduction method is suitable for optimization problems that involve long calculations.

We thus define the Y outputs:

$$Y = HX + SU \tag{11.7}$$

The zone models are then coupled, with the temperatures of the adjacent partitions constituting the outputs of a zone (vector Y) and the inputs of contiguous zones (vector U). The vectors of state X, outputs Y, and driving forces U of the various zones are grouped together to form the global vectors X_g, Y_g and U_g for the entire building, and a matrix F_g is formed by arranging the F matrices in a diagonal.

This system can then be integrated into a time step Δt, so that we can obtain the values of states and outputs in time $(n+1) \cdot \Delta t$, given as X_g^{n+1} and Y_g^{n+1}, according to the values X_g^n and Y_g^n in time $n \cdot \Delta t$. We also note U_g^{n+1} and U_g^n the driving force vectors at time $(n+1) \cdot \Delta t$ and $n \cdot \Delta t$. We then obtain a matrix system as follows:

$$\begin{cases} X_g^{n+1} = e^{F_g \cdot \Delta t} \cdot X_g^n + W_g^\alpha(U_g^{n+1} - U_g^n) + W_g^\beta(Y_g^{n+1} - Y_g^n) \\ Y_g^{n+1} = MGIF \cdot X_g^n + MGIE \cdot U_g^{n+1} - MGID \cdot U_g^n + MGIC \cdot Y_g^n \end{cases} \tag{11.8}$$

The vector U_g comprises driving forces related to climate (temperature, solar radiation, etc.) and occupants (e.g. the heat released from the use of domestic appliances).

Modelling occupancy and climate

The climate data generally used in simulation tools correspond to typical years, constituted by sequences of days from actual years chosen to respect average values over 20 years of temperature, solar radiation, etc. These typical years are adapted to study building design, but meteorological forecasts are better suited to real-time management. These forecasts are generally available at a regional scale, but do not take microclimates into account. A potential line of research involves adapting these data to a local context. In the study presented here, the main objective is to evaluate the feasibility of flattening the heating system's peak electricity consumption. The meteorology is therefore considered as known in advance and represents a very cold week in winter from a typical year.

The occupancy of a building is also considered in a simplified manner. It is modelled in a time form as the number of people present in a residence with the corresponding heat emission proportional to the number of people. The occupancy considered in the residence studied is 4 people, each emitting 80 W, present during the week except for 8 am to 5 pm, and two people present from 5 pm to 6 pm. During weekends, the four people are present all the time. The internal gains in the building correspond to a representative usage of domestic appliances and other energy consumption (lighting, computers, etc.).

Modelling electricity production

This model is used when the objective is to minimize greenhouse gas emissions. The global warming potential (in kg of CO_2 equivalent) is used to build the cost function. To evaluate this indicator, we use the Swiss database, EcoInvent, with the following data for the different means of electricity production.

CO_2 emissions per kWh for the different production means considered.

CO_2	gCO_2/kWh
Nuclear	7.9
Coal	1065.6
Gas	640.8
Fuel oil	882
Hydraulic	3.5

By coupling the EcoInvent data with electricity grid data, we can obtain CO_2 emissions according to the French electricity production mix for each moment of the year. The example studied corresponded to the week of 23 to 30 November 2008, and only counted French national production (imports are added later); the electricity mix and greenhouse gas emissions resulting from this production are presented in Figure 11.1. The CO_2 emissions reach their highest values at the same time as the electricity production of the French facilities, with two peaks at around midday and 7 pm also visible on the CO_2 emissions curve. Carrying out an optimization using this

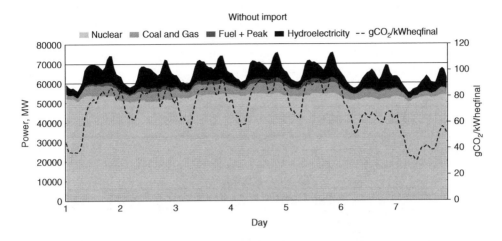

Figure 11.1 French electricity mix without imports and the corresponding greenhouse gases emissions.

cost function should result in flattening not just the 7 pm peak demand but also the peak hours at around midday.

To include imports in this graph, we need to know the electricity mix of the imported electricity, which originates in numerous countries as shown in Figure 11.2. The electricity mix in peak periods is different for each of these countries, but we can see that most imports come from Germany and Switzerland.

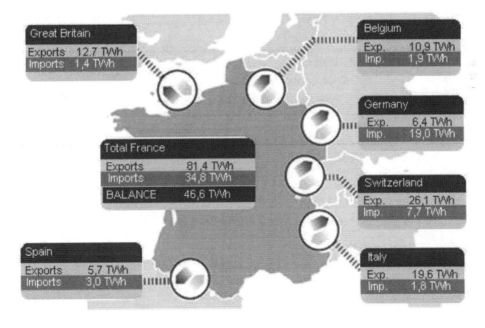

Figure 11.2 French contractual exchange in 2008 with border countries [grid].

Given that electricity production involves high levels of carbon during peak periods in Germany, we put forward the hypothesis that 75% of the electricity imported is thermal, with the remaining 25% nuclear. Of the 75% thermal electricity, we consider that 55% comes from coal and 45% from gas. The electricity mix and greenhouse gas emissions resulting from this production are thus shown in Figure 11.3.

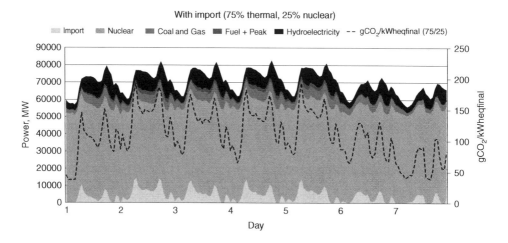

Figure 11.3 French electricity mix with imports and the corresponding greenhouse gases emissions.

When imports are added, the midday CO_2 emissions peak increases and sometimes even exceeds the 7 pm emissions peak. In fact, the CO_2 emissions curve now tends to follow imports more than the maximum production of the French production facilities, with the two curves remaining very close together most of the time, although they sometimes move apart for several hours.

We use the two curves of CO_2 emissions equivalent per kilowatt-hour presented in Figure 11.1 and Figure 11.3 to determine the usage cost of heating power for each time step.

The models representing the different elements of the system having been described, the optimization algorithm used to establish the management strategies is presented below.

Algorithm for optimizing the control

Standard and anticipative control

The object of this work is to devise optimal strategies for energy management in buildings. These strategies are sequences of controls to be carried out at each time step. The optimal strategy obtained is thus carried out in a sequential framework. The objectives can be multiple, e.g. maintaining comfort during summer heat waves, or reducing electricity consumption during peak periods in winter. In both cases, the aim is to obtain an optimal control sequence (e.g. electric heating, blinds or ventilation) over a given period to respond to an objective while satisfying constraints, e.g. comfort. Two types of entity can be distinguished.

- *Physical* entities
  o *The building*, represented in state variables and containing control commands (heating, openings)
  o *The occupants*, who impose comfort constraints that influence the building's thermal dynamics and set optimization targets
  o *The driving forces*, such as the weather, the heat emitted by the occupants and their activities or the site, which determines the local climate and the building's masks.

- *Control* entities
  o *The objective function*: this is the actual objective of the control, to minimize energy consumption, optimize comfort, minimize environmental impacts during a given period.
  o *The command*: this is what will be controlled to reach the previously defined targets. In our case, it involves controlling electric heating, blinds and openings. The outcome of an optimization is a succession of commands minimizing a cost function.
  o *The constraints*: they can either be expressed on the control entities, e.g. limited calculation capacities or a constraint on the command, or on physical entities, where the occupant imposes comfort constraints.

The time step of the dynamic thermal simulation defines the time step of the optimization and thus the discretization of the variables if applicable. A control strategy is established on a finite horizon, i.e. for a limited time period. This limitation is obtained from a stipulation of the local weather forecast, which becomes less reliable as the forecast horizon moves away.

Standard control method

Standard control responds to occupants' comfort objectives. This kind of control can be used to maintain the temperature at 19°C, for example. If meteorological conditions change, causing solar gains to drop, then this decrease can be compensated by an increase in heat gains using heating, to maintain the same 19°C comfort conditions. It is only possible to proceed this way if temperature changes are not too significant. Here, the only objective is to maintain the occupant's comfort, and there cannot be any additional objectives like minimizing energy consumption or totally stopping energy consumption during certain periods while maintaining comfort. However, dynamic control can be used to satisfy these additional objectives.

Anticipative control method

Anticipative control methods are based on anticipating changes in driving forces, constraints or new perturbations. For buildings, these driving forces concern local weather conditions and occupancy. Constraints and objective functions can also evolve, e.g. due to a change in price policy or comfort temperatures.

When a heat wave is anticipated, outside openings can be created during the preceding night, thus generating free cooling that brings down the building's temperature. During very hot periods, comfort can therefore be improved without the use of air conditioning. Standard static control would probably be insufficient to maintain comfort in similar conditions. In the same case of high summer temperatures, an anticipation of the building's occupancy levels can be used to forecast internal heat gains and thus partially compensate using free cooling during the night. These two examples illustrate the benefits of anticipative control compared to standard control.

Control through anticipating local weather conditions and occupancy levels can be used to work on energy storage in the building to minimize energy consumption and improve occupants' comfort. This kind of control can also be used to anticipate significant changes, e.g. a price increase or CO_2 emissions from electricity during peak periods, and thus reduce the costs and impacts of heating and air conditioning.

Dynamic programming is a method for anticipative control that can be used to establish a set of optimal controls over time to minimize an objective function.

Discrete-time dynamic programming

Dynamic programming was implemented by Bellman [BELL57]. The system studied is fully described by a *discretized* state variable at a given point in time t. This variable can be subject to constraints. Dynamic programming studies all of the accessible states at all time steps of the optimization, which means that this algorithm performs very well in a sequential study. Since the whole area of states located in the constraint interval is scanned, the control sequence obtained is necessarily optimal, in the discretization conditions used. On the other hand, it takes a long time to calculate the optimization algorithm, but calculation capacities are making progress.

Variable discretization principle

The optimization by dynamic programming is applied over a discretized time interval at a time step Δt, as are the state variable (known as x) and the driving forces. In other words, for each time step, we take a single value of internal gains, outside temperature and solar radiation. Specifically, the time step t corresponds to state variable $x(t, \cdot)$, internal gains app(t), outside temperature $T_{\text{ext}}(t)$ and overall horizontal solar radiation $R(t)$ with $t \in [0, N]$, N the number of time steps in the simulation.

The state variable x is also discretized considering a number of intervals Ne between the extreme values $x_{\min}$ and $x_{\max}$. This double discretization can be graphically represented on a *state graph*, see Figure 11.4. Unlike other problems that use the graph theory, in our case it is not possible to pass again through a state already covered during a previous test. The study's temporal nature means that we cannot go backwards and can only move from one state to another in a single direction. As a result, all of the policies necessarily involve the same number of steps: the study framework is sequential as well as combinatorial.

The temporal discretization of all the variables used in the dynamic programming involves a multi-step decision process. At the start of each time step – or stage – a choice of commands is made that depends on the system's state (state variable) and the time step characteristics considered (driving forces). The choice of command determines the values of the system variable at the following time step, and thus influences all of the

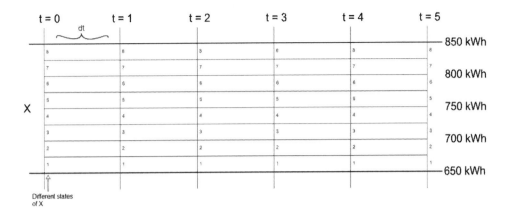

Figure 11.4 Example of state graph with energy as a variable for five time steps.

following time steps. The optimal solution for the optimization is therefore a sequence of decisions (commands) on every time step considered following the objective function criteria.

How dynamic programming operates

An optimization by dynamic programming is carried out over N time steps. The work horizon is thus finite. We are placing ourselves in a *deterministic* framework, i.e. all of the values of future driving forces are known.

We note u as the control variable comprising Nc dimensions:

$$u(t) = u_t \in U_t, U_t \subset \mathrm{R}^{Nc} \tag{11.9}$$

where U_t is the set of admissible commands. The equation of state at each time step t is:

$$x(t) = x_t, x(t+1) = f(x(t), u(t), t) \tag{11.10}$$

with an initial condition $x(0) = x_0$, x being the discrete-time state variable:

$$x(t) = x_t \in X_t, X_t \subset \mathrm{R}^{Ne} \tag{11.11}$$

where X_t is the set of all of the admissible states of dimension Ne. The limits of the set X_t are defined arbitrarily at the start of each optimization.

The equation of state is dynamic since the start of time t is changed at each time step. We use $\Gamma_t(x_t)$ to designate all of the possible successors of x_t, Γ_t being a correspondence of X_t in X_t. By applying correspondence Γ to the initial states and then to their successors and so on, all the admissible states can be defined and thus explicitly the dynamic programming graph.

We can now define a value function that is the price of passing from a state x at time t to another state x at time $t+1$:

$$v_t(x_t, x_{t+1}), \quad x_{t+1} \in \Gamma_t(x_t) \tag{11.12}$$

We work in chronological order, and define a cost function that is the sum of the value functions covered from t_0 to t:

$$V_0^t = \sum_{j=0}^{t-1} v_j(x_j, x_{j+1}) \tag{11.13}$$

where $V_0^t = V_t$ when the cost function starts at t_0.

The aim is to find a policy, or in other words a command vector, to maximize (or minimize) the cost function:

$$J = \text{Max}[V_N] \tag{11.14}$$

In sum, dynamic programming involves finding a sequence of commands $U_N = (u_0, u_1, \ldots, u_N)$ in order to maximize a criterion defined by (11.14), based on a system described by (11.10) with constraints both on the state variable of the system (11.11) and on the command (11.9).

Bellman's principle of optimality

The equation (11.14) can be used to find the optimal policy by comparing all of the possible existing policies. The greater the number of states Ne and the number of time steps N, the higher the number of possible policies. The number of calculations to be carried out thus increases considerably, as do the optimization durations. Bellman's principle of optimality makes it easier to find the optimal policy by limiting the number of calculations: "an optimal policy can only be made up of optimal sub-policies" [BELL 57]. Bellman's equation is the discrete time equivalent of the Hamilton-Jacobi-Bellman equation [BRYS 75]. Applied to our case study, this comes down to saying that an optimal policy from t_0 to $t = N$ can only be made up of optimal sub-policies from t_0 to $t = N - 1$. This optimal sub-policy can in turn only be made up of optimal sub-policies from t_0 to $t = N - 2$ and so on.

$$J = \text{Max}[V_N] = \text{Max}(v_0(x_0, x_1)) + \text{Max}(V_1^{N-1}) \tag{11.15}$$

To illustrate this principle, a simple case is presented in the following Figure 11.5. The state variable is discretized into 5 intervals. After a first time step, five policies move to t_1 through states 1, 2, 3, 4 and 5 with a cost of arriving at these states of respectively c_1, c_2, c_3, c_4, c_5. At the following time step, the five policies have a command from the set U_t taking them to the same state 3. Only the optimal policy is chosen for state 3 at t_2. For this, the value functions are calculated for these five policies, at respectively $v_{t1}(1_{t_1}, 3_{t_2})$, $v_{t1}(2_{t_1}, 3_{t_2})$, $v_{t1}(3_{t_1}, 3_{t_2})$, $v_{t1}(4_{t_1}, 3_{t_2})$ and $v_{t1}(5_{t_1}, 3_{t_2})$.

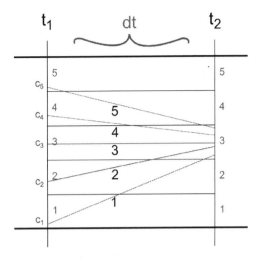

Figure 11.5 Elimination of policies in dynamic programming.

It is now possible to compare the five policies and only retain the optimal policy:

$$J = \text{Max}[V_2] = \text{Max}[c_1 + v_{t1}(1_{t1}, 3_{t2}), c_2 + v_{t1}(2_{t1}, 3_{t2}), c_3 + v_{t1}(3_{t1}, 3_{t2}),$$
$$c_4 + v_{t1}(4_{t1}, 3_{t2}), c_5 + v_{t1}(5_{t1}, 3_{t2})]$$

In this example, the policy starting from the initial state, passing through 1_{t1} and arriving at 3_{t2} is assumed as optimal, and all the other policies are abandoned. At the following time step, policies leaving from 3_{t2} will have an initial cost of $c_1 + v_{t1}(1_{t1}, 3_{t2})$.

Multi-objective optimization

In a multi-objective optimization, two alternatives are possible. The first is to keep a single value function with a weighting factor for each value sub-function:

$$v(x_t, x_{t+1}) = av_1(x_t, x_{t+1}) + bv_2(x_t, x_{t+1}) + \ldots \tag{11.16}$$

In this case, several optimizations are undertaken to find the optimal values of the weighting factors a, b, etc. according to the final objective.

The second method separates each of the value sub-functions, so that the global value function is thus the size of the number of objectives in the optimization:

$$v(x_t, x_{t+1}) = \begin{cases} v_1 & (x_t, x_{t+1}) \\ v_2 & (x_t, x_{t+1}) \\ \text{etc} \ldots \end{cases} \tag{11.17}$$

The cost function thus becomes:

$$V_0^t = V_t = \sum_{j=0}^{t-1} v_j(x_j, x_{j+1}) = \begin{cases} \sum_{j=0}^{t-1} v_1(x_j, x_{j+1}) \\ \sum_{j=0}^{t-1} v_2(x_j, x_{j+1}) \\ \text{etc} \ldots \end{cases} \qquad (11.18)$$

In this case, several policies can be chosen at each state of each time step. In fact, a policy is only eliminated if it is dominated, i.e. if it is less efficient in terms of the objective function for all of the cost sub-functions $\sum_{j=0}^{t-1} v_1(x_j, x_{j+1}), \sum_{j=0}^{t-1} v_2(x_j, x_{j+1})$, etc. than any other policy that reaches this state at time step t. Thus, at the end of the optimization one or several non-dominated solutions exist for each state. All of the non-dominated solutions for **all** of the states considered are optimal and represent a Pareto front.

Application to energy management in buildings

Choice of state variable

The state variable must concisely represent the state of the system studied and its behaviour. For a precise building characterization, there should be as many state variables as nodes in the model. However, for the sake of simplicity and efficiency, we choose a single state variable: the total energy stored in the building (or in a zone), i.e. the sum of the energy of all of the nodes:

$$E = \sum_{i=1}^{nbr\_nodes} C_i T_i \qquad (11.19)$$

where
T_i temperature of the node i [°C]
C_i thermal capacity of the node i [J/°C]

Relationship between energy E and the command used

Dynamic programming works best by calculating the command (e.g. heating power) to reach a pre-determined state. In fixing an energy level to be reached at the instant $t + dt$, the command to be supplied between t and $t + dt$ is calculated to attain this level of energy. The energy stored in a zone (equation 11.19) constitutes an additional output of vector Y (equation 11.7). The totality of the outputs is calculated at each time step using equation 11.8. The solicitation vector includes heating power, possibly air conditioning, the variable part of ventilation and the solar gains transmitted through the glazing. These components are thus directly influenced by the command, so that the command can be identified from the output value corresponding to the state to be attained. This can be done by reversing equation 11.8:

$$MGIE \cdot U_g^{n+1} = MGID \cdot U_g^n - MGIF \cdot X_g^n + MGIC \cdot Y_g^n + Y_g^{n+1} \qquad (11.20)$$

Using this procedure, the dynamic thermal simulation tool is complemented with an optimization module based on the dynamic programming.

The dynamic programming algorithm is applied in the so-called "forward" direction, i.e. the direction moving down in time. A grid of the building's possible energy is set up within an interval $[En_{min}, En_{max}]$ with a number of states Ne. At the initial time step, the building is at a given energy level.

The heating power levels must remain within the interval $[0, P_{max}]$, which constitutes a constraint. If a calculated heating power level is located outside $[0, P_{max}]$, the policy is not stopped, but the power level is fixed at the closest admissible boundary, 0 W if the calculated heating power level is negative, P_{max} if it is higher than the latter.

EXAMPLES OF APPLICATION

The case study is a low-energy building with an initial objective of minimizing energy consumption while guaranteeing a reduction in peak consumption. Thermal comfort is considered to be satisfactory as long as the indoor temperature remains above 19°C with a variation of less than 1.7°C/30 min.

Presentation of the building

The building studied is a modified version of a building on the "INCAS" platform (Institut National d'Energie Solaire, Chambéry). The climate in Chambéry (73) is considered similar to Mâcon (71), for which known meteorological data are more precise. We therefore used the coldest week from the average meteorological data of the town of Mâcon.

The building, which has a total habitable area of 89 m², comprises two levels with a living room on the ground floor and three bedrooms on the first floor. The southern façade includes a large glazed surface (34% of the façade) protected from the summer sun by a roof overhang and a balcony. The northern wall only has two small windows (3% of the glazed surface). The east and west facades, which are smaller, include reasonably sized glazed surfaces of respectively 8% and 14%. All of the windows have low-emissivity double glazing with a U coefficient of 1.35 W/m²/K, apart from the two windows on the north face, which are triple glazed with a U of 0.7 W/m²/K.

The rooms are 2.5 m high on the ground floor and 2.4 m on the first floor.

The building's surfaces are as follows:

	Outside walls	Attic space	Bottom floor	Intermediate floor	Inside walls
Out -> in composition	20 cm of polystyrene 15 cm of heavy concrete	40 cm of glass wool 1.5 cm of plasterboard	25 cm of polystyrene 20 cm of heavy concrete	20 cm of heavy concrete	2 * 1.5 cm of plaster 4 cm of glass wool
U (W/m²/K)	0.15	0.09	0.12	3.13	0.96

We consider an occupancy of 4 people during the week, except for between 8 am and 5 pm, with only two people present between 5 pm and 6 pm. On weekends, the four people are constantly present. Each person emits 80 W of internal heat gain, added to gains from domestic appliances, computer equipment, etc. (from 10 W to 360 W depending on the time of day).

Heating is produced by electric convectors (yield of 1). Mechanical ventilation is maintained at 0.6 vol/h during occupied periods and 0.3 vol/h when the building is unoccupied. This mechanical ventilation includes heat recovery with an overall efficiency of 0.8. During hot periods, when there is no need to heat the building, the exchanger is short-circuited.

The building's thermal model is mono zone in order to simplify calculations for optimization. The attic space and crawl space are taken to be at outside temperature. Annual heating requirements are 14 kWh/m^2, which corresponds to a low-energy house.

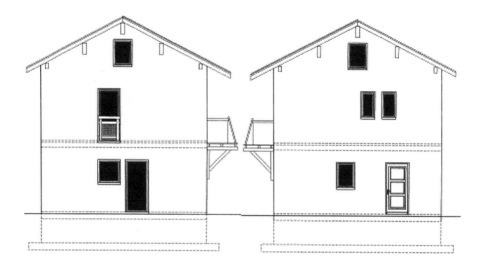

Figure 11.6   East (right) and west (left) façades of the studied house.

Figure 11.7   South (left) and north (right) façades of the studied house.

Management during the heating period

The objective function is to minimize the total cost in euro of heating, constrained by maintaining the temperature above 19°C and the heating power level within the interval [0 W, 5000 W].

The relationship between the total energy stored in the zone and the command is given as seen in equation 11.20. This therefore involves establishing the heating power levels (via the solicitation vector U) in order to obtain the desired energy levels (in the output vector Y).

For a mono-zone thermal model of the building, the heating power level can thus be obtained exactly following the above reasoning. However, for a multi-zone thermal model, the driving force vector at the time step $n+1$, u_{n+1}, is not totally known because we do not know the heating power levels of all of the zones when we calculate the power level in the zone we are interested in. An initial approximate calculation is thus carried out, with nil heating power for the other zones if they have not yet been calculated. When all of the heating power levels have been calculated once for each zone, they are then calculated again iteratively to obtain more precise values. The iteration is stopped when for each thermal zone, between two successive k and $k+1$ iterations, we obtain:

$$\text{abs}(P_{\text{chauff}}^{k+1}[\text{izone}] - P_{\text{chauff}}^{k}[\text{izone}]) < 5\,\text{W}. \tag{11.21}$$

It is thus possible at each point in time to calculate the heating power necessary to reach a predefined energy level at the next time step, based on the energy level of that time step.

Description of the price grid

In the case study, the cost function corresponds to a minimal price. The objective of applying this dynamic programming to buildings during the winter is to maintain an operative comfort temperature – above or equal to 19°C – while flattening heating consumption during a precise time window. This flattened window corresponds to a period of excess electricity consumption in France at the national wintertime level. We consider here a peak period from 5 pm to 9 pm. During these five hours, the price of a kilowatt-hour of electricity is twice that of busy times. Given that in France two different tariffs already operate depending on the time of day, the price grid is as given in Table 11.1.

Table 11.1 New prices used.

	Off-peak hours	Busy hours	Peak hours
Time	1 am–8 am	9 am–4 pm 10 pm–12 pm	5 pm–9 pm
Price per KWh in €	0.0864	0.1275	0.255

Other optimization parameters

The optimization parameters are as follows.

- Number of discretizations of the state variable $Ne = 800$
- Time step of the optimization $\Delta t = 30$ min
- Duration of the optimization: 7 days
- Initialization: 14-hour simulation over a winter period then start of the optimization

Meteorology

In this first study, we do not take into account the meteorological forecast for the optimization using dynamic programming. We consider that outside weather conditions at the building are known in advance (certain future) and thus apply dynamic programming in this "ideal" case.

A very cold week in winter is considered as shown in Figure 11.8.

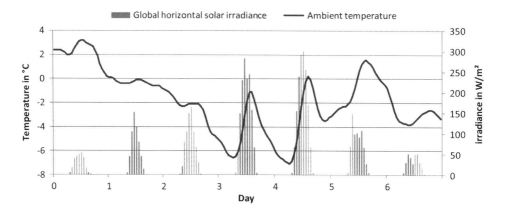

Figure 11.8 Week of very cold weather used for the case study.

Results

An optimization is carried out under the conditions described above. The result of this optimization is a cost of €19.4 (Figure 11.9). The energy level ranges from 520 kWh to 570 kWh, and we observe that it always follows the variations in heating power. The average temperature is 20.1°C and it never exceeds 23°C (Figure 11.10). The average of all of the instant heating power levels used is 1260 W. The total duration of the optimization calculation is approximately 1 minute.

Electricity consumption is flattened during peak times, since the cumulated cost stagnates, indicating that heating is not being used (Figure 11.9). Busy hours are also most often flattened. Overheating the building during the eight off-peak hours combined with solar gains is sufficient to heat the building and maintain the temperature above 19°C for the rest of the day. The flattened electricity consumption extends beyond the five hours initially requested, and in this case can last until 3 pm. The

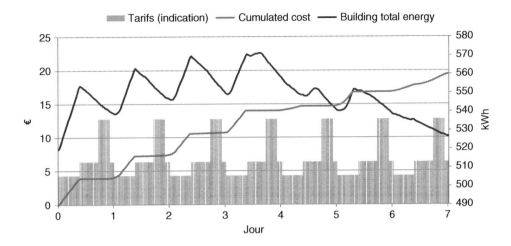

Figure 11.9 Optimal energy trajectory for one week of optimization using dynamic programming.

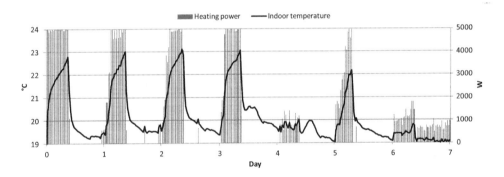

Figure 11.10 Temperatures and instant heating power levels for optimization over a week.

potential for flattening electric heating consumption is therefore highly significant for well-insulated buildings with high thermal mass.

We note that on the last day, the peak hours are not flattened. It is less expensive to evacuate all of the energy stored in the building and thus to use less heating, even though this requires using a small amount of heating during peak hours to maintain the temperature above 19°C. To force the flattening of peak consumption on the last day, we either add one day to the optimization, so that the seventh day is flattened, or increase the price at peak periods.

A simulation using standard control carried out under the same conditions gives a total cost of around €27.3 (Figure 11.11). The optimum regulation obtained using dynamic programming results in a 29% decrease in heating costs for one week of optimization.

On the other hand, the average instant heating power used is 1230 W for standard controls, compared to 1260 W for optimum controls, i.e. a 2% increase in average

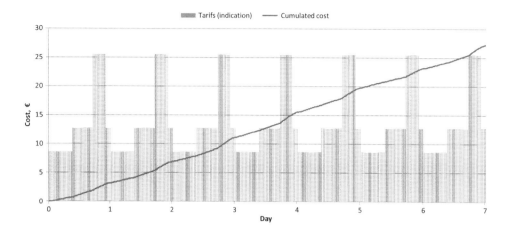

Figure 11.11   Cumulated cost for standard control with peak periods.

heating power. Peak consumption using optimum controls can be flattened by storing energy in the building's thermal mass before the peak period. Even very well insulated buildings lose part of this heat in thermal exchanges with the outside. It is therefore normal that the heating consumption should be higher when peak consumption is flattened in these conditions.

The operative temperature is constrained within the interval [19°C, 26°C] to ensure the building's indoor comfort. It should also respect a constraint on the variation speed for a given period. In Figure 11.10, the greatest variation in temperature is 1.7°C per half hour; thus the temperature variation constraints only just respect the comfort constraints, which impose for example a maximum variation of 2.2°C in one hour or 1.7°C in 30 minutes [ASHR 03].

Control in line with CO_2 emissions

All of the optimizations carried out in this part used the state number $Ne = 800$, a 30-minute time step and weather conditions corresponding to 23 to 30 November 2008 in Mâcon for the outside temperature. This week corresponds to a cold period in a year where the data for the grid's electric mix were available. Since we did not have access to the total horizontal solar radiation for this week, we used data for a similar typical week shown in Figure 11.12.

The meteorology during this week is generally more favourable than that used previously: the temperatures are higher, with similar solar radiation values.

The results of two optimizations with cost functions that include or do not include electricity imports are shown in Figure 11.13 and Figure 11.14.

On the first day of the week, the overall energy level of the building must always be increased to ensure a satisfactory control for the rest of the week. The night from Sunday evening to Monday morning is a period in which CO_2 emissions of the electricity mix are low thanks to the weekend. The increase in the building's total energy is

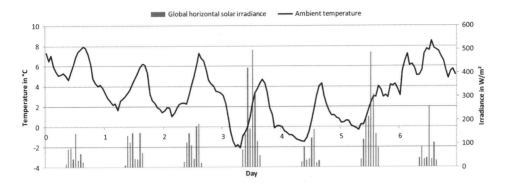

Figure 11.12 Meteorological conditions for week 48 in 2008 in Mâcon.

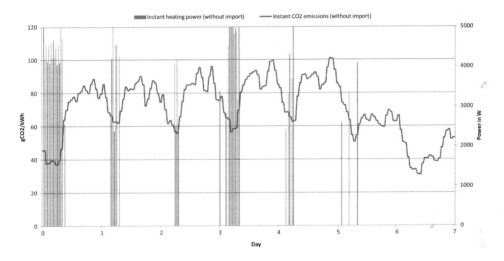

Figure 11.13 Heating power used with optimum control not including electricity imports.

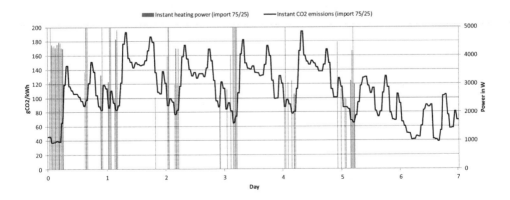

Figure 11.14 Heating power used with optimum control including electricity imports.

thus carried out in a particularly favourable period. Because the cold weather is more clement than previously, heating requirements are low during the week, and practically nil on the last two days. It is therefore fairly easy to flatten the period in which CO_2 emissions are the highest, especially the midday and 7 pm peaks. With CO_2 imports comprising 75% of thermal electricity (Figure 11.14), instant CO_2 emissions increase more rapidly on the first day, and the heating is used less than shown on Figure 11.13. Heating is periodically used during the day at times when CO_2 emissions are lowest. Overall, the optimization algorithm is very well adapted to evolutions in the cost function, and takes advantage of each drop in CO_2 emissions in the electricity mix. Total CO_2 equivalent emissions are 7.4 kg for the week studied. With a standard control maintaining the temperature at 19°C throughout the week, total CO_2 emissions come to 8.9 kg, i.e. 17% more.

Figure 11.15 shows the optimal control for an optimization that aims to flatten the peak electricity consumption period from 5 pm to 9 pm with new meteorological data. Apart from during the first night, when heating is used considerably to increase the building's energy level, heating is used sporadically and only during off-peak hours. Since heating is only used during off-peak hours it also flattens the CO_2 peak hours, even though this was not one of the control's objectives. This explains why the controls are ultimately fairly similar in Figure 11.13 and Figure 11.15, as can be seen in Figure 11.16 which gives the inside temperatures for these two controls.

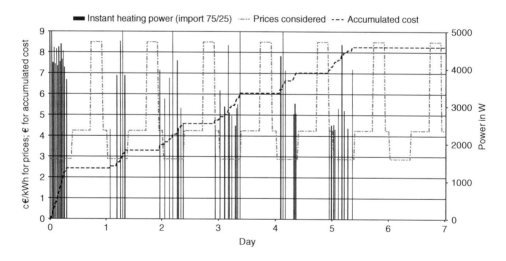

Figure 11.15 Heating power used by optimal control to flatten peak electricity consumption.

Management during heat waves

In very hot summertime temperatures, the objective is to maintain thermal comfort in the building using either forced ventilation or solar protection. Night free cooling is the forced ventilation system used to cool down the building.

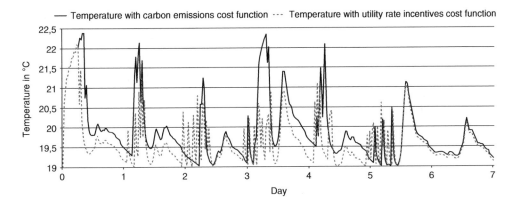

Figure 11.16 Indoor temperature for controls that flatten peak CO_2 emissions and the electricity consumption peak.

Calculating ventilation control

The calculation for controlling ventilation is very similar to that used to control heating, based on equation 11.20.

Ventilation control comes under the driving force vector U. Starting with a fixed energy E_n^{n+1}, we can thus calculate heat gains and losses through ventilation, expressed in Watts. It is possible to connect the power P_{ventil} required to obtain a given energy level to the corresponding ventilation flow. The ventilation command is calculated from this flow. The command is expressed by a percentage of the maximum authorized ventilation.

Definition of the cost function

The objective is to minimize the number of hours during which the temperature exceeds 26°C. In general, to resolve this problem, we must free cool the building during the night or early in the morning when the outside temperature is cool in order to evacuate the energy stored in the building and thus cool it down. Yet during these hours when the building needs to be cooled down, the indoor temperature can be lower than 26°C or very rapidly drop below 26°C. If the cost function is the number of hours during which the temperature exceeds 26°C, then two policies that have different ventilation scenarios and attain the same energy state at a close time step early in the day have the same cost, since neither has so far led to temperatures above 26°C. Given that only one policy can be chosen, one of them is thus eliminated randomly because both result in the same number of hours (0) above 26°C at the time of optimization. However, two different series of commands were used, and the one resulting in the most night free cooling is probably the optimal choice later in the day when solar and indoor gains are higher. The cost function was therefore not a good choice because it does not allow us to associate the different costs with two policies that are only differentiated afterwards.

A cost function could be the operative temperature of the zone. Thus, policies all have different costs when the command is different and the minimization of the temperature encourages free cooling. This kind of objective function results in maximum free cooling when the outside temperature is lower than the indoor temperature. However, in the summertime, outside temperatures can drop below 20°C, and even reach 15°C at night, and free cooling may result in significant discomfort. A cost function corresponding to the difference between inside temperature and a reference temperature T_{ref} for which comfort is ensured is more pertinent.

The value function used is thus:

$$v(E_t, E_{t+1}) = \text{abs}(T_{int} - T_{ref}) \tag{11.22}$$

where $E_{t+1} \in [E_{min}, E_{max}]$.

The reference temperature T_{ref} can be a fixed value throughout the optimized period. However, because indoor comfort depends on the outside temperature (this corresponds to the notion of "adaptive comfort") it is also possible to take a sliding average of outside temperatures as a reference temperature [NICO02]. T_{ref} can thus either be fixed, or obey a law resulting from adaptive comfort:

$$\begin{cases} T_{ref} = 0.049T_{RM} + 22.58 & si\, T_{RM} \leq 10°C \\ T_{ref} = 0.206T_{RM} + 21.42 & si\, T_{RM} > 10°C \end{cases} \tag{11.23}$$

The cost of cooling is not included in the cost function for the moment. The control thus tends to make maximum use of this kind of ventilation to reach its objectives. Two methods are used to limit the use of cooling. The first is to add the cost of using ventilation to the value function: with this method, despite a multi-objective optimization framework, only a single value function is used. Since we do not know the precise electricity consumption of the ventilators, this is included in the value function considering that the maximum use of ventilation corresponds to the cost equivalent of a gap between indoor temperature and the reference temperature of $X°C$. The value function becomes:

$$v(E_t, E_{t+1}) = \text{abs}(T_{int} - T_{ref}) + X^{vent/100} \tag{11.24}$$

where vent is the command used in % of maximum cooling and X is the weighting coefficient of using ventilation. The higher X is, the greater the cost of using the ventilation will be in terms of the first part of the value function.

The second approach to carrying out a multi-objective optimization is to include a value function for each of these objectives:

$$v(E_t, E_{t+1}) = \begin{cases} v_1(E_t, E_{t+1}) = \text{abs}(T_{int} - T_{ref}) \\ v_2(E_t, E_{t+1}) = vent/100 \end{cases} \tag{11.25}$$

A Pareto front can thus be used to visualize all of the non-dominated solutions resulting from all of the energy intervals of the last time step and thus identify the optimal solutions.

Other optimization parameters

Number of discretization points of the state variable $Ne = 800$
Time step of the optimization $\Delta t = 30\,min$
Duration of the optimization: 14 days
Initialization: 14-day simulation over a summer period then start of optimization

Meteorology

We considered one week of heat wave followed by one week of summer weather (Figure 11.17).

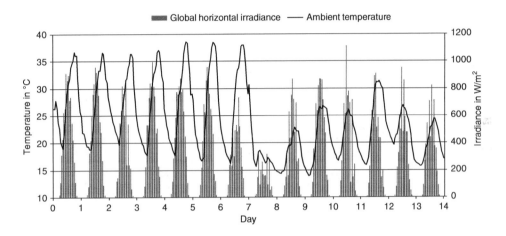

Figure 11.17 One week of heat wave followed by one week of normal summer weather.

The reference temperature T_{REF} evolves throughout the heat wave week, from about 25°C at the start of the optimization to 26.5°C at the end of the first week (Figure 11.18). The reference temperature, which is the comfort temperature, is thus higher than 26°C, the comfort limit generally given for summer periods. Using adaptive comfort allows us to move away from routine conventions and adapt to unusual conditions, such as very high temperatures.

Results of mono-objective optimization

The duration for a 14-day optimization is around 50 seconds. Figure 11.19 shows the optimum control obtained for the two weeks studied. Moderate night cooling is sufficient to reduce the temperature of the zone at the same speed as the reference temperature (adaptive comfort).

Using an AVP approach (average vote percentage indicator) for a building used for habitation, comfort is maintained if the indoor temperature is within 2°C of the reference comfort value [NICO 02]. For the whole of the second week, the optimization algorithm is very efficient because there is an average 0.3°C difference between the temperature of the zone and the reference temperature. The ventilation flow is around 3.3 vol/h for this week. A standard control maintaining the ventilation flow at 0.6 vol/h is also applied. Without free cooling, the indoor temperature rises steeply during the

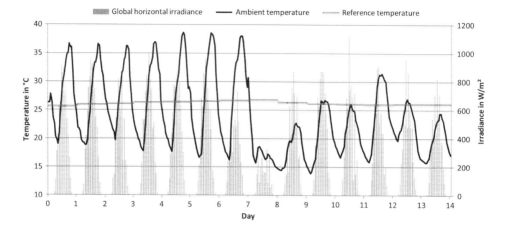

Figure 11.18 Reference temperature according to meteorology for adaptive comfort.

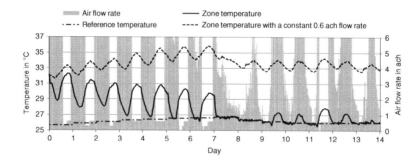

Figure 11.19 Optimum control for optimization with a variable reference temperature resulting from adaptive comfort – 2 summer weeks.

first week to reach 36°C. The inside temperature goes down slightly during the second week but the heat stored in the building's thermal mass is too high for the inside temperature to drop as far as the comfort temperature.

The control resulting from optimization using dynamic programming makes it possible to maintain thermal comfort in a building using significant free cooling up to 6 vol/h. This kind of ventilation requires high electricity consumption from ventilators as well as over-sized ventilation ducts. A new study was thus done to consider solar protections kept closed during the day in periods of high temperature. Flows under 3 vol/h are then sufficient. Research continues on natural ventilation.

Controlling natural ventilation

An aeraulic model used to evaluate air exchanges by natural ventilation was implemented by Maxime Trocmé [TROC09]. An example of results is shown in Figure 11.20. The much longer optimization time can be explained by the iterative calculations necessary in the aeraulic model. As we can see, only one night is needed to bring down the temperature in the building by 10°C. On that night, the outside

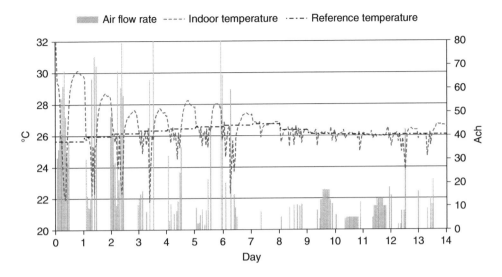

Figure 11.20 Indoor temperature and ventilation flow during the two weeks studied.

temperature is close to 20°C, whereas at the starting point the indoor temperature is almost 32°C. During this first night, the ventilation flows are very high with an average value of 25 vol/h, even reaching 60 vol/h, or one volume per minute. In practice, it is best to avoid draughts and thus flows above 20 vol/h.

During the first nights of the week, the heat stored in the building is rapidly evacuated, so that the window-opening command can be used less intensely at the end of the week when the heat is at its highest. The average absolute difference between the inside temperature and the comfort temperature for this week is 1.45°C, compared to a 3°C difference when only forced ventilation is used. Assuming that thermal comfort is maintained if the temperature is within 2°C of the comfort temperature, only the window command is required to maintain comfort in the building on average over the week. However, the thermal comfort is not respected during every hour of this week, and the temperature sometimes exceeds the 2°C limit. During the first days, the inside temperature is even at times lower than the comfort temperature. However, all of the occupants would probably not be against a slightly cool temperature in periods of high summer heat, especially with easy, fast adaptation possibilities. For the first week, the window-opening command thus leads to a significant drop in the indoor temperature despite the high initial energy level of the building and the very hot outdoor temperatures.

During the second week, the natural ventilation flows are much lower (less than 20 vol/h) making it much easier to keep the indoor temperature close to the comfort temperature. The absolute average difference between the inside temperature and the comfort temperature is 0.25°C during the second week compared to 0.35°C using forced ventilation only.

The flows achieved using the window-opening command (natural ventilation) are much higher than those achieved using forced ventilation. One possible illustration of

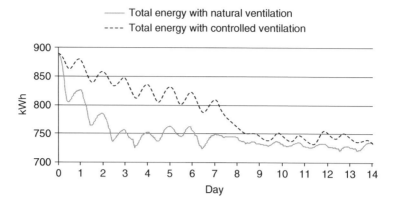

Figure 11.21   Energy in the building using a forced ventilation command and natural ventilation.

this difference is the progress of the state variable in the two case studies (Figure 11.21). The energy drops very sharply during the first three days using the window command to reach an interval [720 kWh, 750 kWh] in which it remains for almost all of the remainder of the period studied. It therefore takes three days to evacuate all of the initial stored energy, despite the very high heat during this period. It takes 8 days for the energy to reach the same interval using the forced ventilation command, thanks to the sharp drop in outside temperature during the second week. Without this weather change, it would undoubtedly take from 11 to 12 days to reach the same energy interval. The natural ventilation command is therefore highly efficient in cooling down a building during very hot periods. For the second week, when the meteorological conditions are more clement, both commands give very similar results. This can be seen not just by the indoor temperature but by the total energy stored in the building. If there is less initial energy, the forced ventilation command could maintain comfort in the building but not as efficiently as natural ventilation.

It is therefore possible to maintain thermal comfort in this type of building, even with very high initial energy and unfavourable meteorological conditions. These controls are efficient because the external temperature drops sharply during the night, creating advantageous free cooling. If this is not available, an active cooling system will be necessary (e.g. air conditioning).

The number of hours of discomfort in the building during the two weeks studied is shown in Figure 11.22. One hour of discomfort corresponds to an hour when the inside temperature is not within the interval $[T_{ref}-2°C, T_{ref}+2°C]$. Natural ventilation is very efficient in improving the thermal comfort in a building, but if we integrate into thermal comfort the notion of a maximum variation of indoor temperature of 2.2°C per hour, then the thermal comfort is no longer respected. A 10°C drop in temperature overnight can be a problem for thermal comfort, especially during a period with few adaptation possibilities. Very high ventilation flows can also create a comfort problem. This limitation results from the fact that we are only taking thermal comfort into account. Similarly for blinds, if they are kept closed, then visual comfort may be significantly reduced.

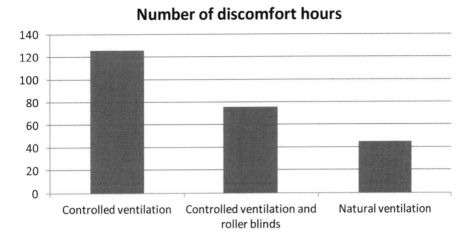

Number of discomfort hours

Figure 11.22 Number of hours of discomfort in a building according to type of command used.

Even though controlling openings gives better results than controlling forced ventilation, there are some limitations to consider, especially in terms of measuring or anticipating supplementary data, such as wind direction or speed. Natural ventilation is more dependent on precise local meteorological forecasts to cool down the building. It also depends on occupants' behaviour: the mono-zone model considered corresponds to a perfect transfer of the air in the building, which thus assumes that all inside doors are open. In practice, this may not be the case, especially at night, though undercut doors allow the air to circulate. Precise forecasts of occupancy and action on windows and doors are thus also important to ensure reliable results from controlling natural ventilation. Window openings also depend on other parameters like rain, wind, noise or neighbourhood safety. In conclusion, controlling the opening of windows can give more beneficial results to cool down a building, but they are more random.

Although the natural ventilation model has room for improvement and greater reliability, this study illustrates the benefits of natural ventilation in cooling down a house in very unfavourable meteorological conditions. We will now continue with this example to look at multi-zone thermal models in terms of natural ventilation.

Multi-objective optimization

The controls resulting from optimization using dynamic programming will now have two objectives: to improve thermal comfort in a building while minimizing the use of ventilation.

Cost weighting

To attempt to reduce the use of ventilation while maintaining comfort, we can integrate the cost of using ventilation. The value function used becomes:

$$v(E_t, E_{t+1}) = \text{ventilation flow rate} + X^*\text{abs}(T_{\text{int}}^{t+1} - T_{\text{ref}}^{t+1}) \tag{11.26}$$

The evolution of the indoor temperature is very similar for *X* values 1 and 9 during the first week (Figure 11.23) because the difference between the inside temperature and the reference temperature is significant due to the initial state of the building. However, for a low *X* value (0.1), the comfort criterion is much lower, and the control strategy does not allow us to reduce the indoor temperature. During the second week, *X* controls with values of 1 and 9 move away from each other with a much higher indoor temperature when this criterion is lower.

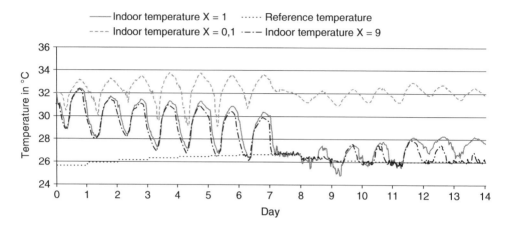

Figure 11.23 Comparison of evolution of indoor temperatures for optimizations with different ventilation costs.

Figure 11.24 shows the evolution of thermal comfort according to average ventilation for different given values of *X*. This study allows us to a posteriori choose the best value of the weighting coefficient of the value functions in line with users' objectives.

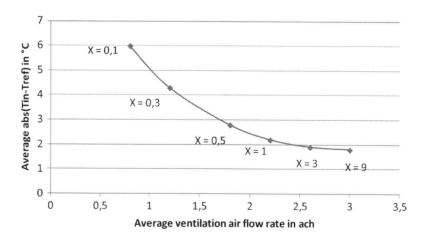

Figure 11.24 Correlation between indoor thermal comfort and average ventilation according to cost of using ventilation.

If, for example, the occupant's thermal comfort is an important datum, an X value of 1 will guarantee the use of reasonable ventilation with relatively good thermal comfort in the knowledge of the conditions of the first week simulated.

Taking ventilation cost into account is therefore highly efficient to reduce the use of forced ventilation while maintaining comfort in the building. Comfort is maintained except during very hot periods when the indoor temperature exceeds 26°C, due to the building's initial conditions and the very hot period.

This multi-criteria optimization method is interesting but requires a high number of optimizations to obtain the results shown in Figure 11.24. A multi-criteria optimization with two separate value functions can give a Pareto front with results similar to those from a single optimization.

Pareto front

The cost function comprises two components:

$$v(E_t, E_{t+1}) = \begin{cases} v_1(E_t, E_{t+1}) = \text{abs}(T_{\text{int}}^{t+1} - T_{\text{ref}}^{t+1}) \\ v_2(E_t, E_{t+1}) = \text{ventilation\_flow\_rate} \end{cases} \tag{11.27}$$

At each time step and for each energy value of the state variable, several strategies can be retained. Only those strategies that are less efficient than all other strategies for both criteria together are eliminated. The number of strategies calculated thus increases rapidly, resulting in memory problems for the computer.

To reduce the number of combinations calculated, some strategies are eliminated even if they are not dominated for both criteria. We assume that it is not useful to retain two strategies that give very similar results for the same state and time step. This avoids keeping too many similar combinations for each time step and thus allows us to increase the number of discretization intervals Ne to 800 and the number of simulated days to 7. Without reducing the number of strategies retained, the maximum discretization number would be 400 for 5 simulated days. A control strategy is eliminated, even if it is not dominated for both criteria, if it is located too close to another strategy. In other words, when for a given time step and given state variable value, we have:

$$\text{abs}(v_1(\text{reg 1}) - v_1(\text{reg 2})) + \text{abs}(v_2(\text{reg 1}) - v_2(\text{reg 2})) < \text{Epsilon} \tag{11.28}$$

where
v_1 (reg 1) the cumulated cost of the first control on the first criterion
v_1 (reg 2) the cumulated cost of the second control on the first criterion
v_2 (reg 1) the cumulated cost of the first control on the second criterion
v_2 (reg 1) the cumulated cost of the second control on the second criterion
Epsilon the minimum "distance" between two strategies

The epsilon value is taken as 0.3 in the optimization presented below, carried out with a 30-minute time step.

Figure 11.25 shows the Pareto front resulting from an optimization by dynamic programming that lasted less than 19 minutes. The same graph shows the result obtained in the previous paragraph for the multi-criteria optimization including a single value function in the objective function. The maximum number of strategies that

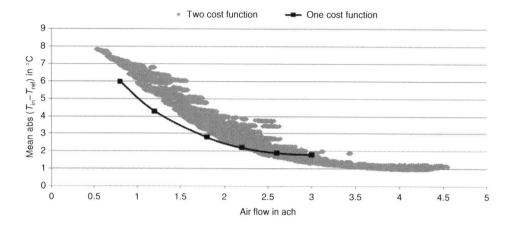

Figure 11.25 Pareto front for thermal comfort/ventilation flow.

reached the same state at the same time step and were retained is 381. This means that up to 381 strategies are retained per state with 800 states in total for each 30-minute time step for an optimization lasting 7 days.

The results obtained by the Pareto front are slightly different: we can see that for a low average flow, there is around one additional degree Celsius compared to when the objective function comprises a single value function. On the other hand, for high average flows, the results obtained using multi-criteria optimization correspond to improved thermal comfort. The range of controls calculated is also much greater, from an average ventilation flow of 0.5 vol/h to almost 4.5 vol/h. With an optimization using two separate value functions, building occupants can choose a posteriori the control that suits them best. Dynamic programming can thus be used to develop both mono- and multi-criteria controls in reasonable times.

CONCLUSIONS

A dynamic thermal simulation tool was supplemented with an optimization module based on dynamic programming. The state variable considered, i.e. the total energy stored in the building, was calculated from the temperatures of each grid cell taken from the finite volume model. This algorithm was studied for a low-energy house with the aim of flattening the electricity consumption peaks in wintertime, and controlling the thermal comfort during hot weather. In both cases, the optimal command proved advantageous in comparison with standard controls: the result was reduced heating costs in winter, and the maintenance of acceptable indoor temperatures in summertime with limited use of mechanical ventilation.

Future research will include a case study of buildings with different insulation and thermal mass levels, in order to understand the limits of these kinds of management strategy based on energy storage. The optimization could also be applied to different criteria, in particular concerning environmental impacts.

Optimization necessitates significant calculation time. To devise real-time management strategies would probably require using model reduction and anticipative control techniques, whose performances could be compared to those of the optimal strategies identified by the method presented above.

REFERENCES

American Society of Heating, Refrigerating and Air-Conditionning Enginners (ASHRAE), "Thermal Environnemental Conditions for Human Occupancy", *Third Public Review*, 2003.

R. Bellman, "Dynamic programming", *Princeton University Press*, Princeton, USA, 1957.

T.B. Hartman, "Dynamic control: fundamentals and considerations", *ASHRAE*, vol. 94 (11.1), n° 3112, pp. 599–609, 1988.

B. Peuportier, I.B. Sommereux, "Simulation thermique simplifiée appliquée au bâtiment", Centre d'Energie de l'Ecole des Mines, 1992.

A.A. Argiriou, I. Bellas-Velidis, m; Kummert, P. andré, "A neural network controller for hydronic heating systems of solar buildings"; *Neural Networks*, vol. 17, pp. 427–440, 2004.

M. Bauer, "Gestion biomimétique de l'énergie dans le bâtiment", *Thèse n° 1792*, Ecole Polytechnique Fédérale de Lausanne, 1998.

R. Bellman, "Dynamic programming", *Princeton University Press*, Princeton, USA, 1957.

J.E. Braun, K.W. Montgomery, N. Chaturvedi, "Evaluating the Performance of Building Thermal Mass Control Strategies", *HVAC&R Research*, vol. 7, n°4, pp. 403–428, 2001.

F. Calvino, M. La Gennusa, G. Rizzo, G. Scaccianoce, "The control of indoor thermal comfort conditions: introducing a fuzzy adaptive controller", *Energy and Buildings*, vol. 36, Issue 2, pp. 97–102, 2004.

T.Y. Chen, "Real-time predictive supervisory operation of building thermal systems with thermal mass", *Energy and Buildings*, vol. 33, pp. 141–150, 2001.

"Les chiffres clés du bâtiment 2009", *Agence de l'Environnement et de la Maîtrise de l'Energie*, 2009.

G. Escriva-Escriva, I. Segura-Heras, M. Alcazar-Ortega, "Application of an energy management and control system to assess the potential of different control strategies in HVAC systems", *Energy and Buildings*, vol. 42, pp. 2258–2257, 2010.

Favre B., Étude de stratégies de gestion énergétique des bâtiments par l'application de la programmation dynamique, thèse de doctorat, Ecole des Mines de Paris, septembre 2013.

E.M Greensfelder, G.P Henze, C. Felsmann, "An investigation of optimal control of passive building thermal storage with real time pricing", *Journal of Building Performance Simulation*, vol. 4, n°2, pp. 91–104, 2011.

P.O. Fanger, "Calculation of thermal comfort: Introduction of a basic comfort equation", ASHRAE Transactions, 1967.

M. Hamdi, G. Lachiver, "A fuzzy Control System Based on the Human Sensation of Thermal Comfort", *Fuzzy systems Proceedings*, IEEE World Congress on Computational Intelligence, 1998.

P. E. Hart, N. J. Nilsson, B. Raphael, "A Formal Basis for the Heuristic Determination of Minimum Cost Paths", *IEEE Transactions of Systems Science and Cybernetics*, vol. 4, n°2, pp. 100–107, 1968.

G.P Henze, C. Felsmann, G. Knabe, "Evaluation of optimal control for active and passive building thermal storage", *International Journal of Thermal Sciences*, vol. 43, pp. 173–183, 2004.

M.A. Humphreys, "Thermal comfort temperature and the habits of hobbits", *Standards for thermal comfort: indoor air temperature standards for the 21st century*, pp. 3–12, 1995.

K. Le, "Gestion optimale des consommations d'énergie dans les bâtiments", *Thèse de doctorat*, Institut Polytechnique de Grenoble et de l'Université de Danang, 2008.

J. Liang, R. Du, "Thermal Comfort Control Based on Neural Network for HVAC Application", *Procedings of the 2005 IEE Conference on Control Applications*, 2005.

P. Malisani, B. Favre, S. Thiers, B. Peuportier, F. Chaplais, N. Petit, "Investigating the ability of various building in handling load shiftings", *IEEE Power Engineering and Automation Conference*, Wuhan, China, 2011.

E.H. Matthews, D.C. Arndt, C.B. Piani, E. van Heerden, "Developping cost efficient control strategies to ensure optimal energy use and sufficient indoor comfort", *Applied Energy*, vol. 66, pp. 135–159, 2000.

K.J. Mc Cartney, J.F Nicol., "Developing an adaptive control algorithm for Europe. " *Energy and Buildings*, vol. 34, pp. 623–635, 2002.

F.B. Morris, J.E. Braun, S.J. Treado, "Experimental and simulated performance of optimal control of building thermal storage", *ASHRAE Transactions*, vol. 100, n°1, pp. 402–414, 1994.

P.M. Narendra et K. Fukunaga, "A Branch and Bound Algorithm for Feature Subset Selection", *IEEE Transactions on Computers*, vol. 6, n°9, 1977.

T. Nagai, "Optimization method for minimising annual energy, peak energy demand, and annual energy cost through use of building thermal storage", ASHRAE Transactions, vol. 108, part 1, paper number 4496, 2002.

J.F Nicol, M.A. Humphreys, "Adaptive thermal confort and sustainable thermal standards for building", *Energy and Buildings*, vol. 34, pp. 563–572, 2002.

A.M. Nygard Ferguson, "Predictive thermal control of building systems", *Thèse de doctorat n°876*, Ecole Polytechnique Fédérale de Lausanne, 1990.

F. Oldewurtel et al., "Energy efficient building climate control using Stochastic Model Predictive Control and weather predictions", *American Control Conference*, 30 Juin 2010 – 2 Juillet 2010.

B.W. Olesen, O. Seppanen, A. Boerstra, "Criteria for the indoor environment for energy performance of buildings – a new european standard", *European Organisation for Standardization (CEN)*, 2006.

B. Peuportier, I.B. Sommereux, "Simulation tool with its expert interface for the thermal design of multizone buildings", *International Journal of Solar Energy*, vol. 8, 1990, pp. 109–120.

S. Poignant, B. Sido "Groupe de travail sur la Maîtrise de la pointe électrique", Rapport parlementaire Poignant – Sido, Avril 2010.

B. Polster, B. Peuportier, I. B. Sommereux, P.D. Pedregal, C. Gobin, E. Durand, "Evaluation of the environmental quality of buildings – a step towards a more environmentally conscious design", *Solar Energy*, vol. 57, n°3, pp. 219–230, mars 1996.

"Arrêté du 26 octobre 2010 relatif aux caractéristiques thermiques et aux exigences de performance énergétique des bâtiments nouveaux et des parties nouvelles de bâtiments", *Journal Officiel de la République* Française, Octobre 2010.

H. Sane, M. Guay, "Minmax dynamic optimization over a finite-time horizon for building demand control", *American Control Conference*, Seattle, Washington, USA, 2008.

P. Somol, P. Pudil et J. Kittler, "Fast Branch and Bound Algorithms for optimal Feature Selection", *IEEE Transactions on Pattern Analysis and Machine Intelligence*, vol. 26, n°7, 2004.

M. Trocmé, "Aide aux choix de conception de bâtiments économes en énergie", *Thèse de doctorat*, Mines ParisTech, 2009.

V.V. Tyagi, D. Buddhi, "PCM thermal storage in buildings: A State of art." *Renewable and sustainable Energy Reviews*. vol. 11, 2007, pp. 1146–1166.

Chapter 12

Eco-design in practice: How does urban biodiversity fit in?

Alexandre Henry & Nathalie Frascaria-Lacoste
AgroParisTech, Laboratory for Ecology, Systematic and Evolution, UMR CNRS/UPS/AgroParisTech, Paris Sud University, ORSAY cedex

We live in an increasingly urban world. Towns already cover 2% of the Earth's surface and are responsible for consuming over 75% of natural resources (Müller & Werner, 2010). Today, more than half of the world's population live in a town (CBD 2007). At the same time, the trend is a move towards massive globalization. This is particularly the case in Europe and North America. Faced with depleting natural resources and deteriorating natural environments due to urban sprawl, sustainable development has become a major issue for many of those involved in urban planning, in particular local authorities and businesses.

Towns are extremely artificial places and it is thus difficult to define biodiversity in such a context. Urban biodiversity is profoundly determined by the organization, planning and management of the built environment, which is itself influenced by economic, social and cultural values. This biodiversity is complex, and results from an assembly of horticultural species, species that have spontaneously migrated from their natural habitats into towns, and species resulting from natural hybridizations between native and non-native species in urban contexts. In addition, the composition of species is highly controversial, especially exotic species that dominate ecosystems once introduced (Williams & Jackson, 2007; Dearborn & Kark, 2009).

Paradoxically, in Europe, towns often accommodate more species than rural areas (Hope *et al.*, 2003). Nevertheless, this specific wealth is relative since it only concerns angiosperms, and in terms of animals, mostly birds (Wittig, 2010). In addition, due to biotic homogenization, most of the species found in urban areas are generalist rather than specialist (Julliard *et al.*, 2006). This particular specific wealth, even for native species, can be explained in a number of ways: towns have developed in very heterogeneous landscapes (Kühn *et al.*, 2004); they are highly structured themselves (Niemela, 1999); their high temperatures encourage different species to colonize these areas (Knapp *et al.*, 2008). Potentially invasive exotic species are introduced into urban environments (Kühn *et al.*, 2004) and escape from them to create colonies outside the town.

With their very particular specific wealth, towns have developed ecosystems that are profoundly different from those organized before humans existed. We call these "new ecosystems", "emerging ecosystems" or "non-analogous ecosystems" (Williams & Jackson, 2007; Hobbs *et al.*, 2006). The key characteristics of these ecosystems are the emergence of new combinations of species presenting a potential that can change the way ecosystems work, and previously unseen human influence resulting from deliberate and non-deliberate action (Hobbs *et al.*, 2006).

Today, very little information is available on the significance of new ecosystems in urban biodiversity evaluations and on the significance of the services they already provide. It is now essential to include them in reflections on tomorrow's urban environment because they already play a part. They may represent very useful biodiversity. Numerous questions arise: Will these new ecosystems increase in number, totally wiping out all native species? What do they signify compared to ecosystems built up around native species? How do they work? Do they function in new ways or do they optimize existing functions? Should we be thinking about new ways of managing them? How should we take on these new assemblies? Which should we privilege? Should we connect them? If so, how? What socio-economic aspects should be considered in relation to them? How can we develop a management system that maximizes the beneficial changes they bring about (in looking for the actual benefits) and reduces the negative impacts?

This complex reflection cannot take place without strong partnerships between all urban stakeholders, decision makers, urban planners, landscape designers, ecologists and town residents, explaining the importance of recreating spaces where a more diverse and operational type of nature could regain lost ground. This movement needs to be built on different spatial and temporal scales. Humans built these new ecosystems; it is therefore up to us to devise a management system to guide their development. The task is not simple, given that it will be very difficult to return to a more natural state, in terms of effort, time and money. As a result, we must totally reconsider towns as a whole, accept these assemblies for what they are and for the benefits that they provide, and set up genuine "adaptive management" based on partnership and "learning by doing".

In addition, towns, as places of knowledge concentrating financial and human resources, can be much more powerful places than rural zones for educating about biodiversity, resulting in a genuine awareness and acceptance of the challenges of living diversity and the potential acceptance of a reworked biodiversity that may not always be physically attractive. Towns can also be places for experiments to understand the impact of socio-economic change on ecosystems. New ecosystems bring new challenges, initiating a variety of new ways of thinking and managing that can be a genuine asset for urban areas. Towns need to be convinced and take on this opportunity, which makes a concrete participation to reconciling man with nature.

Towns are centres of economic, political, financial and social power, as well as culture and innovation. Urban areas, with their "new ecosystems" that are potential drivers of expected ecosystem services and rich with new biodiversity, provide a public place that can give rise to new fields and varied discussions, thus promoting totally new, original places with a different type of nature, and developing new local languages with specific words and management tools for each area. The time has come for urban stakeholders to focus on these new ecosystems and to question their relevance in towns, bringing functional, ecological, and socio-economic added value. If ecosystems are more efficient and sustainable, new management modes should integrate this crucial dimension and limit the incessant turnover of "tissue gardens" (plants that are rapidly changed before they have had time to wilt) (Blanc *et al.*, 2005) and the voluntary reintroduction of native species that can no longer adapt. This debate is digging deeper and looking closely at our way of thinking and knowledge. We need exchanges between all urban stakeholders to move towards towns that have profoundly altered relationships with nature. Ultimately, it is these reconsidered towns that will initiate these changes and guide those confronted with the same issue.

Over the last few years, awareness of climate change has led urban planning practices to change and start integrating the notion of sustainable development. Thus, new measures have been developed, including thermal regulations, standards and energy-saving certification. However, these measures have had little positive impact on preserving biological diversity. Can we thus really talk about sustainable buildings and eco-neighbourhoods?

In the space of several years, and in particular since the "Grenelle de l'Environnement" round table in 2007, the number of eco-neighbourhood projects in France has risen from a handful to several hundred. This can be seen as urban planners' response to the environmental problem linked to climate change and urban sprawl. Nevertheless, the eco-neighbourhood concept remains unclear. For some, it involves constructing a set of buildings with excellent energy performance compared to former standards. For others, it illustrates a much deeper reflection on how neighbourhoods fit into the environment and their impact in terms of producing waste, water pollution and clean transport.

The general definition of an eco-neighbourhood established by the Minister of the Environment stipulates that it must respect the principles of sustainable development and be designed to offer accommodation for all in a quality living environment, with a limited ecological footprint. More precisely, the design of an eco-neighbourhood should promote responsible management of resources. In addition, this type of neighbourhood should fit into the existing town and surrounding territory. It should also participate in local economic development. Social diversity and a "living together" approach result from housing built for all people in a range of types (from one-person studio flats to family apartments, as social housing with the potential for ownership) and with a shared vision between development stakeholders and residents starting from the design of the neighbourhood.

In the face of this general definition of the concept of eco-neighbourhoods, regulations are still unclear, and no certification or accreditation systems exist for this type of neighbourhood. However, the MEDDTL (Ministry for Ecology, Sustainable Development, Transport and Housing) is planning on creating an "EcoQuartier" label to stabilize their status by making them part of the French law on urban planning. Henry *et al.* has illustrated the disparity between the various projects and the insignificant place allotted to biodiversity.

It would therefore be worth regulating this practice so that only those neighbourhoods that respect precise standards, and in particular those that maintain or promote ecological operations of the site on which the project is built, receive "eco-neighbourhood" accreditation. These neighbourhoods would also need monitoring to control their management and decide whether or not they can retain the label. Nevertheless, obtaining accreditation should not be a neighbourhood's goal, but rather the outcome of an approach. Although a respectable goal, if the primary motivation is to obtain a label, then the resources put in place to create an "eco-neighbourhood" will not go any further than what is required, potentially creating an obstacle to urban innovation.

As well as observing poor diversity in urban developments, we note a lack of appropriate tools available to urban planners to tackle this problem. LCAs are a perfect example. This decision aid tool, which is frequently used in the building field to calculate a product's environmental impact (from design to end of life), does not efficiently integrate biodiversity into its calculations. It would be worth improving this part to estimate actual impacts on biodiversity as best as possible. A solution to make this indicator more effective would be to consider the impact on ecological functions or ecosystem services, rather than simply the number of species destroyed.

Beyond its current function, we are also keen to point out the potential deficiencies of using this type of tool. A very recent example is using LCAs to compare different types of green space so as to install the one with the lowest environmental impact. Not only are LCA calculations approximate from a biodiversity point of view, but this new way of using the tool could bring the risk of homogenizing practices and thus diminishing biodiversity and altering the way the ecosystem works even more systematically.

Henry & Frascaria-Lacoste have developed a new tool to help urban planners in their work. Their first decision aid tool (BioDi(v)Strict) is simple enough for non-specialists to understand, inexpensive and easy to use, so as to encourage as many urban planners as possible to take better consideration of biodiversity in their practices. The tool is based on the diversity of residents in connection with groups of bio-indicator species (plants, butterflies, amphibians and nesting birds). For its first application, we chose Cité Descartes as a pilot site (see chapter 14).

The development of this type of new tool to aid decisions and cooperation is a first significant step in the process of the consideration of biodiversity by companies.

This is all the more relevant today, as companies find themselves in a paradoxical situation in the face of ecological crisis. Intentionally or otherwise, they contribute to the growing erosion that in numerous circumstances constitutes the base of their competitive advantage and value creation, i.e. biodiversity. As a major source of innovation, biodiversity often conditions a significant share of business turnover, in particular via its dependence on ecosystem services. For construction and land planning companies, their dependence on ecosystem services resides in their need for renewable raw materials (wood) or non-renewable ones (minerals, sand), the physical properties of the ground (stability), regulations to combat natural risks (floods, landslides, earthquakes), climate (temperatures, rainfall) and proximity to cultural services (green spaces, landscapes). Despite this strong dependence, the pressures on these services are numerous, such as the destruction, mineralization and fragmentation of environments and the introduction of exotic species.

Faced with risks and opportunities linked to deteriorating ecosystems, companies are confronted with questions in terms of ecosystem services when devising and implementing their strategies. Yet for most companies, maximizing their return on investment and developing or maintaining their competitive advantage do not involve investments and innovations in favour of maintaining biodiversity and the associated ecosystem services.

This consideration by the company is thus partly linked to a cost/benefit analysis and partly to its own motivation. Four types of policy are often observed in companies (Houdet, 2008). (1) Status quo or inertia: the company does not feel concerned by biodiversity and does not change its practices. (2) Reactive policy: the company is conscious of biodiversity-related risks for its activity; to justify itself this policy tends to be based on standards, certification and communication. (3) Proactive policy: a policy that implements actions in favour of biodiversity without challenging its activity. (4) Win-win: a policy that sees a genuine economic opportunity in biodiversity leading to deep-seated reshaping of the strategic decision process.

Thanks to tools developed as part of the ParisTech-VINCI chair, we can move from a policy of inertia, often observed in the domain of construction and spatial planning, to a reactive policy (such as accreditation), or even a proactive policy (such as ecological optimization of urban developments).

In a company in the construction field that significantly impacts its environment, these two types of policy are an essential first step, but they are still insufficient. This is because a reactive policy uses existing certification (HEQ, BREEAM or LEED), but does not necessarily go any further; in addition, standards and certification do not guarantee efficiency, which is closely linked to user behaviour. A more ambitious, innovative step would be to move to a "win-win" policy in which biodiversity becomes a genuine driver for the company.

In fact, taking biodiversity into account can be an excellent way to increase the status of a company and the environment it impacts, by making its activities more sustainable and boosting its reputation vis-à-vis the public. To do this, the company will need to make profound changes by adapting its technologies and production methods and integrating problems linked to living systems into its management, performance assessment and innovation systems. This therefore involves looking for new products, processes and organization modes in nature and the living environment, taking inspiration from natural loops to rethink the metabolism of companies and territories and obtain added value from them that can enter into the company's accounting system (Houdet, 2008). The goal of this model would be to maintain and restore the broadest possible ecosystem diversity, while allowing companies to develop and prosper.

The crucial idea is a return to nature in the shape of a formidable business challenge for the future, through which companies can find new forms of competitiveness while applying a "win-win" approach. This would represent a strong, spontaneous affirmation that biodiversity is genuinely and effectively considered within business. Instead of a "greenwashing" veneer, this would be a move towards a real business/nature partnership in which biodiversity, like a model, would at last be respected and integrated as it deserves.

REFERENCES

Blanc, N., Bridier, S., Glatron, S., Grésillon, L. & Cohen, M. 2005. Appréhender la ville comme (mi)lieu de vie. L'apport d'un dispositif interdisciplinaire de recherche, in Mathieu, N. & Guermond, Y. (Ed.), La Ville Durable, du Politique au Scientifique, Paris, Cemagref/Cirad/Ifremer/Inra, pp. 261–281.

CBD – Convention on Biological Diversity 2007. Cities and Biodiversity Engaging Local Authorities in the Implementation of the Convention on Biological Diversity. UNEP/CBD/COP/9/INF/10, 18 December 2007.

Dearborn, D.C. & Kark, S. 2009. Motivations for conserving urban biodiversity. Conservation Biology. 24: 432–440.

Henry, A., Roger-Estrade, J. & Frascaria-Lacoste, N. The eco-district concept: effective for promoting urban biodiversity? Submitted in Cities Henry, A. & Frascaria-Lacoste, N. Biodiversity in decision-making for urban planning: Need for new improved tools. Submitted in Journal of Urban Planning and Development.

Hobbs, R.J., Arico, S., Aronson, J., Baron, J.S., Bridgewater, P., Cramer, V.A., Epstein, P.R., Ewel, J.J., Klink, C.A., Lugo, A.E., Norton, D., Ojima, D., Richardson, D.M., Sanderson, E.W., Valladares, F., Vila, M., Zamora, R. & Zobel, M. 2006. Novel ecosystems. Theoretical and management aspects of the new ecological world order. Global Ecology and Biogeography. 15: 1–7.

Hope, D., Gries, C., Zhu, W., Fagan, W.F., Redman, C.L, Grimm, N.B., Nelson, A.L., Martin, C. & Kinsig, A. 2003. Socioeconomics drive urban plant diversity. Proceedings of the National Academy of Sciences of the United States of America. 100: 8788–8792.

Houdet, J. 2008. Intégrer la Biodiversité dans les stratégies des entreprises: Le Bilan Biodiversité des organisations. FRB. Paris: Orée. 393p.

Julliard, R., Clavel, J., Devictor, V., Jiguet, F. & Couvet, D. 2006. Spatial segregation of specialists and generalists in bird communities. Ecology letters. 9: 1237–1244.

Knapp, S., Kühn, I., Schweiger, O. & Klotz, S. 2008. Challenging urban species diversity: contrasting phylogenetic patterns across plant functional groups in Germany. Ecology Letters. 11: 1054–1064.

Kühn, I., Brandl, R. & Klotz, S. 2004. The flora of German Cities is naturally species rich. Evolutionary Ecology Research. 6: 749–764.

Müller, N. & Werner, P. 2010. Urban Biodiversity and the Case for Implementing the Convention on Biological Diversity in Towns and Cities, in: Müller, N., Werner, P., Kelcey, J.G. (Eds), Urban Biodiversity and Design, Wiley-Blackwell, Oxford, UK.

Niemelä, J. 1999. Ecology and urban planning. Biodiversity and Conservation. 8: 119–131.

Williams, J.W. & Jackson, S.T. 2007. Novel climates, no-analog communities and ecological surprises. Front. Ecol. Environ. 5: 475–482.

Wittig, R. 2010. Biodiversity of urban-industrial areas and its evaluation – an introduction, in Müller, N., Werner, P. & Keley, J.G. (Eds), Urban Biodiversity and Design, Wiley-Blackwell, pp. 37–56.

Part 3

Application to a case study, Cité Descartes

Chapter 13

Extended Cité Descartes: a territory to eco-design

Fabien Leurent

Paris-East University, City, Mobility and Transportation Laboratory, Ecole des Ponts ParisTech, Ifsttar, UPEM

INTRODUCTION

Cité Descartes is an urban fragment located in the east of the greater Paris urban area. Urbanization started in the 1980s to build housing and in particular accommodate major scientific and technical organizations, like Ecole Nationale des Ponts et Chaussées, ESIEE, Paris-Est Marne la Vallée University and CSTB, which have since been joined by IFSTTAR and the Paris School of Urbanism, etc., along with several companies including innovative start-ups. At a time when urban renewal is under study for some of the most ancient buildings, the Cité Descartes neighbourhood continues to inspire ambitious urban development projects. In 2010, the local public development body, *Etablissement Public d'Aménagement* (EPA) in Marne la Vallée, launched an urban development competition called "Cœur de Cluster Descartes", or in other words, Cité Descartes at the core of a bigger area. The competition was won by Ateliers Yves Lion (AYL), with their project for a neighbourhood bringing together Infrastructure, Town and Nature. The Eco-design Chair has chosen this urban development territory, which is a hotbed for design projects, as a study field to be shared by its various research themes.

This chapter presents the main lines of Cité Descartes, as a project territory and an eco-design field. It is divided into three parts. In the first, we describe the territory and trace out over the history of its urban development. We then consider the urbanization project of EPA Marne and Ateliers Yves Lion, before setting out the range of themes studied by the Chair.

A TERRITORY UNDER DEVELOPMENT

In 1965, the regional planning and development plan, or *Schéma Directeur d'Aménagement et d'Urbanisme*, for the Paris region established the territory of Marne-la-Vallée as a new town, to accommodate and polarize the demographic and economic development of the French capital towards the east. This territory gathers 26 municipalities, sharing an *Etablissement Public d'Aménagement*, EPA Marne, for the west and central sectors, with a twin body, EPA France, for the east sector. The area covers 150 km² and it is spread out along 4 km to 8 km north to south and around 20 km from east to west, ranging from 14 km to 35 km from the centre of greater Paris

Figure 13.1 Aerial view of Cité Descartes (2011).

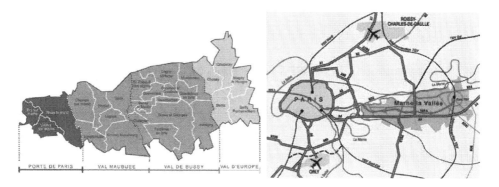

Figure 13.2 (a) the new town, (b) regional location.

(Paris Notre Dame). One of these municipalities, Champs sur Marne, accommodates a 2 km² scientific site, Cité Descartes.

Four distinct sectors are identified from east to west within Marne la Vallée: the first was urbanized before 1960, the second mostly between 1980 and 2000, and the 3rd and 4th mainly since 1985. Cité Descartes is located in the 2nd sector, west of the Champs sur Marne municipality, with Noisy le Grand immediately to the west and Emerainville to the south. In its statistic definition (INSEE's IRIS Cité Descartes) it covers 90 hectares in a rhombus shape.

Cité Descartes originated in 1983 as a response to the significant lack of research and higher education establishments east of Paris. Land premises were thus acquired

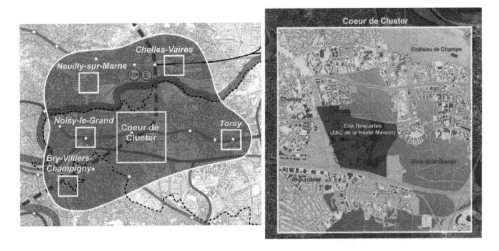

Figure 13.3 (a) the Descartes Cluster and (b) its "cœur de cluster" (cluster core).

and used to locate major organizations: engineering schools (ESIEE, ENPC, ENSG set up from 1992 to 1995), Marne la Vallée University established in 1999, and recently IFSTTAR in 2010. The latter, along with ENPC and CSTB located close by, are public organizations affiliated to the Ministry of Sustainable Development, and together they supply the site with scientific and technical capacities of international rank in the field of Sustainable Cities ("Sustainable City" Scientific and Technical Cluster). This capacity is reinforced by a competitiveness cluster focusing on "Sustainable Cities and Mobility", Advancity, and confirmed by the presence of young innovating companies that have come to join those already established on the site, benefitting from the available land and an efficient transport system: the A4 motorway and lines A and E of the RER regional express railway.

The site's history involved different stages. The competitiveness cluster was created in 2005 at the time of a national policy on territorial competitiveness and attractiveness. A few years later, in 2008, the concept of the Descartes Cluster emerged: a territory of around 70 km² (Figure 3a), comprising several activity hubs and in particular Cité Descartes, purported to make the cluster core (*cœur de cluster*) (Figure 3b). The cluster core is centred on the Cité Descartes and extends it by integrating the adjacent neighbourhoods of Noisy le Grand: Champy to the west, Richardets to the south; and those of Champs sur Marne: Nesles to the north, Château to the north-east and Bois de la Grange to the east.

Along with this major territorial project for economic development, in 2009–2010, EPA Marne boosted the local urbanization project with an "extended Cité Descartes" comprising Cité Descartes itself and the adjacent urbanized neighbourhoods (figure 3b); covering around 1500 ha of which 500 ha woodlands, with 700 ha in Champs sur Marne and 350 ha in Noisy le Grand.

At the same time, the regional development of greater Paris received new impetus with the Grand Paris Express project: an automatic, high-capacity railway network

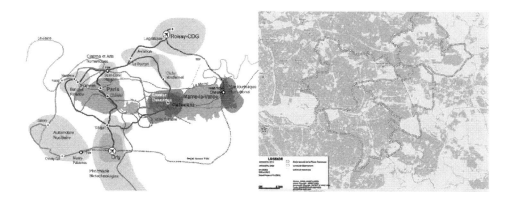

Figure 13.4   (a) Descartes Cluster and Grand Paris Express, (b) "Descartes Territory".

linking the main population and economic activity hubs within the Paris urban area. The decision was made to link up the Descartes Cluster with two stations: one at Bry-Villiers-Champigny and the other at Noisy-Champs joining up with line A of the RER (Figure 4a). In 2012, some regional planners (the State office for regional infrastructure – *Direction Régionale de l'Equipement*) observed a slow-down of the "Descartes territory" extending well beyond Cité Descartes in all directions (figure 4b), and recommended boosting the project, with Cité Descartes playing a key role, to encourage innovations in products and processes and foster the industrial development of specialized companies.

AN AMBITIOUS URBAN PLANNING PROJECT

In 2010, the organizer of the urban planning competition (EPA Marne) and the winner (Ateliers Yves Lion) together devised a sustainable urban planning project for the extended Cité Descartes: the first one defined the project, the other deepened and expanded it. This project is highly ambitious due to the range and dimension of actions to be taken, combining: the *urban dimension* in terms of functions and intensity as well as overall identity and infrastructural facilities with an emblematic role; the *environmental dimension* in order to preserve resources, especially energy, emphasize water features and wooded areas, preserve biodiversity and allow the site's users to enjoy them; and the *transport dimension* (transport function in the broad sense), to connect the territory to the rest of the urban area and in particular the internal layout of different modes, different networks and local insertion into the city and its environmental heritage.

a) The order from the client EPA Marne

The consultation document clearly stipulates the stakes and general orientations. After establishing the metropolitan context of Grand Paris and the general stakes of sustainable development, it sets out the particular stakes of continuing to develop the extended Cité Descartes: namely, cultivating excellence in eco-technologies and promoting and supporting lines of economic development, to create a dynamic tertiary and economic hub in the eastern part of greater Paris.

The project's framework comprises "the terrains of the Haute Maison special planning district (ZAC); the land premises of road routes (formerly designated as A103 and A 199) belonging to the State, EPA Marne and the public domain; private terrains; business parks (Richardets park, Malnoue ZAC); commercial zones including the commercial centre in Champy and a number of wooded areas (La Grange, Grâce, Parc de Champs, La Butte verte)".

Four main orientations are defined:

1 *Develop the Cluster's economic services* in nine duly identified industries, by developing activities in designated areas (a tertiary hub very close to Noisy-Champs multimodal station and some points with available land), taking advantage of eco-designed buildings for organizations depending on the ministry of sustainable development (the Bienvenue and Coriolis eco-buildings, near the ENPC's Carnot building), extending the business incubator that offers property suitable for innovating small- and medium-sized firms.

2 *Promote natural areas*: this involves: using wooded areas to satisfy a demand for nature from residents and other users of the site; "giving water back its place in the landscape" by enhancing and interconnecting water features present on the site; encouraging access to the landscape by landscape-friendly transport means (foot or bicycle); and fostering the ecological continuity of habitats for animal, plant and fungal species.

3 *Make mobility more sustainable*, by developing a diverse multimodal range of services that combines accessibility to the territory with the demand for personal services from users, and also with environmentally efficient means of transport.

4 *Affirm urban identity*, i.e. the "mix of density, flows and diverse functions" . . . . "in order to increase the occurrence of meetings, which is crucial to clusters". The orientation is stipulated in terms of (a) a busy, dense, mixed urban axis (boulevard du Ru de Nesles), (b) a multi-modal, multi-functional hub at Noisy-Champs railway station, (c) hosting of new residents and jobs, (d) urban reorganization around the Noisy-Champs hub, and a few other target locations, (e) infrastructural facilities to be contextualized.

We have summed up the general orientations, but the main part of the consultation document takes the details further, e.g. by stipulating specific infrastructural facilities or means of transport. In addition, appended documents set out the client's expectations: some of these documents are preliminary feasibility studies.

b) Ateliers Lion's proposal

The project put together by Ateliers Yves Lion responds point by point to the issues and general orientations established by the client. The proposal gives substance and body to the urban planning project: it identifies functions and facilities, places them individually in space, fitting in with each other and with the existing urban setting and the environmental heritage, in order to give the site an overall urban structure, with the form of a genuine city. The architects have proposed the title Infrastructure-Town-Nature as a motto for the neighbourhood, to give it an identity.

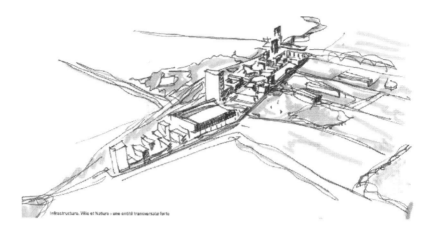

Figure 13.5 A composite identity, Infrastructure, Town and Nature (source Ateliers Lion).

Beyond the diverse set-ups of the particular themes, the proposal brings functions and infrastructure together in the space and with respect to major environmental challenges, especially energy and the wood cycle. Thus, the project has a high-level function of integration, borne by several pillars, and further reinforced by the imagination and visualization of the proposed layouts. In the architects' drawing that symbolizes the project, a major urban axis is featured out (Figure 5) that is purported to link north to south and to overcome the major fracture lines associated with east-west roads, passing above the A4 motorway with a terraced superstructure, where an emblematic urban building will allow drivers to identify the urban site. The figure also shows a significant green belt and several indications of a blue belt.

The project package comprises six parts covering respectively: (1) Nature, (2) urban identity, (3) programmatic approach, i.e. spatial projection of urban infrastructural facilities, along with the functions and treatment of spaces, (4) the environment, (5) mobility and movement, (6) phasing and assessment principles. This organisation makes a clear distinction between Nature, to which residents want access, and the Environment, a set of resources requiring sustainable management and impacts to be controlled. In the following sections, we reorganize these proposals into three main sections, respectively City, Environment, and Mobility, then discuss the phasing of governance, finally commenting on the project in terms of content, cooperative process and methodology.

c) Town: urban identity, atmosphere, functions, facilities and place on the site

The urban planners have designed an urban identity that is targeted primarily to the residents, then to the other individuals using the site – including people working there and thus the companies employing them. The key principle is to create a harmony between Infrastructure, Town and Nature: rendering the countryside omnipresent in the town, offering spacious housing of different types, ensuring a local atmosphere

Figure 13.6 Housing, nature and transport atmosphere on the boulevard du Ru de Nesles (source AYL).

that fosters conviviality, and organizing places into belts in line with north-south and east-west axes that harmoniously fit in with the local fabric.

These principles concerning the quality of urban life aim to respond to the town-living aspirations of inhabitants and the client's requirements (Figure 6). The urban planner proposes to increase the size and comfort of housing to attract well-off residents; to house students close to the multi-modal station to facilitate access to the entire built-up area; to vary the types of building available to businesses to take advantage of micro-local green and blue potential; and to organize local groups of buildings that foster community life and cooperation.

The spatial programming of activities is designed to go further than the client's instructions in terms of new surfaces to allot to housing and especially to businesses, in order to reinforce densification as much as possible and thus the town effect.

In addition, the spatial organization is proposed as a series of centres (Figure 8) structured around major facilities (multi-modal station, terrace over the A4 motorway, nautical facilities associated with the campus, main square on the campus – cf. Figure 7) and the tram, which should anchor the residential Yvris neighbourhood in Cité Descartes.

d) Environment: interact with Nature

The architect-urban planner has conceived fertile, multiform interaction with Nature: making it omnipresent in the town (cf. above), and creating local blue and green reserves organized into continuities that create belts:

- A network of green belts linking wooded areas with planted corridors along the side of the roadway, also including pedestrian and cycling lanes so that users can remain constantly in contact with nature and access wooded areas more easily.
- A network of blue belts featuring the two ponds, two streams and valleys present on the east-west axes and on the Boulevard du Ru de Nesles.

Le Maison Descartes / Cité Internationale du Développement Durable : la vie urbaine au cœur de la Cité Descartes

Figure 13.7 The main square on the campus (source AYL).

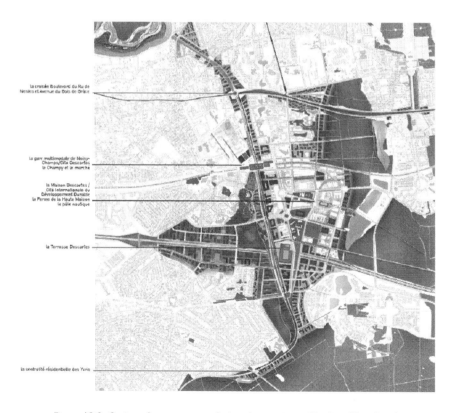

Figure 13.8 Series of centres to polarize the territory (Ateliers Yves Lion).

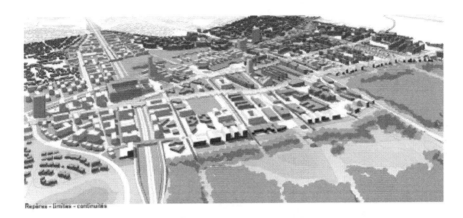

Figure 13.9 A town penetrated by Nature (source Ateliers Yves Lion): marks, boundaries and continuity.

These proposals respond to the client's environmental specifications (cf. Figure 9). However, the urban planner goes further, proposing a series of measures to constructively interact with the environment:

- State-of-the-art energy performance for buildings, via their construction or auxiliary measures (ventilation, revegetation),
- Use of geothermics, with district heating systems,
- Local collection of renewable energy from the sun or wood for biomass,
- Cyclic exploitation (circular management) of wood as a building material and biomass.

e) Transport: modes, networks and their local insertion

In terms of transport, the urban planning project translates the client's orientations into space. Soft traffic, on foot or by bicycle, would be made easier by specific lanes encouraging short journeys. The idea is that the multifunctional development of activities would encourage this kind of exchange, still all too rare in 2012. Soft traffic would involve individual modes with easy access for users. The plan also involves the installation of self-service vehicle systems for bicycles and cars (cf. Figure 6). For private cars, the policy is to restrict usage by not increasing roadside parking capacities (along with the eviction of commuters' cars) and orienting public parking towards high-rise car parks that require less land and ground waterproofing.

As well as local stopping points and liaisons, the project focuses on organizing high-capacity liaisons. In public transport, the highest available capacities are on lines A and E of the RER, both travelling east-west, and on the future Grand Paris Express, which will provide a north-south tangential liaison. These will be supplemented by smaller yet quite high capacities: an extension of the T4 tram line to serve the neighbourhood on the north-south axis, high-performance bus routes on the RD 199 and express lines on the A4 motorway, both travelling east to west. The network would also include

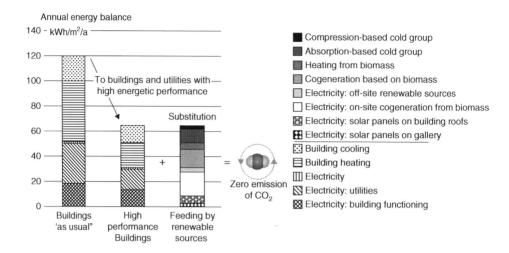

Figure 13.10 Improve the energy balance (source Ateliers Yves Lion).

standard bus routes that would travel over the main infrastructure lines rather than remaining within grid blocks.

The local insertion of transport means in the town is devised to work on several levels: at micro-local scale, by putting parking places primarily in buildings and public high-rise car parks; at location level, through urbanizing the Noisy-Champs multi-modal station, where shops and services are planned (cf. Figure 12); at road section level, by transforming the Ru de Nesles axis from its original state resembling an old inter-urban road, into an urban boulevard in the true sense of the term, i.e. a wide lane bordered by trees then buildings, open to a variety of traffic serving local activities as a priority (Figure 6). A fairly similar vocation is proposed for the RD 199 and A4 main roads, with significant redevelopment of the latter's cross-profile.

f) Phasing and governance

Lastly, the winning project proposes to carry out the development in four phases: first a launch phase involving two monumental installations and bus lines prefiguring the tram lines and high-level-of-service buses; then an implementation phase for the tram, high-level-of-service buses on the RD 199, an express bus on the A4, and the development of an urban boulevard on the Rue de Nesles axis; next a re-development of the area between the multi-modal hub and the RD 199, linking the Richardets and Malnoue neighbourhoods into the urban fabric; and lastly a conditional phase to urbanize the still underused terrains along the A4 motorway.

This phasing recognizes a difficulty in the timing: it takes a long time to implement a tram line -thus the need to prefigure it by a bus line. However, it omits a much greater difficulty: the Grand Paris Express will not exist before 2020, and yet the reconstruction of the Noisy-Champs multi-modal station is a crucial element that conditions the tram link-up.

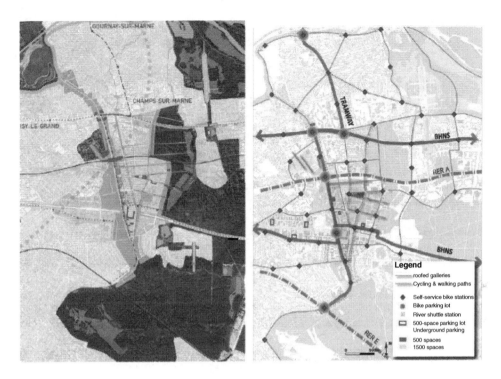

Figure 13.11 Transport lines and their networking (source Ateliers Yves Lion).

This brings us to identify a crucial governance issue: the entire urban project is organized around the tram, and therefore the multi-modal hub; this involves major transport investments that will not be contracted by EPA Marne, which focuses on land and property. This aspect will therefore require to mobilize the two major stakeholders for the public transport system in Ile de France, i.e. STIF, which is the organizing authority and the Société du Grand Paris, which is responsible for installing and implementing the Grand Paris Express.

g) Comments on the project's design

We end this presentation of the urban planning project with three sets of comments: first, the actual content of the project; second, the design process; and third, the design methodology.

Concerning the *project's content*, the composition of elements constituting the site constrains the spatial layout, yet without saturating the project's potential functionalities: as proof, circular management has been proposed for energy and wood resources. Also, the project is still virtual: it identifies the potentials and makes them manifest; it indicates and recommends directions; but its register is that of proposals, or even suggestions, that are based more on intuition than assurance – as opposed to a simulation based on an explanatory model. The result is a degree of fragility, in particular when it

Figure 13.12 Make Noisy-Champs station multi-modal and multi-functional (source AYL).

comes to material and financial feasibility. On a material level, we have pointed out the planning difficulty, which should probably be tackled with the priority reconstruction of the multi-modal station, without waiting for the Grand Paris Express tunnelling machine, but anticipating its passage. On a financial level, the urban planning project sketches out an evaluation of the cost of developing the surfaces, not including all of the super-structural or infrastructural public facilities. However, funding makes a difficult issue for a project of this size. At this stage of the project, the following issues remain unresolved:

- The efficiency of the Town-Infrastructure-Nature identity concept. A motto like "Marne la Vallée Green Campus", or "East Paris Eco-Campus" would target expected clients more efficiently and make the location easier to interpret within greater Paris.
- The residential attractiveness for well-off populations. In principle, the environmental measures are favourable (although beware of road traffic pollution), but the existing social neighbourhood of relatively modest residents would clearly require specific handling to attract a first wave of new residents.
- The urban project expressed by a general layout plan does not guarantee a concrete neighbourhood life to flourish. This would require sufficient users carrying out more than one activity after another on the site, in a sequence, taking advantage of a local urban fabric that is sufficiently dense in diverse opportunities. The corresponding spatial setting acts on a finer scale than the general layout plan, i.e. the elementary pairing between functions and buildings and thus the bottom-up constitution of an urban fabric.
- The in-depth, diverse treatment of the environment does not guarantee its actual performance. This will depend on sufficient numbers of clients in order to yield

the ecological benefits of massive systems, such as high level of service bus routes and district heating systems.

On the *design processes* level, we can mention the progressive, converging and collaborative characters, as well as the incompleteness.

Progressiveness: the consultation document followed by the response of the winning application constitute two stages in the definition of the project. The first has already made good progress, in specifying the facets concerned and the designation and individualization of the main elements to be integrated. The second is more accomplished still, since it fixes the different items in space and in doing so sets out the neighbourhood relationships, continuities, corridors, networks of belts and an overall shape.

Convergence: the second stage of the design process continues with the same desire to integrate as the first stage, and even reinforces it by basing it on several pillars. Space is the project's corner stone, its crucial integration factor, but it is not the only pillar and lever. Thus, the project's maturation is a consolidation of the original ambition. The urban planner fully assumes the stakes assigned to the project as a whole, and achieves them through the composition of its elements. He goes even further, by designing each element to serve the overall stakes, not just as a piece of the whole, but to fulfil the outcome in itself. In other words, this kind of element is transcended by the project. These elements include circular management resources ([1]), such as major super-functionalized facilities ([2]), and sections of road that work not just within their own modal network, but act as a support to collect energy and accommodate habitats and activities.

Collaboration: the client and its urban planning consultant are clearly both confirmed planners. The client establishes the challenges and major orientations to match the needs and potential of "its" territory. Its urban consultant, who is more of a composer than a manager or operator, projects the project into space. He or she also brings an informed technical capacity on recent progress: technical solutions for building or transportation, management of environmental resources. He or she, in this case, brings knowledge of the surrounding area encompassing the target territory: which makes backing and even mutual support easier, with larger-scale projects – here the "Dorsale Est" green belt to the east of greater Paris, and the redevelopment of the Seine River banks at the edge of the central city – just downstream the junction with the Marne Valley.

Incompleteness: at project stage, the board of territorial decision-making is incomplete, and the transportation planner(s) need(s) to be integrated into a cooperation circle. Promoters to deal with elementary property operations are also lacking.

Lastly, on the *design methodology*:

• The project documents play a vital role: whether on paper or digital, whether prescriptive (originating with the client) or indicative (with the urban consultant), the document expresses the project and determines the lay-out.

---

[1]Sub-systems rather than elements.
[2]Or trans-functionalized: transformed in the function rendered.

- Graphics are essential to represent objects in their particular form and context and boost the imagination. Digital images are now widely used.
- Designers use a directory, a referential of types: architectural types for buildings and other urban items (roads, squares, etc.), types of technical objects (mode of transport element, such as a vehicle or station), and types of corridor. They adapt each object of their project to the function to be fulfilled and the place to be occupied, by searching for properties on several levels (the trans-functionality mentioned above). This adaptation of the composition is crucial to endow the place with an intelligence specific to the service of the assigned function, an intelligence via design that is thus artificial.
- Although designers compose the elements, they also know how to hybrid types. For them, buildings and transport elements are also environmental resources in one or several ways. Reciprocally, a particular environmental device (like a planted roof) can naturally integrate into a technical component. Their composition breaks away from and disregards conventional boundaries between technical specialities.

COMMON FIELD OF STUDY FOR THE ECO-DESIGN CHAIR

In 2009–2010, the Chair held discussions with EPA Marne to choose a current eco-neighbourhood project to use as a common application field for its studies. Cite Descartes was chosen following a suggestion by EPA: with the advantage that Ecole des Ponts was already located on the site, well informed about its urbanization plan and a contributor to it with the construction of two new high-performance green buildings.

a) The place of engineering in design

In the professional world, an engineer's position fits into a specialized scientific and technical domain: engineers have sectorial rather than general skills. An engineer working in his or her speciality is sensitive to fundamental issues (like solidity for a construction or profitability for a business model) and to the composition and operation of technical objects to be used or to design. Engineers possess scientific models of these objects and, increasingly, simulation tools and assessment methods. The technical toolbox of simulation models and evaluation methods is very useful in design projects, to devise alternatives, simulate each of them, and assess the advantages and disadvantages: based on assessment results, it is possible to take retrospective action on a scheme alternative's design in order to gradually devise a solution that is satisfactory for all of the issues at stake. Thus, a project's design fits into a decision aid cycle that goes through four stages making up a circle: design, simulation, assessment, comparisons and feedback.

In architecture and urban planning projects, basic simulation involves situating the elements of a project in space in relation to each other: the typical result is a map of the overall layout. Architect-urban planners can be satisfied with this result, but increasingly they work with specialized engineering to study more closely how a specific aspect works. Calculating the mechanical behaviour of civil engineering installations has scientific bases dating from the 19th century. Thermal behaviour and the energy performance of buildings are much more recent scientific domains – perhaps half a century old. The study of ecosystems is an emerging field for modelling that is still far from mature. Somewhere in the middle, modelling land use and transport is fairly developed for the spatial scale of a large territory (part of a built-up area or greater), but not yet for the local scale of a neighbourhood.

As a reminder, other engineering roles in projects are: financial engineering, organizational engineering of project management, and increasingly, information systems engineering to manage digital information.

b) The Chair's specific objectives

Given the state of the art, the Eco-design Chair decided to tackle a common case study with a set of clear targets:

- By theme dealt with by the Chair, the case study serves as a demonstration application: to illustrate the application of a mature model, or the application of an emerging model, or the application of the theme's characteristic indicators.
- Share a marked-out territory as a common application context: which facilitates mutual comprehension between the teams working on (a) particular aspect(s).
- On a given theme, the ambition is not to broadly explore the wide set of possibilities offered by the study field, but rather to select a few as typical cases. Even though the theme has been studied by specialized engineers working with EPA Marne or Ateliers Yves Lion, the Chair's approach markedly differs from a professional design study: it involves the classroom projects of high-level students as well as research activities.
- Similarly, although the various themes studied by the Chair constitute a broad range, they are no match for an integrating project devised by an urban planner, which goes beyond sectoral engineering skills.

These targets have been fulfilled by application chapters in most of the doctoral theses supported by the Chair, as well as projects by engineering students. In particular, the study field provides a guiding thread for the ATHENS course, which brings together ParisTech schools and lasts a week, that the Chair has created on its general theme. Nevertheless, each theme has been treated in a fragmentary way, and bringing together all of the themes covered creates above all a juxtaposition, not deep-rooted integration.

c) Range of themes covered

The following chapters include:

- A territorial diagnosis for the extended Cité Descartes, at the metropolitan scale of greater Paris. The team at Ponts, working with several engineering student groups, characterized the local urban functions by quantities of residents and jobs, closely interacting with the rest of the built-up area, and in particular through the main transport networks. Land use and transport scenarios were devised and simulated.
- Once again involving the Ponts team, but this time at local neighbourhood scale, a detailed analysis of the layout in space of urban functions, buildings and transport means: by evaluating accessibility indicators and simulating the use of car or public

transport modes. The car parking theme was studied in more depth, creating scenarios of supply and demand and simulating their usage relationship (Houda Boujnah's thesis directed by Fabien Leurent).

- The design of buildings for residential, tertiary and school usage around the boulevard du Ru de Nesles. Several alternatives of architectural form and positioning in space were put together by the Mines team (thesis by Grégory Herfray and work by Eric Vorger directed by Bruno Peuportier). The alternatives are the object of an LCA, with simulation of thermal and energy behaviour.
- A characterization of biodiversity in the extended Cité Descartes: the team at AgroParisTech (Alexandre Henry's thesis directed by Nathalie Frascaria) explored this them along five lines, with specific indicators for each line. A scenario of restoring this fundamental environmental capacity was also designed and assessed.

CONCLUSION

In this chapter, we have presented the Cité Descartes urban planning project and discussed its contents. The urban planner's detailed expression of the urban project is a great deal more persuasive: the documents presenting it provide an agreeable, gratifying and stimulating read. In addition, we have highlighted the design activity as a process: setting out the project's formation and consolidation stages and explaining the roles played by those involved.

Lastly, we have situated the studies that this concrete field inspired at the Chair. The range of themes implements items into space on four overlapping scales: firstly the infra-building scale for all inside or superficial layouts; then the micro-local scale of a group of neighbouring buildings, the public transport station, the parking lot, or the local habitat of an ecological taxon; then the local neighbourhood scale, for its overall functions and constitutive sub-systems such as transport corridors, or green and blue belts; and lastly the whole territory scale, for transport and energy networks, the use of local urban functions whose users mainly come from the rest of the territory, and reciprocally for environmental impacts caused on the outside by users of the neighbourhood.

The local context has not however allowed us to maintain long-term close links between researchers and those out on the field. To strengthen the linkage, an instance of collaborative action could be to share sustainable development objectives in order to boost the coherence of a project carried out by a large number of stakeholders (including EPA, local elected representatives, urban planners, promoters, architects, engineering firms, companies, managers, users, etc.): does everyone have the same vision of what needs to be done to "meet the needs of the present without compromising the ability of future generations to meet their own needs"? What is more, it is vital to gather professionals' opinions and comments on the researchers' work.

ACKNOWLEDGEMENTS

This chapter borrows extensively from the project documents drawn up by EPA Marne and Ateliers Yves Lion; we thank them for granting us permission to use text and images. We also thank Eric Alonzo, professor at the "Ecole d'Architecture de la Ville et des Territoires", for supplying us with valuable keys for informed reading.

REFERENCES

Ateliers Yves Lion. Cité Descartes cœur du Cluster Descartes. Dossier de présentation. Avril 2010, 15 pages A3.

DRIEA. Portrait de territoire Descartes. 2011.

EPA Marne. Cité Descartes, cœur du Cluster Descartes Ville Durable: consultation internationale de programmation urbaine. Plaquette de présentation, novembre 2009, 15 pages.

EPA Marne. Cité Descartes, cœur du Cluster Descartes: Programme du concours de maîtrise d'œuvre urbaine. Janvier 2010, 9 pages.

EPA Marne. Réseau de plate formes multimodales pour la mobilité sur Marne la Vallée. Document de travail, mars 2010, 10 pages.

Chapter 14

Territory and transport in a metropolitan context

Thierno Aw, Nicolas Coulombel, Fabien Leurent,
Sergio Millan-Lopez, & Alexis Poulhès
Paris-Est University, City Mobility Transport Laboratory, Ecole des Ponts ParisTech,
Ifsttar, UPEMLV

INTRODUCTION: THE TOWN AS THE MATRIX OF THE NEIGHBOURHOOD

At the scale of the built-up area, a neighbourhood is an urban fragment: although its internal complexity is generally great, its functions and the way it operates are dominated by its interaction with the rest of the built-up area, or in other words with its metropolitan context. The context's influence and domination are achieved through the built-up area's territorial extension both in space and as a set of activities, as an urban mass of inhabitants and jobs: this mass puts pressure on the neighbourhood to contain, on a local level, urban functions, inhabitants, jobs, and various amenities (e.g. services, shops, green areas). In addition, the urban mass in space, in order to operate locally and globally, equips itself with major resources: dense built areas to use the land efficiently (cf. building height), high-capacity transport networks such as express roads and in particular rail lines for public transport. These major resources affect the local level; they determine the operation of the neighbourhood that must deal with them: the situation in relation to major facilities, clusters or the most intense urban hubs and in relation to modes of transport that provide accessibility, conditions the contents of the neighbourhood and its own facilities with a local scope. Lastly, the geographical influences that are salient at a given time have taken form gradually: the weightiest facilities involve long-term investments, creating significant inertia in the built-up area and making planning incremental. To sum up, the *fundamental coupling of a neighbourhood with its metropolitan context occurs when it is at one with the built-up area, as a receptacle of activities and a user but also a bearer of major facilities.* Planning can add more elaborate couplings, like specialization in certain functions – cf. the scientific vocation assigned to Cité Descartes. This kind of specialization makes the neighbourhood attractive to the activities concerned, and its attraction will be all the greater in scope and intensity if the whole built-up area contributes with its mass that generates flows, transport facilities to channel the flows, and an arrangement of similar polarities: this is the notion of a marketplace, a feeding and catchment area.

This contextual dependence should be integrated into the design of a neighbourhood, which is a piece in the puzzle of the built-up area. The neighbourhood is a location opportunity for a large set of stakeholders already present in the built-up area and likely to re-locate: this opportunity for territorial development influences the area's performance on social, economic and environmental levels. The effect is no doubt marginal at the scale of the built-up area, but impacts that are low for the whole zone can be high on a neighbourhood scale. As well as the performance on the inside, there is also therefore an effect on the outside, which can be just as great and thus needs evaluating. Mutual impacts between the neighbourhood and the rest of the built-up area constitute a *commercial relationship* whose *equitable character* raises questions. It is necessary to clearly establish residents' exposure to local conditions, including transit flows; and at the same time, establish the impacts produced by these residents on other local residents. The equitable character depends on the *balance of impacts* respectively endured and produced.

This chapter treats the Extended Cité Descartes as a neighbourhood of the Paris metropolitan area, which we also refer to as greater Paris and which is one of the foremost large urban metropolises in Europe. We start by presenting the metropolitan context: the gradual formation of greater Paris, in the geographical area that it has structured significantly through its internal urban polarities and major transport facilities. On this basis, we study the urban functions assumed by the Extended Cité Descartes in the reference state (its recent situation) or that it would assume in a virtual future state (a development scenario): in terms of land use and facility needs to be conceived. We then study the physical and economic conditions of operating the neighbourhood in a steady regime: its physical operation in terms of energy consumed by users in buildings and their everyday movements, and its economic operation in financial terms imposed by the property market in particular. We then focus on transport as a structuring facility for interaction between the neighbourhood and the rest of the built-up area: we make a diagnosis of the state and performance of car and public transport modes in a reference state (from 2009), and study major projects to develop transport networks.

FORMATION OF THE METROPOLITAN TERRITORY

Originally, the site of Paris resulted from the confluence of rivers: the Marne and Oise Rivers join the Seine and together create the boundaries of an island of land, "île de France". This situation at the heart of a major basin suitable for farming led to the emergence of a town around a core protected by two branches of the Seine River: the Ile de la Cité.

a) Affirmation of territorial functions

Over the centuries, the town took on political and administrative functions in addition to its initial commercial function when the kings of France moved there, starting with the Merovingians and Capetians. In modern times, the city's economic and financial preponderance have reinforced these functions. In the industrial era, Paris accommodated and stimulated various industries, in particular the mechanical construction of machines, parts for engineering works, cars and aeroplanes and other more traditional types of production. These activities contributed to the area's economic and demographic development, aided by the arrival of modern techniques, along with efficient

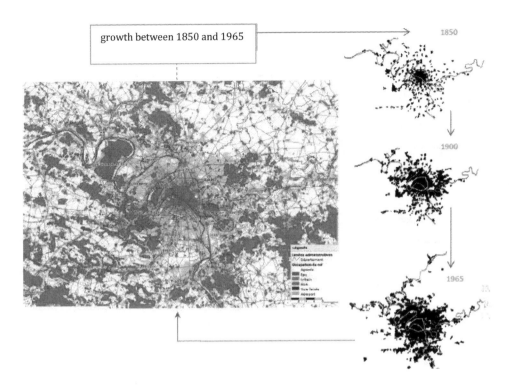

growth between 1850 and 1965

1850

1900

1965

Figure 14.1 Greater Paris at three points in time: 1850, 1900, 1965.

modes of transport[1] that allowed the built-up area to operate as a whole and move onto a higher plane in space and demographics.

Figure 14.1 shows the growth of the Parisian area in three successive eras; from 1.5 M inhabitants in 1848, to 8 M in 1956, along with 4 M inhabitants in 1900 interconnected by the metropolitan network – compared to 12 M inhabitants in 2010 (Figure 14.2). Technical and demographic developments have participated in the same historic process of joint evolution: with the progressive transformation of economic production, social activities and life styles, the area has become a highly significant centre in terms of population, diverse production and intense consumption. This significance in terms of people, intense production and exchanges is also illustrated by average income, and thus by the large size of its own urban markets, i.e. domestic markets: employment and labour, for industry and services, construction and property, and transport.

Over the course of this general evolution, the relationship between Paris and the rest of the world has transformed. The original relationship of proximity with a gradually expanding agricultural area has strengthened in absolute terms, but moved into the background in comparison with the industrial, commercial, financial, political and

---

[1] As well as other technical networks, like water supply and sanitation, and energy supply (gas then electricity).

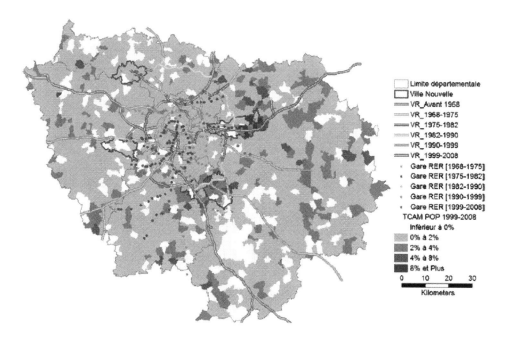

Figure 14.2 Major facilities, new towns and demographic dynamics 1999–2008.

cultural functions played out on a national, continental and even international scale; and the same can be said for service functions of a domestic nature, i.e. those that respond to a local demand that has increased as a result of demographic and economic developments and a more service-based economy. In terms of flows of financial value, the main relationships between the built-up area and the Parisian Basin are now service-based: the built-up area's high-level functions are used on site or from a distance by the Basin's inhabitants; in addition the latter acts as a labour pool, or housing overspill – with same outcome but from different starting points.

b) A short history of an unusual urbanization

In Ile de France, human presence initially took the form of a scattering of village buildings. Paris only gradually stood out from other medieval towns, mainly in its role as royal place of residence and so political and administrative centre. However, the national power only permanently settled in Paris at the turn of the 17th century, when Henri IV organized the original urban centre, Ile de la Cité, in an architectural form that remains to this day. Along with the installation of political power came the court nobility: a privileged class whose consumption stimulated the development of luxury goods in terms of construction and craftsmanship. In the mid-19th century, the fortunate conjunction of Napoleon III's urban vision, the intersectoral administrative activity of Baron Hausmann, and technical progress, conferred the city of Paris with the main features of today's urban form, i.e. land use rationalized between dense islands of homogeneous architecture and broad avenues of traffic, forming a network connecting railway stations that developed at the same time towards the provinces and abroad.

This road and rail network proved simple to use for trolley bus and tram traffic, and then for the construction of the Métro. In the early 20th century, public transport efficiently connected the city's 4 million inhabitants so that individuals could work and live in different places. The built-up area then expanded along the rail lines, absorbing other old urban centres. In the second half of the century, the car progressively became the main means of transport and took over the network, which was partially renewed with motorways and urban expressways. The combination of roads and cars, simple and flexible for urban planners and efficient for users, led to rapid local developments that were not limited by corridors and could therefore be located anywhere – wherever there was a need or an opportunity, in the absence of regulations.

After the Second World War and reconstruction, populations arriving from the former colonies and the baby boom boosted and accelerated demographic development. In 1960, the population of greater Paris exceeded 7 million inhabitants, a third in the city of Paris, and the other two thirds in the suburbs. In reaction to a demographic forecast predicting 14 million inhabitants by 2000, in 1965 the French government passed a Schéma Directeur d'Aménagement et d'Urbanisme (SDAU – regional planning and development programme) for the Parisian region, based on a polycentric principle and the idea of new towns on the outskirts accommodating a significant share of new inhabitants and jobs, and new transversal lines on the rail network, one running from east to west and two from north to south. The demographic forecast and the regional development programme were only partially achieved. In 2000, "only" 11 million people lived in greater Paris, and the additional rail lines were reduced to the east branch of the RER A, the RER E line also running from the east to the centre, line 14 of the Métro, from northwest to southeast but confined to the centre, and extensions of some Métro lines: for the most part, the additions involve *interconnections in the city centre*, which create a regional rail network. In parallel, cars have become more widespread, a network of urban expressways has been created, and new towns have only absorbed *one quarter* of the rising numbers of inhabitants and jobs. The historical centre of the city of Paris has mainly spilled over into its suburbs, losing 1 million residents since 1900 and 160,000 jobs since 1965.

In 1976, the new SDAU produced by the Ile de France region reiterated the major principles of polycentric development, but adjusted the demographic forecast to 12 million inhabitants by 2000. It redefined the major transport projects (i.e. motorways and RER, radial and peripheral liaisons) and made a place for social and environmental issues, i.e. maintain the social and spatial balance of the area, protect the environment and heritage. In 1994 the demographic projection was revised to 11 M inhabitants for 2015: a new regional development plan (SDRIF) counted on a hierarchy of centres to limit urban sprawl, and on the protection of natural areas; in terms of transport the programme identified the need for tangential liaisons and predicted that roads would respond to the rising need for journeys within the suburbs. The next regional development plan, in 2007, maintained the challenges and insisted on striking a new balance between residential and economic functions. It focused on a densification at the heart of the area and the creation of a rail bypass, Arc Express, in the close suburbs.

In fact, the 2000 decade was marked by a re-densification of the heart of Paris, gradually extending from the centre in a spiral, combined with a transport policy encouraging the use of public transport and restricting car capacity (traffic and parking). Congestion rose sharply in the backbone networks of these two modes at rush hour, thus reducing accessibility to territories and increasing discomfort for users. In addition, property prices went up constantly from 1996, further reducing inhabitants' margins for manoeuvre. The French government's reaction was to devise the Grand Paris Express project to significantly increase rail capacities and efficiently connect major existing centres with future ones, thus reducing the pressure on property and anticipating future urban development. The project, negotiated with local governments from 2008, was progressively adopted by all urban planners in Ile de France: around the eighty or so stations concerned, Territorial Development Contracts exist to organize and boost local development of urban, economic and social programmes.

Comments. This overview illustrates that the function of a national capital has significantly contributed to the social and economic development of Paris, on a multisecular timescale. More recently, Paris has become the focal point of main national transport networks: road, rail, motorway, high-speed trains, and a major centre for flight transfers thanks to its two international airports. These modern, efficient connections on a very large spatial scale are structural elements that necessarily feature on regional development maps. Since 1965, the Parisian region in the broad sense has been the object of overall development through regional development plans that determine land use and major transport networks on a regional scale, and anticipate at least the next two or three decades ahead.

Each programme is only partially achieved, and every new programme revises its predecessors and adapts its planning in line with major political issues. In 1976, protecting the environment mainly involved protecting natural areas, with little focus on improving air quality and even less on reducing greenhouse gas emissions.

The need for railway bypass liaisons was identified late in the day; due to a lack of foresight, the infrastructure will be mostly deep underground, involving very high construction costs.

Lastly, local planners, promoters and individual stakeholders (households and companies) each play an important role in urban development. General planning dating from 1965 only centred on a small area of urbanization. In the 2000 decade, this "spontaneously" adapted itself (i.e. mostly autonomously) to new congested transport conditions, focusing once again on the heart of the city.

c) Portrait of metropolitan dynamics

Figure 14.2 shows the boundaries of the new towns in the region and how road and rail networks developed between 1968 and 2008, along with local demographic dynamics from 1999 and 2008.

Figures 14.3 and 3 bis respectively concern population and jobs: each shows the progress in size from 1968 to 2008 in relative terms (annual average growth rate evaluated geometrically). Demographically speaking, the new towns and the Roissy centre have seen much higher relative growth in population, but starting out from a much lower level. Growth has been largely disseminated in the outskirts of the built-up area, while the population at the heart of the city has gone down slightly, mainly due to a reduction in average household size. Economically speaking, the relative progress

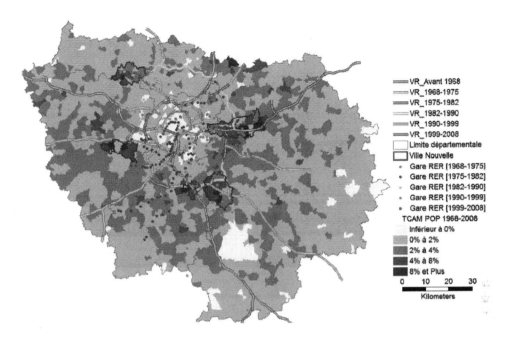

Figure 14.3 Demographic dynamics Ile de France from 1968 to 2008.

Figure 14.3bis Economic dynamics in Ile de France from 1968 to 2008.

Wooden areas
Farming
Water
Rural (other)
Open urban
Single-family housing
Multi-family housing
Activities
Equipments
Transport
Construction site

Figure 14.4 Types of land use east of Paris in 2008.

of employment also shows the role played by the new towns, and by diverse towns on the edge of the built-up zone.

d) Marne la Vallée, new urbanization hub

The new town of Marne la Vallée, covering around 15 thousand hectares, was designed in an angular sector focused on the centre of the built-up area to the east, ranging from 15 km to 35 km from the centre, in a 6 km circular arc. In 1965, the first sector on the west side was already urbanized, as well as several areas along the Marne River (Noisiel, Torcy, Lagny). From 1982 to 2008, urbanization gained over four thousand hectares on former farming land (fig. 4); woods and forests were preserved. In 2012, the new town accommodated 300,000 inhabitants and 131,000 jobs. The ratio of active workers to jobs was 0.9, which is respectable in a regional context with relatively low unemployment but a persistent polarization of employment in the very centre. The ratio exceeded 1 in sectors 2 and 4, where the state deliberately planned the location of Cité Descartes and business zones, along with a tourism cluster around Eurodisney.

The size and sustained pace of urban growth can mostly be explained by the dimension of the Parisian region, and its colossal population, for which Marne la Vallée is one of several overspill outlets. Impetus also comes from the gradual extension of major transport capacities running east to west, with the A4 motorway in 1980 and the eastern branch of RERA A beyond Torcy in 1987; from north to south, the main liaisons are the A86 and A104 (called "La Francilienne") bypasses built in the 1980s.

Prior to the Grand Paris project, the central scenario of Marne la Vallée's development was a continued trend of urbanization, filling in the empty spaces while preserving wooded areas: the urban clusters would progressively join up to create a general urban area. By 2030 the forecast was that 50% of the working population in the new town would work there, illustrating an internal coherence.

The Grand Paris project completely altered the trend scenario: the anticipated urbanization would be more intense, faster, and more focused on the Descartes cluster and the tourism cluster. The relationships with the rest of the built-up area would be relatively less focused on the very centre and more on those centres connected by the Grand Paris Express, which should encourage economic development through more diverse opportunities, and a modal shift from the car to public transport for bypass connections. From this history and future perspectives, it is worth noting the crucial role and dominant influence that the overall territorial project of greater Paris has on Marne la Vallée's urban development. The built-up region imposes the demography, economy, major facilities and overall layout of the new town. The new town is a territory under influence: the influence feeds into it, stimulates it and helps to equip it. However, it makes it more complicated for the town to create its own identity, specific urban atmosphere and operate autonomously or self-sufficiently.

URBAN FUNCTIONS AND TERRITORIAL ATTRACTIONS OF THE EXTENDED CITÉ DESCARTES

a) Dependent, long-term urbanization

Greater Paris imposes its dominating influence on Cité Descartes in particular through: delimiting the site, assigning it with particular functions, locating clustered activities and major installations there. Recent history has shown a succession of decisions that are mostly consistent but sometimes subject to revisions and no obvious logic in their order. Thus, the major A4 installation existed prior to the site's delimitation, the Noisy-Champs RER A station was installed shortly afterwards, and then the site was redefined as Extended Cité Descartes, which transformed and increased the intensity and functions of the urbanization project, as well as the role of the railway station.

If this joint evolution of the various aspects is to make efficient use of resources, it is important to maintain a logical coherence: new measures should make the best of the situation created by previous measures. The progressive revision of the project's diverse facets one after the other is simply the mark of an adaptive process of evolution and development.

b) Essential functions and their functional indicators

Urbanization has supplanted the pre-existing agricultural function and replaced it with three explicit functions: firstly, to lodge households, secondly to accommodate economic activities and so jobs, and thirdly to educate students. This latter function is coupled with a particular specialization of activities in technological research and development, which are two crucial characteristics of Cité Descartes.

The respective functional indicators are: for housing, the number of people or households; for economic activity, the floor surface area per type of activity; and for education, the number of students. As an alternative to floor area for economic activity, the number of jobs can be used as a functional indicator: for tertiary activities we can convert 30 m² into one job. In a life cycle assessment, the corresponding functional units would be the product of the functional indicator for a length of time, let us say one year.

A fourth function is largely, but in fact implicitly, borne by the territory: transit traffic, in particular on the A4 motorway and RER A but also on other roads (Boulevard du Ru de Nesles on the RN 370).

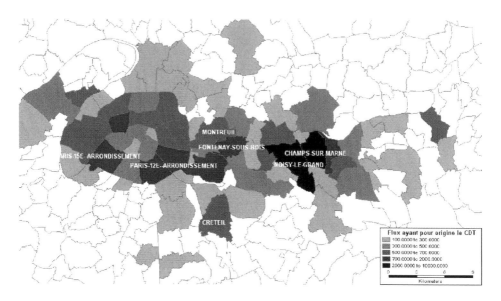

Figure 14.5 Employment catchment areas in Champs sur Marne and Noisy le Grand in 2010.

c) Relationship to space

Cité Descartes is located on the borders of the centre of the built-up area, inside which the density generally exceeds 5,000 inhabitants per km². It is still significantly covered by wooded areas to be preserved. The intensity of land use is quantified by the spatial density of functions: in 2010 these were 21 inhab/ha and 70 jobs/ha.

In addition to this static situation, other spatial relationships complement the activities established by individuals using the site. For the two towns of Champs sur Marne and Noisy le Grand, in 2010, Figure 14.5 shows the labour catchment area (or recruitment spots) for local employment, and Figure 14.6 shows the employment catchment area that attracts resident workers. Clearly, the configuration of greater Paris and the situation that the Cluster Descartes core occupies within it determine these two areas and explain their differences. The employment catchment area leans strongly to the west due to the hypertrophy of activities in the central zone as far as La Défense and even beyond, and opens out to the north because of the Roissy business centre. The labour catchment area recruits mainly close to the site: it extends far to the west along the RER A axis and opens out along the east side on the edges of the built-up area. The characteristic dimension of each area is the average distance between work and home for the workers concerned: around 13 km for the employment catchment area and 10 km for the labour catchment area.

d) Attracting stakeholders

The two towns of Champs sur Marne and Noisy le Grand developed progressively, especially from 1975 to 1999, when most of the urbanization took place in the Marne

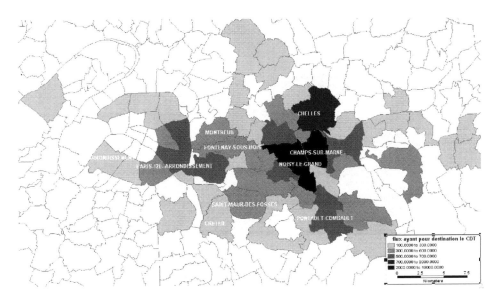

Figure 14.6 Labour catchments areas in Champs sur Marne and Noisy le Grand in 2010.

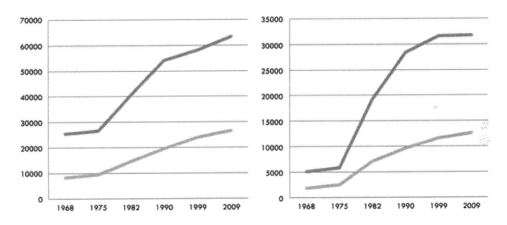

Figure 14.7 Household supply and inhabitants: progress from 1968 to 2009 (upper curve: Noisy le Grand, lower curve: Champs sur Marne).

la Vallée sector (Figure 14.7). This residential population is younger than in the built-up area as a whole; 15–29 year olds make up the biggest age group (Figure 14.8).

The social composition includes a lower share of retired people than in general, slightly more labourers and intermediate-level employees, and slightly less managers and professionals (Figure 14.9). Thus the household demography shows that the supply of housing has attracted households, but relatively more modest ones, since the most solvent households prefer to live more centrally or to the west of the city (Figure 14.10).

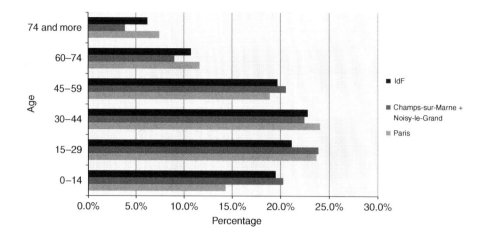

Figure 14.8 Demographic composition.

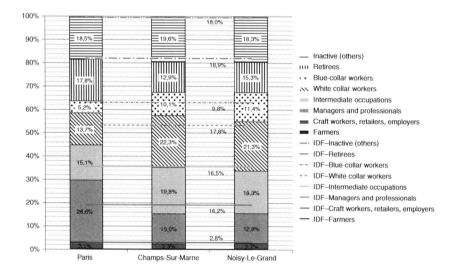

Figure 14.9 Social composition.

Concerning companies, for our two towns, in a broad zone around the A4 motorway, the dominant activities are research and teaching, government services and social work, as well as wholesale and retail commerce, and transport, hotel and catering services (Figure 14.11). A survey of companies[2] indicates that the crucial factors for their location are distance to the A4 motorway (and so access to the main road network)

---

[2]Devised and carried out by a group of engineering students at ENPC on a TUSMUR course in 2012.

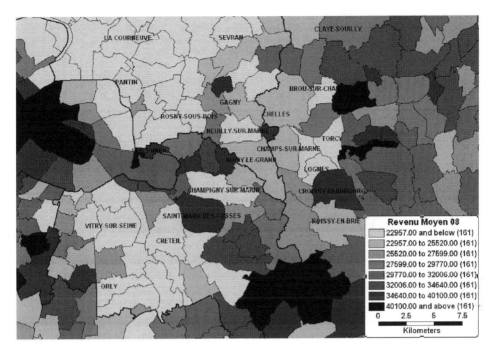

Figure 14.10   Household income: averages by town in Ile de France, 2008.

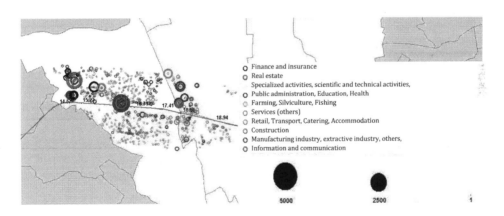

Figure 14.11   Business activity layout.

and to the centre of the city, ahead of rail access and the price of land, followed by the local presence of clients and suppliers (Figure 14.12). Thus the site's attractiveness mainly depends on its location in the built-up area and in relation to major transport networks.

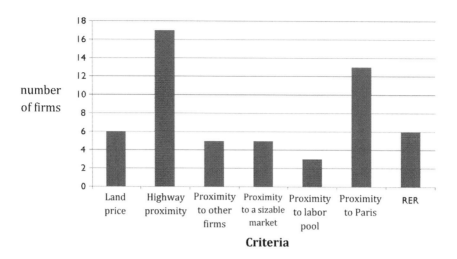

Figure 14.12 Attraction factors according to a sample of companies in Marne-la-Vallée.

HOW THE TERRITORY OPERATES

The territory's composition and operation take shape over time. Factors that seem permanent in the short term are subject to evolution, transformation and renewal in the mid and long terms. Land is obviously the most stable, durable component, followed by installations for particular functions. Buildings and infrastructure constitute one type of land use among others. In the very long term, counted in decades, the main evolution of an urbanized territory is in fact its urban development. This is evaluated using the functional indicators mentioned above, not yet associated with environmental impact indicators. The mid term, counted in years, concerns the evolution of facilities and their usage. This is the time scale of population and employment trends, which evolve through a set of particular events of a demographic and economic type the accumulation of which progressively transforms the territory on a human and social level. The short term is the scale of the territory's everyday operation, marked by flows on regular circuits: traffic of passengers and goods, and flows of raw materials and energy.

a) Mid-term flows and major equilibriums

We have observed that housing and jobs progress by several percent per year (Figures 14.3, 3bis & 14.7). These are the territory's main mid-term evolutions.

The Extended Cité Descartes urban planning project highlights a balance between the Town, Nature and Infrastructure. This involves a local balance within the site. On the scale of the entire built-up area, the relationships between the site and its environment present a major equilibrium challenge, i.e. a balance between the working active and employment. In the current context, at 2.3 people per household and an activity rate of 45% among inhabitants, this balance is almost the same as the balance between

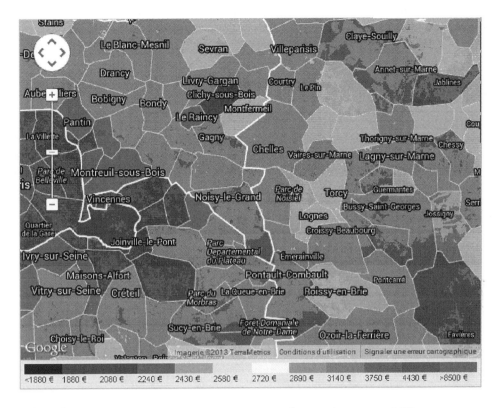

Figure 14.13 Property purchase prices (€/m^2) in local averages in 2012.

housing and jobs. In 2010, the number of dwellings and jobs came to 15,900 and 9,800 respectively, with a ratio between jobs and working people of 62%. According to the Ateliers Lion project, by 2020, the numbers could be 17,200 dwellings and 21,800 jobs; the relationship between jobs and working people would thus be 127%. This kind of progress would mean an extensive transformation of the functional hierarchy: work would significantly predominate housing.

The indicators for the other functions are 13,000 students and 120,000 vehicles/day for road traffic on the A4 at the level of the Champs sur Marne interchange.

The residential attractiveness is reflected by property prices. In 2012, their situation was fairly high in local terms (Figure 14.13), relatively high in national terms, but quite low on a regional scale. The latter illustrates the relative modest incomes of households that pay a higher price to live there rather than in other more sought-after sites. The tension in relation to the local level shows a degree of relative preference. Lastly, for the most part the site attracts people on average incomes. To attract residents with higher incomes would require ensuring a high-quality living environment and significantly developing local amenities.

It would seem that property prices are suitable for local investments, as long as land prices are controlled by the development planner (EPA Marne). This means that buildings could be eco-built and facilities financed (site preparation, roads). On the other hand, there is no margin to finance urban amenities or major facilities: given the regional situation, there is hope of solidarity at this scale or higher, to aid amenities and facilities, and for social housing, including for students.

b) Everyday flows: energy

We roughly modelled the everyday usage of the Extended Cité Descartes to estimate energy consumption. On the basis of a disaggregated description of buildings and activity establishments (cf. next chapter), we divided floor areas between residential and professional use, distinguishing several natures of professional activity. Applying standard ratios, the consumption of energy due to building usage was evaluated at 14 GWh/year for housing and 71 GWh/year for the other activities, for 2010.

In addition, using the Modus road traffic model (cf. annex), we assessed energy consumption from journeys entering the zone or leaving it. The car journey restriction is essential, since most journeys are made by car; they cover the longest distances and are much greater consumers per unit of distance than the other modes. At the evening rush hour in 2008, consumption was estimated at 9.6 MWh for entering traffic and 9.9 for leaving traffic. On this basis, we estimated daily consumption, counting five hours in the day as rush hours and the others as given proportions of the rush traffic, then calculated annual consumption by multiplying by the number of days in the year. The total needs to be divided between the zone and the rest of the built-up area since each journey has two extremities – let us say by half, which gives around 37 GWh per year.

From these rudimentary and very approximate calculations, we can see that energy consumption from building usage dominates consumption from transport usage by a factor of 2. We can relate this consumption in the usage phase to that of other phases in facilities' life cycle. For buildings, taking a technical life span of 40 years and 80% consumption over the entire cycle for the usage phase, i.e. 2 points per year, when fully operational, renewal counts for 2.5% of buildings, multiplied by the 20 points of non-usage phases, i.e. 0.5 points. The usage phase represents 100% times 2 points, thus 2 points, in a logical relationship with other phases for a total of 2.5 points per year. However, if urbanization develops by 5% over the year, then in the transition regime, the re(construction) phases would represent (5% + 2.5%) times 20 points, or 1.5 points, thus the same amount as usage and practically double the energy consumption estimate for the building.

Similarly, for cars we could break down the energy consumed into life cycle phases; once again, an 80% usage share is justified, but for a much shorter technical life span – let us say 10 years, therefore the 20 points of non-usage incur 2 points per year. To everyday consumption, 8 points out of 100 in the life cycle, we add two points for amortization, coming to a total of 10 points. Thus, the consideration of other phases only adds a quarter to the estimate of consumption linked to transport, with the same car equipment.

This summary assessment in a reference situation illustrates that consumption from using buildings significantly dominates that of transport. In the future, when ecobuilding and energy renovation of buildings will have had their effect and divided by 2, 4 or 6 the consumption of everyday building usage, then this total will drop sharply and transport will represent a significant share, whereas for buildings the usage phase will correspond to the construction phase, whose own eco-design will become a major issue.

TRANSPORT AS A TERRITORIAL FACILITY

In a metropolitan context, local functions of a professional character (including higher education) are mostly employed by users coming from elsewhere. Bringing users to local functions is an essential function of the passenger transport system, and it is fulfilled using the technical, financial and environmental qualities of diverse modes of transport. This section presents both the functions fulfilled and the technical means used to connect the Extended Cité Descartes to the rest of greater Paris. It makes a diagnosis of the existing situation: this kind of diagnosis necessarily precedes the design of major transport facilities because of the complexity of the system at the scale of the major territory that the built-up area constitutes.

a) Functions and multi-modality

A transport system fundamentally takes on two functions: crossing (or straddling) space, and accessing each transit end location. Within the territory as a major system, the transport system is a sub-system that connects places: its fundamental performance is the quality of service offered by the origin/destination pair ("O-D relationship"). Other impacts, such as on the environment, are added to this fundamental performance, and they can be dealt with based on traffic volumes per modal network section ("local traffic loads").

By focusing on a neighbourhood as the extremity of an exchange transit with the rest of the world, service quality for passengers is generally worked out from the time required to make the journey to the other end of the transit. Figure 14.14 shows the locations served from Cité Descartes at a given time by car or public transport (PT), during the morning rush hour on a weekday in 2012. It illustrates that the car mode operates well below its nominal performance, since the network of rapid roads covers the territory well and the maximum speed limit always exceeds 70 km/h, which means that when there are no traffic jams, it takes less than an hour to go from one end to the other. In these conditions, PT is time-competitive with the car on a broad east-west axis, but not from north to south. This anisotropy is less pronounced for the car mode.

This multi-modal situation of transport service quality influences users' choice of transit mode: cars were used more than PT to reach the extended Cité Descartes in 2008 (Figure 14.15).

b) Situation of the car mode

The road network is used to transport freight as well as passengers: heavy goods vehicles and other delivery vehicles contribute to the traffic on the road network. In general, they travel outside rush hours because the bad traffic conditions would reduce their productivity and thus profitability. Passengers restricted by work or study have less time flexibility and are more exposed to congestion. At Cité Descartes, the road network channels the biggest flows: the A4 motorway serves a vital function in the

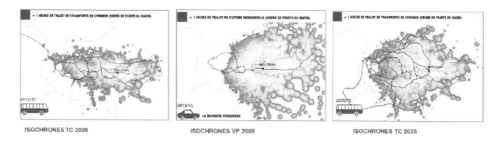

Figure 14.14 Time taken to access locations from Cité Descartes in the morning rush hour.

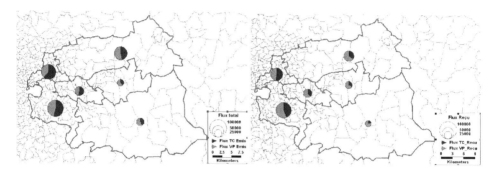

Figure 14.15 Modal shares leaving and entering the Descartes territory in 2008.

east-west axis, with highly degraded speeds on the sections that it shares with the two bypass motorways, the A86 and A104 (Figure 14.16).

Most of the A4 motorway's clients come from the region (Figure 14.17a): its corridor function is clearly shown by its "traffic arborescence" (Figure 14.17b), which superposes the routes of its users between their respective starting points and destinations. The high load generates significant environmental impacts, especially since it is congested, at relatively low speeds that mostly generate higher emissions.

c) Public transport situation

The prioritization of lines is even greater in PT, from bus lines with relatively low frequency to the RER railway with high frequency and very high capacity, than for the car between the street to the motorway (factor of 10). The RER A plays a major role east of Paris, connecting up to the city centre with movements between its north-east and south-east branches (Figure 14.18). All stations on the line are subject to intense passenger flows with the outside or the rest of the network (Figure 14.19).

However, passengers who want to cross north to south on either side of the RER A corridor must take two bus journeys and change at the multi-modal RER station, i.e. a double transfer and potentially long waits since frequency is fairly low. As a result, this kind of journey is much harder using PT than a car (Figure 14.20).

This situation inspired a project to extend line T4 of the tramway from Clichy-Montfermeil to the Yvris-Noisy le Grand station on RER E, south of Cité Descartes.

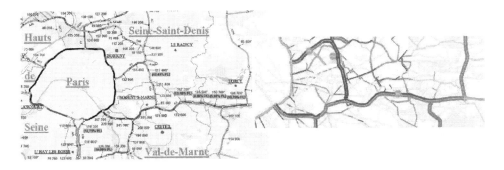

Figure 14.16 Daily traffic loads (a) in congested state (b) on the main road network between the edge of Paris and Cité Descartes, morning rush 2008.

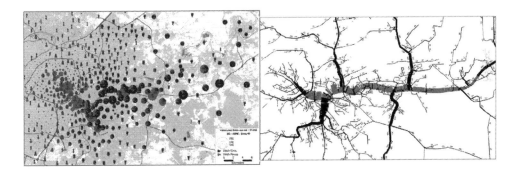

Figure 14.17 Inflow (a) and outflow (b) of road traffic on the A4 motorway, morning rush 2008.

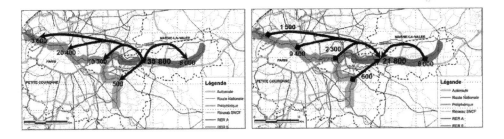

Figure 14.18 Flow relationships served by RER A, morning rush 2008 (source EPA Marne).

DESIGN OF METROPOLITAN TRANSPORT DEVELOPMENT

A diagnosis of connections between our neighbourhood and its environment shows the extent of the built-up area's influence, which imposes its masses of inhabitants and jobs, its internal distances and major transport facilities. It also imposes its regional planning and development programmes: in this case, the Grand Paris Express project

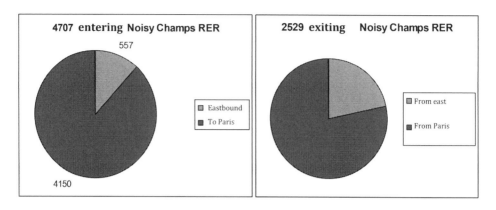

Figure 14.19 Traffic of passengers accessing RER A at Noisy-Champs station, morning rush 2008.

would radically transform the PT network, especially at Cité Descartes with the Noisy-Champs station.

a) Become a junction in a main PT network

On Figure 14.14, the right side shows a simulation of accessibility using PT from Noisy-Champs in 2035, assuming that Grand Paris Express (GPE) has been built. The time performance of PT would thus be similar to that of private cars on the north-south access, and exceed it in a westwards direction! The logical outcome is an increase of PT's modal share in movements of neighbourhood users, cf. Figure 14.21.

Without waiting for Grand Paris Express to be delivered, we simulated the use of the T4 tramway extended to the Yvris station and going through Chelles and Noisy-Champs in 2020: usage levels would be very high during the morning rush hour, fully justifying the project (Figure 14.22). The arrival of GPE would compete with the extended T4 between Noisy-Champs and Chelles, however the station spacing is planned to be much closer for the tram, so that it should still to be used as the main mode for short and average distances, and as a feeding mode for longer distances.

b) Sublimate the road network?

The network of "express" roads in Ile de France is unlikely to change much by 2030; only a few links are due to be developed locally. Technological progress in vehicles is expected to reduce environmental impacts per unit of traffic: appearance of electric vehicles, and in particular reduction in fuel consumption of combustion vehicles. We might also expect increased road security and improved handling of traffic disruption. However, by 2030, it is very unlikely that the flow capacity in vehicles per lane and per hour will significantly differ from the standard value. Nor is it very likely, unfortunately, that traffic will be intensively regulated on the network of urban expressways to restore free-flowing traffic, even though the access and price controls exist to do so.

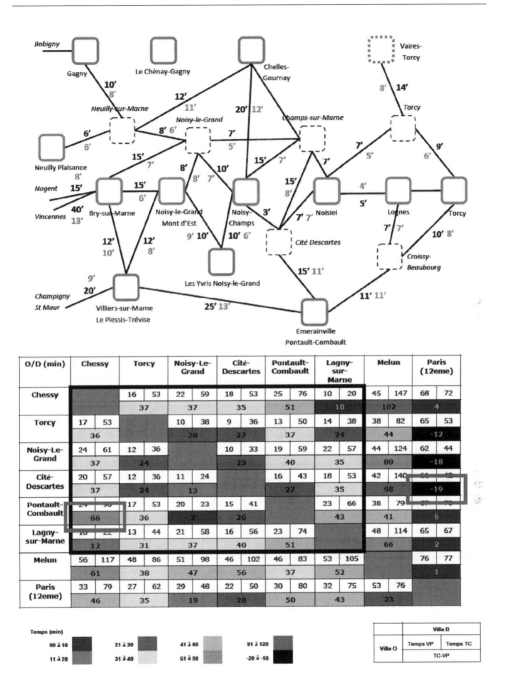

Figure 14.20   (a) Comparison of journey time in public transport (black) and cars (red), morning rush 2008; (b) Time by origin-destination.

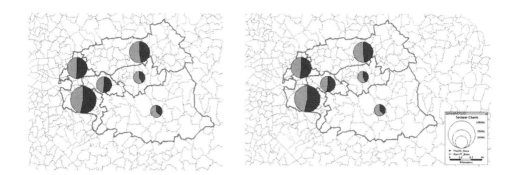

Figure 14.21 Modal shares leaving and entering the Descartes territory in 2030.

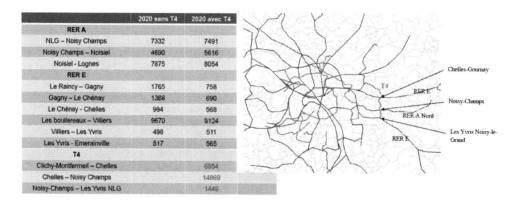

	2020 sans T4	2020 avec T4
RER A		
NLG – Noisy Champs	7332	7491
Noisy Champs – Noisiel	4690	5616
Noisiel - Lognes	7875	8054
RER E		
Le Raincy – Gagny	1765	758
Gagny – Le Chénay	1388	690
Le Chénay - Chelles	984	568
Les boullereaux – Villiers	9670	9124
Villiers – Les Yvris	498	511
Les Yvris - Emerainville	517	565
T4		
Clichy-Montfermeil – Chelles		6954
Chelles – Noisy Champs		14869
Noisy-Champs – Les Yvris NLG		1449

Figure 14.22 Extension of T4 to Yvris: (a) route, (b) traffic simulation morning rush 2008.

Overall, the performance of the car mode should stay the same or worsen, which will make PT more attractive and increase the need to develop it.

One interesting development option does however remain: to use the existing tram rails to carry efficient public transport and increase passenger capacity thanks to the capacity of each PT vehicle. Tram lines already follow this principle: the RN/RD 370 is to be re-used to extend the T4 tram line running north to south. On the east-west axis, the road route already integrated into the urban environment is ill-suited for the tramway; an alternative solution is Buses with a High Level of Service (BHLS), due to their commercial speed and comfort, and in particular high frequency and regularity. The urban project at the core of the cluster includes two BHLS lines: one on the RD 199, i.e. to the north, and the other on the A4 motorway to the south. This second project is the most difficult since the motorway accommodates a very high level of traffic, and is thus not just more expensive, but an emblematic development, implying that urban quality should be put ahead of car travel. This proposition of functional

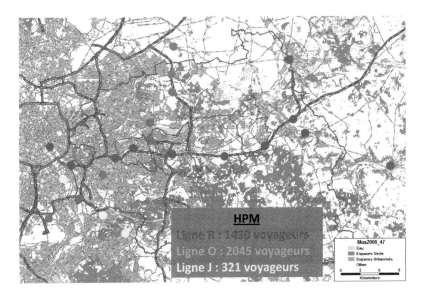

HPM

Ligne R : 1430 voyageurs
Ligne O : 2045 voyageurs
Ligne J : 321 voyageurs

Mos2008_47
Eau
Espaces Verts
Espaces Urbanisés
Other

0   3   6   9
Kilometers

Figure 14.23 Clientele of BHLS lines using the A4 motorway, simulation of early rush 2008.

prioritization has not received consensus from all stakeholders since the A4 motorway was originally designed to efficiently ensure long-distance connections.

We simulated several BHLS routes using the A4 motorway: either express routes directly linking a major origin to a major destination (busway principle), or with intermediate stations to allow efficient connections (common route principle). The latter solution has difficulty competing with lines A and E of the RER, which are already very efficient on an east-west axis. In fact, other projects connecting an east-west passage with a north-south function would attract higher numbers of clients (Figure 14.23, orange line): the major need to improve transport east of Paris is above all in the north-south axis, and could even justify new routes to cover main east-west areas.

c) Certitudes and uncertainties

The neighbourhood's subordination to the whole territory's planning project necessarily restricts the scope of local planning, which must imperatively work in a metropolitan context. At regional level, although the GPE project has found consensus and become a reference for local planners, its overall feasibility is not yet guaranteed financially, which makes it uncertain, and all the more so given the distant deadlines. Uncertainty shrouds some of its lines and, for most lines, the timing of production. This funding issue is dealt with by setting up phasing, which results in a series of lines and renders the planning modular (Figure 14.24).

However, independently from organizing financial resources, population and activities continue to grow in the Paris region, and requirements for facilities are constantly increasing. If the urbanization of Cité Descartes takes shape faster than the Grand Paris Express, then it is likely that transport access will become saturated. A

Figure 14.24 Lines of the Grand Paris Express.

solution would be to anticipate a mid-term situation, let us say in 2020, involving rein-forced transport capacities to prefigure and accompany the GPE project; probably with a redesigned, reinforced network of bus routes, connected to the RER A and E lines which will certainly have been increased in capacity, and thus extended west for line E.

CONCLUSION

Cité Descartes' urbanization is pervaded by its metropolitan context, which has pro-voked, driven, boosted it, and determined its size and structure ... and the story is far from over. We have highlighted the specific functions assigned to the site and shown how this major new urbanization project reinforces and reorders them, so that the Extended Cité Descartes can become a driver of economic development in the Grand Paris. This major project aims to redevelop the neighbourhood, intensify its activities and even extend its area, which will substantially, perhaps radically, transform it.

In this context of major transformation, it is particularly useful to simulate scenarios in order to select efficient measures, and even more useful to evaluate a pertinent set of aspects and impacts. However, it is already difficult to specify a case study, represent its content, and define scenarios (i.e. place them and characterize them). The simulation models available were designed to work on the scale of a metropolitan area and not that of a neighbourhood. They deal with the formation of movement flows and local traffic loads based on land use and major transport facilities (cf. annex); they are increasingly completed by specific impact models for particular environmental themes (i.e. noise, air quality, energy) or socio-economic themes (e.g. accessibility, cohesion).

In the current state of the art, land use is specified exogenously. This involves specifying a priori not the content of a planning policy, but its concrete effects, and without knowing the reactions of the territory's stakeholders, who however autonomously decide the location of their activities. Models exist that make land use endogenous and deal with location decisions and even land and property operations: to date, their application remains experimental and is the subject of research. The Société du Grand Paris is backing three integrated transport and land use models: we hope that they will become usable by planners at local scale.

The interface between the two spatial scales of the neighbourhood and the metropolitan area also needs to be modelled, to "virtualize" everything that constitutes this interface, thus making it easier to design, and even eco-design. In particular, public transport routes or journey routes are territorial objects that interact with the two space scales. In the CapTA model, public transport routes are treated individually. This facilitates detailed design, which is already far-reaching for the transport function but still superficial in terms of fitting the route into the territory: the actual route is only represented by a geometric line, with no attention to the urban fabric or the physical conditions of the support terrain. This kind of abstract treatment of the support space creates an even greater difference between the design of a transport connection and that of a building. For a building, the terrain is fixed and thus invariable, whereas the line of the connection can be the object of different alternatives, routes or zones, each with specific local and micro-local conditions.

ACKNOWLEDGEMENTS

This chapter borrows from the recent chronicle of greater Paris urbanization in a study by Sétec International for the Société du Grand Paris, and the study of transport from the city of Paris to Marne la Vallée from another Sétec International study for the Association de Collectivités Territoriales de l'Est Parisien (ACTEP). We would like to thank these organizations for giving us permission to use the text and images. In addition, "our" diagnosis and simulations were produced with help from engineering students at Ecole des Ponts, as part of a course project: we thank these students for their active contribution.

REFERENCES

Aw, T. (2010) La ville nouvelle de Marne-la-Vallée et son insertion dans la dynamique francilienne: Evaluation des enjeux du renforcement de la structure polycentrique sur les systèmes de déplacements. Thèse de doctorat de l'Université Paris-Est, soutenue le 6 décembre 2010. (Direction J Laterrasse).

Aw T., Laterrasse J., Leurent F. (2011) "Une prospective 2030 de l'usage du sol et du transport en Ile-de-France". In B. Duplessis and C. Raux, eds, Economie and développement urbain durable. Presse des Mines, Paris.

Duby, G. (1987) Atlas historique: l'histoire du monde en 317 cartes. Larousse.

Enpc (2011) Réaménagement de l'A4: conversion en boulevard urbain. Rapport de projet du cours TUSMUR 2010–2011, 34 pages.

Enpc (2012) Réaménagement de l'A4 en avenue métropolitaine. Rapport de projet du cours TUSMUR 2011–2012, 166 pages.

Enpc (2013a) Etude de l'usage du sol, de l'économie territoriale et du transport pour la Cité Descartes dans le cadre du CDT Cluster Descartes. Rapport de projet du cours TUSMUR 2012–2013, 87 pages.

Enpc (2013b) Reconception des transports dans la ville nouvelle de Marne-la-Vallée. Rapport de projet du cours TUSMUR 2012–2013, 36 pages.

Hillairet, J. (1956) Connaissance du Vieux Paris. Editions Princesse, Paris.

Sétec International (2012) Lot 6 des études d'évaluation socio-économique du Grand Paris Express. Rapport pour la SGP.

Sétec International (2012) Etude de trafic pour une ligne de bus à haut niveau de service sur l'autoroute A4 entre Paris et Marne la Vallée. Rapport pour l'ACTEP.

ANNEX: SIMULATION OF LAND USE AND TRANSPORT

Here we clearly set out the method for simulating land use and transport, extending the description sketched out in the chapter on territorial facilities.

A.1 Application conditions

This method is known under the name of "four-stage diagram" for studying travel demand. It is based on sharing responsibility between the analyst (study engineer) and the model:

- The analyst specifies land use in terms of local demographic variables, e.g. population, jobs by type of activity. He or she does this by temporal state: typically a year in a particular scenario.
- The analyst also specifies the means of transport, e.g. network structure and traffic operating laws, composition of services in public transport, individual motorization, individual possession of driving licence or subscription to a particular mode of public transport.
- The model simulates the behaviour of those in transit, focusing on individual trips. Mobility decisions are broken down by sequence: travel or not (trip generation stage), choose a destination for a certain trip purpose (spatial distribution of traffic stage), choose a mode of travel with a fixed origin and destination (modal choice stage), and choose a route on a modal network with a fixed origine, destination and purpose (route choice stage, traffic assignment model on a modal network).

These principles describe mobility mostly as a set of individual trips. Each trip is characterized by its end locations, at the origin and destination. The study area is broken down into "demand zones". On its boundary, each point of passage is modelled by an interface zone with the outside, called the "barrier zone" or "outside zone". The model is applied for a set time condition: often a rush hour for a set of working days in an urban setting, in order to size networks for the busiest periods; or an "ordinary" working day for an inter-urban network.

A.2   Basic models in the "four-stage diagram"

At the *generation* stage. Per category of population and per activity purpose, the number of trips emitted (or received) is modelled for each zone as a linear combination with positive coefficients of local land use variables. These coefficients translate statistic regularities observed from a survey of residents in the territory (one or several categories of population) or of "specific generators", such as airport travellers or tourists attracted by a particular centre (each with its category(ies) of population).

At the *spatial distribution* stage. Once again per category of population and per activity purpose pair at the respective origin and destination (e.g. home-work purpose), emissions from the origin zones and receptions at the destination zones are related to the travelling conditions by all of the modes. The result is a origin-destination (O-D) matrix of trip flows during the period concerned. In general, the relationship is the "gravity" model: the O-D flow is proportional to the emission at its origin, to the reception at its destination, and to a decreasing function of the O-D generalized cost. Some gravity formulae are compatible with a micro-economic description of individual behaviour to choose a destination option from its origin, based on the "discrete choice" principle described below for the mode choice.

At the *modal choice* stage: Once again by category of population and by O-D activity purpose, as well as by O-D relationship, the multi-modal flow is shared between a set of transport modes specific to its category (depending on the capacity to access the mode, e.g. motorization for the car mode). In general, the share is based on a microeconomic "discrete choice" model, i.e. that selects one option from several modal options. A utility function is associated to each mode option, and the individual chooses the maximum utility option for him or herself.

At the stage of *traffic assigment* to routes on the modal network. Once again by category of population, by O-D activity purpose, by O-D relationship and also by mode, the modal flow is shared between the routes supplied on the modal network. In general, sharing is based on a microeconomic model of choice: each route is characterized by its cost in time, distance and financial expenditure, brought together into a "generalized cost", weighting each term by a sensitivity specific to the category and purpose. The axiom of rational economic behaviour is that the user chooses the route with the minimum generalized cost.

In the assignment model, we "bring together" the routes chosen for the various journeys by superposing the individual trajectories travelled on the network. For each element of the network, the result is a local traffic load during the period concerned. The element's journey time can be linked to this load based on a physical congestion law that takes specific local capacity into account (e.g. around 2,000 cars per hour for a motorway lane in steady, non-disrupted regime). The *traffic equilibrium* is accomplished when the mutual dependencies between the traffic loads and the travel conditions are all made compatible with each other.

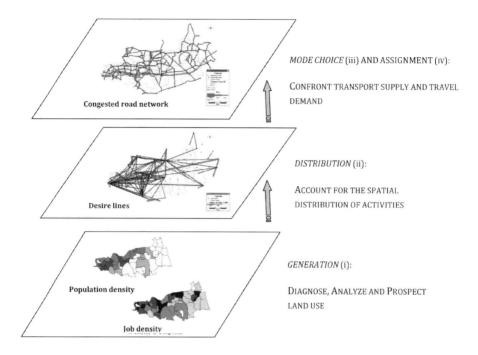

Figure 14.25 Formation of journeys based on local land use conditions.

A.3 Using the results

For network planners, local traffic loads and the associated qualities of service (e.g. average speed by road section) constitute the key result of the transport supply-demand model. The results by O-D relationship are also important to characterize flows in space ("desire lines") and the quality of service per geographic liaison: the O-D time is greater for users than the O-D distance that appears more intuitively on a map.

Diverse socio-economic and environmental impacts can be assessed based on these result indicators. From the quality of service conditions, we can deduce the user welfare (their benefit taken from the trip). From local road traffic conditions, adding a statistical description of the car population, we deduce local emissions of noise and various pollutants, as well as energy consumption and greenhouse gas emissions.

Figure 14.25 shows the bottom-up formation of local traffic loads based on land use variables in Marne la Vallée.

Figure 14.26 illustrates modelling stages, the use of databases, and a comparison with reference observations.

A.4 Limitations of the model

The model is based on statistic regularities as much as "explanatory mechanisms": the microeconomic behaviour of trips, along with the physical laws of flow conservation and local traffic condition formation.

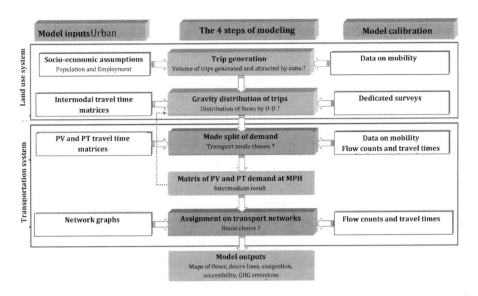

Figure 14.26 Models and data in the application process.

Some statistical regularities can be replaced by specific explanatory mechanisms:

- A discrete choice model for the destination, in compatibility with or to replace a statistic spatial distribution model.
- By a time choice model to establish the portion of trips made during a particular period.
- A residential location model to render endogenous the split of residences (location of households) and jobs (location of firms) between the demand zones.
- A trip frequency choice model, to determine emissions from a zone.
- A motorization model to explain equipment with a vehicle as a choice among several options. Similarly, models for possessing a public transport travel card or a driving licence, etc.

In addition, dynamic models exist for road traffic, in a macroscopic or microscopic (micro-simulation per individual) form.

However, the most significant improvement in terms of mobility behaviour is combining all of the trips of one particular person, by reconstituting loops of activity and movement (e.g. a return trip from an individual's permanent base for a main activity purpose) or even better, daily schedules of activities and trips. This is because travellers think mostly on these higher scales when choosing their modes of movement: they anticipate their return and the time spent at destination, in particular considering local car park conditions, the price of which can depend on the time spent on site.

Other extensions concern interpersonal constraints, typically interactions between individual members of the same household (e.g. the need to accompany a child) or the composition of a travel group (which will strongly influence the individual cost for the car mode).

Chapter 15

Activities, accessibility and mobility at neighbourhood scale

Houda Boujnah, Nicolas Coulombel, Natalia Kotelnikova-Weiler, Fabien Leurent, Sergio Millan-Lopez & Alexis Poulhès
Paris-Est University, City Mobility Transport Laboratory,
Ecole des Ponts ParisTech, Ifsttar, UPEMLV

INTRODUCTION

a) Context: the neighbourhood as a design object

What qualities are necessary or desirable to make a neighbourhood sustainable, in all of its social, environmental and economic facets? But first, what is a neighbourhood? According to the Hachette dictionary, it is, "Part of a town presenting distinctive traits in terms of its functions, attendance and population". For eco-design, this is our definition: "A delimited area offering a range of activity opportunities to inhabitants in a proximity relationship". It is therefore an area characterized by (1) demarcation vis-à-vis the exterior and proximity on the inside, (2) the human presence of users and the performance of activities, (3) a functional composition of diverse types of activities, offering a potential for local activity combinations by the same user. Ecodesigning a neighbourhood means designing its composition, organization and operation while respecting the environment, society and the economy.

Real estate developers and builders tend to eco-design neighbourhoods with major material components, i.e. for habitation, sustainable buildings; for the living environment, spaciousness and access to nature; management of energy and material flows devised on a local level; buildings allocated to activities chosen so that the spatial layout encourages local interaction; walking and cycling facilities for internal circulation; and for connections with the outside, car parks creating vestibules to "contain" cars, as well as access stations and public transport lines. This thus means that neighbourhoods should be planned as sites fitted with a whole series of virtuous measures.

However, are these virtues sufficient to attract inhabitants and other users, services and businesses? The location within the built-up area is a crucial factor that has already been covered. Focusing on local aspects, a neighbourhood's attractiveness to social stakeholders is a qualitative notion that needs to be accounted for by one or several indicators, each with a degree scale (qualitative or quantitative). This kind of indicator can be used to compare project alternatives in terms of attractiveness and potential usage, and thus, integrate sensitivity to social issues into the design engineering.

b) Objectives: recognize the composition and equip the design

The Chair's aim is, for its objects, to pinpoint the functions, develop sensitive indicators, and equip design engineering with simulation models and evaluation methods.

At the Chair, the notion of neighbourhood is tackled from several angles associating physics, ecology and geography. This chapter looks more specifically at the geographer's point of view, while seeking connections with the other domains.

To begin with, our two objectives for a neighbourhood were: firstly, to make an initial investigation of the internal composition in terms of buildings, activities and transport means; secondly, on a methodological level, to explore the resources available for design engineering, including databases and spatial and statistical analysis methods, in order to account for the neighbourhood's different facets, with the composition and operation of each facet.

Our exploratory research was based on the Cité Descartes neighbourhood: we used national information sources available (IGN and Insee databases) and additional information collected locally, some via surveys that we devised and carried out with the help of student groups at the Ecole des Ponts.

c) The chapter's content and organization

The body of the chapter comprises three parts respectively devoted to: activities in relation with buildings; micro-local accessibility for potential users; and local transport conditions, with a special focus on car parking.

URBAN FUNCTIONS AND BUILDINGS IN THE NEIGHBOURHOOD

Traditionally, territorial planning treats neighbourhoods mainly as receptacles of inhabitants and jobs. However, over the last twenty years, diagnosis methods have emerged, some of them to link polluting emissions to the resident neighbourhood of the emitters (Gallez and Hivert, 1995), and others to locate energy consumption within a town (Antoni et al., 2009). In this section, we present a totally disaggregated model to match activities and buildings in order to synthesize the elementary components of the town, as genuine urban particles. Using this model, we can locate more precisely, and potentially assign the emission and reception of impacts to individuals.

We start by presenting the matching model; we then treat energy consumption in building usage as an impact caused, and exposure to noise as an impact endured. We finish with a methodological overview.

a) Highly disaggregated land use model

Very detailed national information bases are now available. One of these is Insee's GéoSirene database, which locates each "firm" establishment by postal address, "firm" in a broad sense including authorities and associations: the total was 7.5 M of this kind of establishment for 5M firms on the national territory in 2010.

In addition, IGN's topographic database (BD Topo) locates each built item in the geographical area with metric precision: the item is described by a series of 3D vectors that outline the contour, and thus the area on the ground and the height. A link between these two bases is provided by the IGN's postal address database, which identifies each postal address by the geographic coordinates and a point along a road.

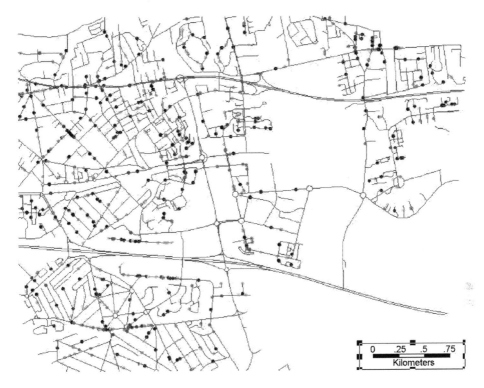

Figure 15.1 Firm establishments around Cité Descartes in 2010, source GéoSirene, Insee.

Our matching model summarizes the urban functions fulfilled by each building. The treatment process comprises three steps:

1   The floor surface area, which indicates the building's capacity to accommodate activities, is calculated by estimating a number of floors based on the building's height multiplied by its area on the ground.
2   Activity establishments located at the building's address are connected to it, and through them the number of agents and the nature of the activity. By applying standard ratios of floor area to job, according to the nature of the activity, we deduce the surface area occupied by professional activities inside the building.
3   The residual surface of each building is assigned to the residential function; we attribute the building with a number of occupants proportional to its residual surface. The coefficient of proportionality is established by applying the Iris Insee population (the basic statistical information grid square).

The end result is that each building is allotted with one or several natures of activity, each with a number of concerned individuals (resident or agent by nature of activity).

Figure 15.2 Topography of built items around Cité Descartes in 2010, source BD Topo, IGN.

b) Impacts caused: energy consumption

Knowledge of "urban particles" allows us to assess impact indicators, if we possess ratios per surface area or per occupant according to nature of activity. We thus estimated the consumption of energy caused by using buildings, with a distinction according to residential or professional function and nature of activity (Fig. 15.5).

c) Impacts endured: exposure to noise

Reciprocally, the building's position allows us to estimate the exposure to impacts endured by its occupants, in particular noise or a type of pollution. We modelled the noise emitted by road traffic in Ile de France (Modus model) and individual building's exposure (Figure 15.6). The evaluation of the impact could be extended by calculating the number of occupants exposed to this level of noise in each building.

d) Methodological assessment

In terms of methodology, modern geographic databases are information-rich, allowing a highly disaggregated analysis of functions and buildings in space. The matching

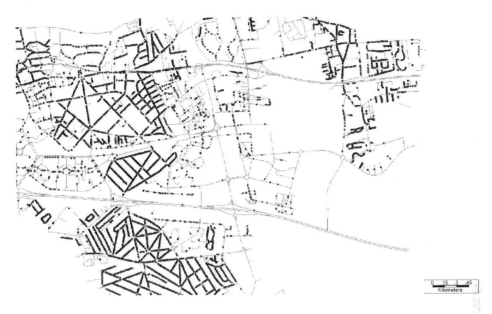

Figure 15.3 Postal addresses of establishments around Cité Descartes in 2010, source BD Adresse, IGN.

Figure 15.4 Activities in buildings around the Cité Descartes.

model devised here is based on simple principles: it combines basic information of a specific type with typical ratios established elsewhere. This simplicity may make the results too crude if the site dealt with differs too widely from the average conditions represented by the typical ratio.

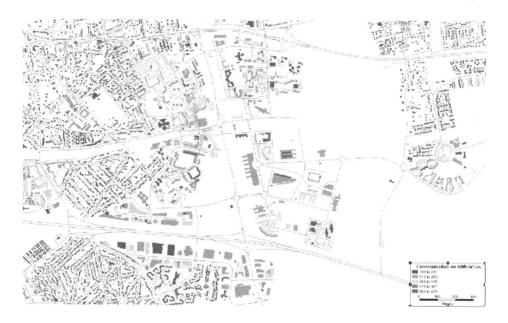

Figure 15.5 Energy consumption in buildings around Cité Descartes.

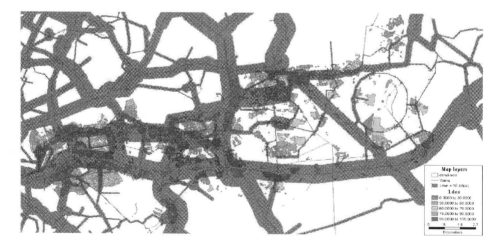

Figure 15.6 Strip exposed to 70+ dB in Marne la Vallée during morning rush hour, 2008.

Notwithstanding this risk, disaggregation should significantly improve the precision and robustness of the estimations. A priori, indicator estimates based on an enumeration of basic objects should prove to be a lot more precise than evaluation using aggregates – in particular for impacts that vary in space.

Thus, we modelled the mode choice of an individual travelling from home to work according to his or her address, which conditions feeding into PT (Figure 15.7).

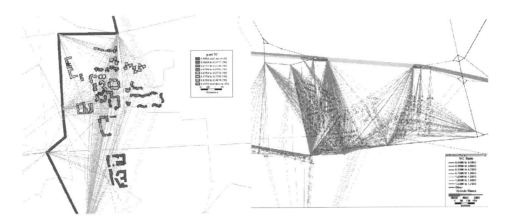

Figure 15.7 Disaggregated access to PT and private cars, morning rush in 2008.

For the buildings concerned in the Nesles-Sud neighbourhood, with the disaggregated model we simulated a 65% mode share for PT, whereas the aggregated model only simulates 35%.

Lastly, we must emphasize the crucial underlying role played by the Geographic Information System: GIS is the indispensible kingpin for truly interacting with geographic databases.

ACCESS TO AMENITIES: QUESTIONS OF SCALE

A household settling in a town looks for a set, or "basket", of opportunities: employ-ment, services, shops, leisure activities, etc. The town offers a set of opportunities called "amenities" in an urban economy. The household must then access these amenities: easy access genuinely determines the potential of the opportunities that the household disposes of according to its place of residence. This is the notion of accessibility, which can be broken down into local accessibility to amenities close by, or accessibility to the entire built-up area using a transport network.

In this section, we develop indicators to illustrate accessibilities at local level: accessibility to activities, and accessibility to the main transport network. We start by showing the need to refine the analysis to an elementary, micro-local level. We then develop "functional diversity" indicators, of the functional richness available close by, one by small grid square and the other by micro-local item. We then tackle accessibility to the built-up area. Lastly, we establish a methodological assessment.

a) Refine the grid to analyze urban intensity

National statistical information is available to users at the Iris scale, i.e. a block grouped for statistical information: a zone containing at least two thousand inhabitants. The example of Cité Descartes shows that this grid is not fine enough to analyze neigh-bourhoods (Figure 15.8). In particular, the density indicator is not significant for a

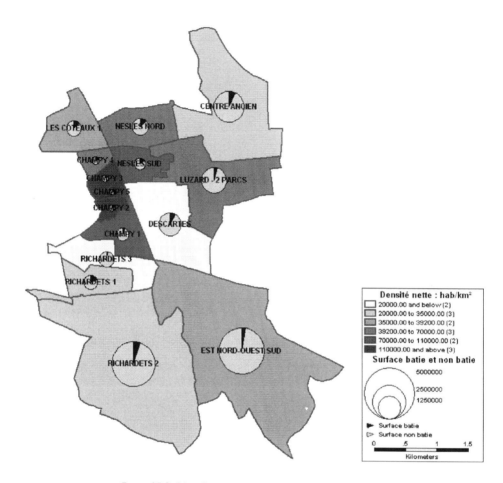

Figure 15.8 Net density around Cité Descartes in 2008.

heterogeneous zone (cf. the Iris Centre of Champs sur Marne, which includes the castle and its extensive gardens, which brings down the average density, although strictly speaking the urban centre is fairly dense). The scale of the Iris makes it possible to focus forecasts of inhabitants and jobs without having to locate them in close detail in homogeneous urbanization-type zones (in the sense of a local urbanization plan (PLU) which includes the land use coefficient).

For Ile de France, the Institut d'Aménagement et d'Urbanisme Régional (institute for regional development and urban planning) has developed a database of land use modes (Modes d'Occupation du Sol – MOS) by land parcel (Figure 15.9). This base allows us to locate demographic projections more finely. We did so for the Descartes site, anticipating the MOS map at the final horizon of the urban planning project (2030), revising the 2008 MOS according to the block plan drawn up by the urban planning consultant (Ateliers Yves Lion). We identified the following transformations: suppression and creation of housing parcels, transformation of housing parcels into

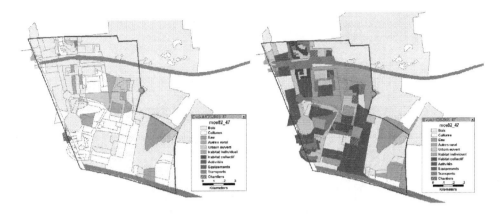

Figure 15.9 Land use modes at Cité Descartes in 1982 and 2008 (source IAU-IDF).

activity parcels and conversely, suppression or creation of activity parcels. Based on estimated templates of (existing or planned) buildings in the modified parcels, we estimated the number of jobs or inhabitants added or deleted. Through this analysis, we deduced the net variation in population and jobs for each Iris from the reference state in 2009 to the anticipated state in 2030.

The project zone is mainly residential. Its total population was 45,400 inhabitants in 2009, which according to the urban planning project analysis should only increase by around 9% to reach 49,400 inhabitants in 2030. At the same time, the number of dwellings is set to rise from 18,070 in 2009 to 19,410 in 2030 (+7%). Lastly, the number of jobs should increase from 19,720 in 2009 to 30,760 in 2030 (+60%). The ratio of jobs to housing should increase from 1.07 in 2009 to 1.58 in 2030, further accentuating the site's professional specialization.

Urban intensification should mainly be concentrated in the Iris Champy 1, Descartes and Richardets 3, at the heart of Cité Descartes, where the project creates housing and jobs (Figure 15.10).

b) By basic grid, an indicator of functional diversity

In between the land register parcel and the statistical block, a 300 m square grid constitutes an intermediate scale that is sufficient to characterize accessibility by foot to an activity establishment. Pedestrians can even access neighbouring units easily from their starting unit: this extends their accessible area by foot to a square grid measuring 900 metres long. On this grid, we have defined an indicator of functional diversity, which quantifies the diversity of functions locally available: the functional types present are identified from a section taken from the following 11 sections: Woods and Forests, Crops, Other rural, Water, Open urban, Individual habitat, Collective habitat, Activities, Facilities, Transport, Worksite (*not counted*). The grid indicator totals the indicators of the units that make up the grid, and relates the total to the number of units, i.e. 9.

Figure 15.10 Population and jobs around Cité Descartes (a) in 2009, (b) in 2030.

For the Extended Cité Descartes, we note that on the vast majority of the territory, at least 7 different categories of MOS are accessible by foot (Figure 15.11).

The indicator discriminates between zones that are rich and poor in amenities, but remains too aggregated for characterizing individual situations.

c) Accessibility to amenities at micro-local scale

We made a more detailed characterization of the functional diversity, measuring the minimal distance (as the crow flies) between each housing parcel and a range of amenities close by. The following range was chosen: shops, primary schools, green areas, local facilities.

The housing parcels considered are strictly located within the 15 Iris of the project. The facility parcels can also be found outside of these 15 Iris. We first measure the distance between each MOS "housing"-type parcel, and all the parcels of a given type of facility (e.g. all of the "shop"-type parcels). For each housing parcel, we determine the minimum distance as the crow flies to each type of facility: shop, primary school, green space, local facilities. These characteristics are represented in accessibility maps for each of the 4 types. To constitute an index of total accessibility, for each housing parcel we calculate the average of the 4 specific distances.

In general, for all the housing present on the site, accessibility is fairly high (Figure 15.12) since the four categories of amenities are located less than 500 m away as the crow flies (average of 480 m for all housing). The only exception is the Iris of Emerainville, East North-West South, were the various facilities are on average over 600 m away as the crow flies. It is also worth noting that the distance as the crow flies is only an indicator of accessibility. In fact, the actual path covered by a pedestrian

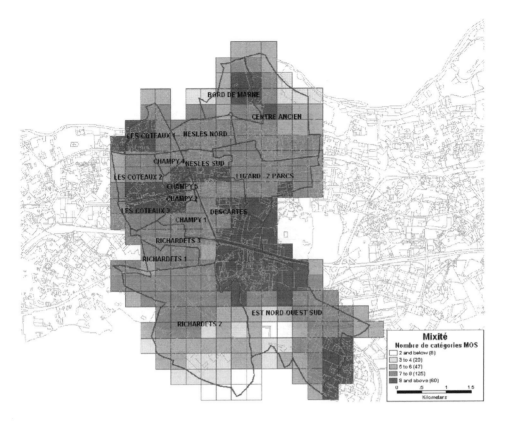

Figure 15.11 Functional diversity around Cité Descartes in 2008 (source IAU-IDF).

depends on the infrastructures available and their accessibility to pedestrians. Thus, crossing over a motorway, following the position of flyovers, can require taking a much longer route (relatively) than the direct line between the starting point and the destination.

In addition, the type of facility considered must be distinguished. Although all housing enjoys good access to green areas (this is in fact one of the assets of this site, identified and preserved by the urban planning project), the same is not true, for example, for access to shops (Figures 15.13 and 15.14). Whereas the distance from green areas does not exceed 120 m on average for all housing, the distance from shops is greater, around 1 km on average, with a significant disparity between the East North – West South Iris of Emerainville (distances of over 1.2 km) and the rest of the site.

Distances to local facilities and primary schools are similar (respectively 400 m and 370 m on average) but present different spatial configurations (Figures 15.15 and 15.16). Access to local facilities is better in the north-east part of the site and less good in the south-west part and the opposite for access to primary schools.

To sum up, the site's green spaces are a great asset and widely present. This presence is "natural". Local facilities and primary schools are intentionally located to respond

Figure 15.12 Accessibility to all amenities in 2008 (source IAU-IDF).

to inhabitants' needs and are distributed around the territory in a more homogeneous manner, which guarantees good access for the whole site. Shops are more subject to land market laws, demand from customers on the site, available property, means of supply and access, and the catchment areas of shops already on site, which results in competition and polarity. Their distribution over the site is thus more heterogeneous and access to them is more variable.

In total, the analysis of accessibility to amenities drastically refines the analysis of diversity. Diversity is high on the whole site, although accessibility results show that disparities exist in terms of service supply. Diversity does not thus guarantee good access to amenities.

d) Accessibility to the metropolitan area

A city offers many more amenities than a neighbourhood, but in the form of a large catchment area that is unsuitable for evaluating overall accessibility. A Hansen-style (or Poulit) indicator of accessibility, which aggregates from an originall activities of a certain type in the metropolitan area, amortized according to their respective distance from the starting point, counts the abundance of the nature of activity in the area, but does not count either the quality of each opportunity or the diversity of the natures. This latter gap can be rectified by aggregating the

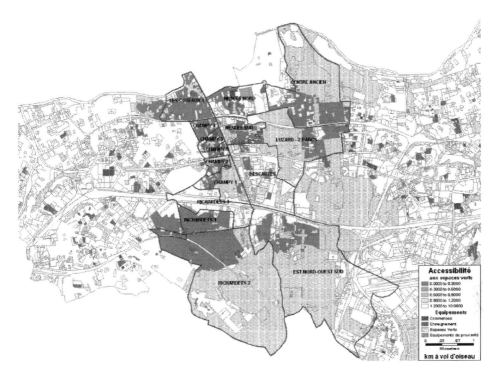

Figure 15.13 Accessibility to green spaces in 2008 (source IAU-IDF).

same number of indicators as natures, but the first shortcoming is fundamental. This kind of enumeration indicator can nevertheless be useful for stimulating the imagination.

In addition, an index of the access time to the main network in the metropolitan area also "captures" the possibility of accessing diverse amenities: it is evaluated on a quantitative scale without ambiguity, and its interpretation is simple.

For the Extended Cité Descartes, based on the MOS 2003 and the bus network in 2009, we evaluated the access time to an RER station. This was split into two parts: the access time to the bus stop, and the journey time in the bus itself. The first was tackled as follows: calculate the distance as the crow flies between the MOS housing parcel and the closest bus stop then convert into access time taking a walking speed of 4 km/h. The bus journey time was calculated from existing bus network characteristics. The waiting time was not included.

Figure 15.17 shows a lack of connections between the north and northeast of the site and the mid- and long-distance network.

e) Methodological assessment

In this part, we matched urban resources: a house at the origin, one or several amenities at the destination. The basic information contained in databases is suited to this kind

Figure 15.14 Accessibility to shops in 2008 (source IAU-IDF).

of evaluation, thanks to the Geographic Information System, which opens up other possibilities.

Maps are irreplaceable supports for representing a collection of individual situations: each with its location and specific characteristics illustrated by a graphic symbol or colour. Maps give an overall restitution and prepare the summary without taking its place. In the previous section, location was simple, of order 1, since each place is treated individually: a map is sufficient to reproduce the whole thing. In the present section, the location is crossed, of order 2, by matching places: this is the origin-destination relationship, which is difficult to reproduce on a map apart from for flows.

Lastly, as a reminder, accessibility to amenities and accessibility to transport do not exhaust the question of access to the city, which also includes access to housing, employment, social solidarity, etc.

PARKING A CAR IN CITÉ DESCARTES

Activities in Cité Descartes are mainly carried out by users from the outside, and most of these come by car. This means of transport is the main choice for exchange trips between the neighbourhood and the outside and is also used for internal trips. This

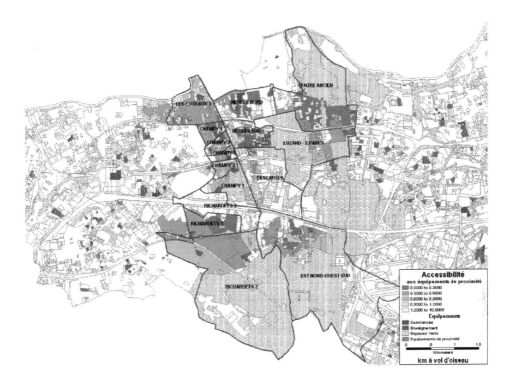

Figure 15.15 Accessibility to local facilities in 2008 (source IAU-IDF).

leads to a general need for parking and on-site functions, strengthened by the possibility of feeding into Noisy-Champs RER A station by car. The development project at the core of Cluster Descartes will reinforce urbanization and thus the demand for activity and movement; it relies on the significant development of heavy public transport, which should retrospectively act on the use of modes of transport and reduce the car's share by journey unit. The combined effect of these two influences on the demand for parking is uncertain in the long term. In the mid term, if urbanization progresses faster than the main PT lines are set up, then it is likely that cars will retain their 2010 modal share and that parking needs will increase in the neighbourhood.

The eco-design issues for a parking system are set out in a specific chapter. For a particular neighbourhood like Cité Descartes, it is important that the capacity installed is sufficient to satisfy local demand, to avoid parasite transfers to adjoining neighbourhoods and limit journeys spent looking for a place.

In this section, our objective is to identify parking requirements at neighbourhood scale, so as to size the required capacity and propose location sites and management modes. We start by modelling parking requirements based on local functions and mobility conditions. We then examine supply and usage in a 2010 reference situation. Next, we create a parking scenario for 2030 based on the Atelier Lion's urban planning project, and discuss whether supply and demand are locally matched.

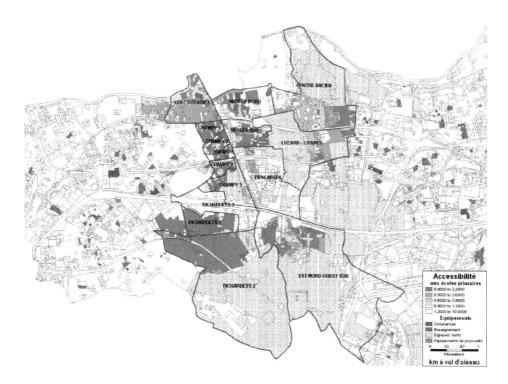

Figure 15.16 Accessibility to primary schools in 2008 (source IAU-IDF).

a) Parking requirement model

Each urban function motivates parking, according to its local intensity and mobility conditions. In the case of Cité Descartes, the main functions to consider are home, regular work activity (workers employed on the site) or occasional activity (business visitors), university, whose clients are students, and local services – here reduced to shopping.

We modelled the formation of parking demand using a five-stage diagram:

1   *Functional generators.* Each function motivates specific individual activities, in terms of a specific frequency, called the generator. The "resident population" generator, given as X_H motivates inhabitants' presence at their homes or at an activity close by. The "local work" generator, given as X_W (if necessary distinguishing several types) motivates workers' presence, as well as business visits, and the attendance of clients for a Shopping motive. The "university" function is specifically generated by the number of student places.

2   *Local activity demand.* Through an activity purpose, demand is proportional to its local generator according to a daily factor given here as ρ_A for activity of type $A \in \{H, W, B, S, U\}$ for Home, Work, Business, Shopping and University. The daily demand to locally carry out activity A is $D_A = \rho_A \cdot X_A$. For the home motive, ρ_H

Figure 15.17 Access time from housing to a station.

complements the unit of factors for other purposes that are likely to attract a resident elsewhere – e.g. work in another neighbourhood.

3   *Parking place demand.* Mobility conditions impose on a neighbourhood, per activity motive, a certain modal share for the Car mode as a Driver. We call $\tilde{\rho}_A$ the product of this modal share (a proportion) and the generation rate ρ_A. The daily demand for parking places is $\tilde{D}_A = \tilde{\rho}_A \cdot X_A$ for activity A, to be added together for all activities to obtain the overall daily demand, $\tilde{D}_\Sigma$.

4   *Parking occupancy load.* We then need to temporalize the demand for places according to the time and duration of each activity. We focus here on a daytime period from 9 am to 5 pm, which reflects the continuous occupation of a place by a "commuter", i.e. an occupant who comes for the day to work or study. A student with a car counts as a commuter, and so does a resident with a car who does not use it and leaves it parked in the neighbourhood. However, we count one hour spent on a shopping activity, with a coefficient of relative time $\delta_s = 1/8$ in commuter equivalent, and 4 hours for a business visit, with $\delta_B = 1/2$. The overall occupancy load of the daytime period comes to $C_\Sigma = \Sigma_A \delta_A \tilde{\rho}_A X_A$ commuters.

5   *Modal and spatial breakdown of the load.* The neighbourhood's functional generators can be disaggregated by spatial element, such as parcels or urban blocks, along with their entire chain of consequences on activity demand, parking demand and occupancy load. If the micro-local capacity is sufficient and free or cheap enough, then users find it worthwhile parking right at their destination. Capacity constraints can lead to transfers, cf. ParkCap model, inside our outside the neighbourhood, and between private, semi-public and public modes.

This sequential model is macroscopic. It reproduces statistic regularities: generation factors, the share of car driver mode. Its logical order is strongly inspired by the four-step diagram to study travel demand.

Feeding into the Noisy-Champs RER A station constitutes a specific purpose, the generation of which is complex. In the absence of a metropolitan model to simulate the use of the station, we can directly measure the associated parking load.

b) Supply inventory in 2010

Enumerating the parking places in a town is often a challenge due to the abundance of places, the diversity of forms, and the difficulty in recognizing private forms incorporated into buildings. We made an inventory of existing supply in 2010 at the scale of Cité Descartes (the Descartes Iris had 1,900 inhabitants in 2008 according to Insee), extended to the north by the Nesles-Sud neighbourhood of Champs sur Marne (Nesles-Sud Iris had 4,700 inhabitants in 2008). On the site, private places and areas are visible on the surface and did not prove difficult to identify (this would not have been as easy for building courtyards). The enumeration was carried out in two steps involving: first, examination of a high-resolution aerial photograph, then a field visit to precisely confirm the locations and identify garages. The information was recorded on MapInfo GIS, based on an innovative conceptual data model that distinguishes road sections and areas treated as polyline objects. A comprehensive digital model of the road network was used, taken from the OpenStreetMaps web server, a free collaborative service put together by internet users.

The places were described according to three characteristics: situation, form and operating mode. The *situation* is along a road or in a dedicated area: a public area that could be connected to a shop or station; or a "private" area for a household or business. The *form*, which could be parallel, perpendicular or an angle, determines the space taken up lengthwise and across in relation to an access axis, and conditions parking manoeuvres. The *operating mode* concerns access conditions (possibly reserved for the disabled or deliveries), time conditions (possibly limited duration) and price conditions (free, 'temporary' payment or season ticket).

In 2010, the site included 7,560 parking places, of which 40% on the road. Road parking was 54% parallel, 43% perpendicular and 3% at an angle. Usage was 95% free, with the remaining 5% subject to a season ticket. Off the road, 75 areas (or lots) were counted, with a highly variable number of places per lot: median 30, average 60. Only the shopping centre car park was paying, for stays lasting longer than 30 minutes.

We divided the study territory into 11 zones (Figure 15.18). Table 15.1 identifies the zones, characterizes their occupancy and indicates their parking capacity.

c) Knowledge of users in 2010

Still in the spring of 2010, we investigated parking demand by intercepting and interviewing car drivers parking their vehicle. Almost 300 interviews were carried out in four zones with varied characters (2, 3, 8 and 11), at target points (including the Noisy-Champs RER station and the Champy shopping centre), on a Friday between 7.30 am and 7 pm. The conditions of the survey obviously restrict the sample's representativeness and the results are essentially qualitative in scope.

Figure 15.18 Parking zones and local capacities on the Descartes-Nesles site.

Table 15.1 Parking inventory at Cité Descartes in 2010.

Zone	Surface (ha)	Land use	Road places	Car park places	Density (places/ha)
1	8.2	East: residential. West: offices, workshops, restaurants & shops in groups	54	379	53
2	9.2	Non-residential: RER A station (Noisy-Champs) and proximity, car parks & office buildings. Champy shopping centre	164	644	88
3	16	Residential with a few small shops and a secondary school	408	163	36
4	12.4	Residential and ede of Grâce Woods	102	334	35
5	13.4	Mainly residential, plus 6th form college, grocer's, bank, pharmacy & doctor's surgery	276	346	46
6	22	Offices, Esiee, Ibis Hotel, residential buildings (including student accommodation), restaurants, pharmacy, bank	262	653	42
7	19	Universities: Ecole des Ponts ParisTech and Ecole d'Architecture des Villes et des Territoires	197	259	24
8	9.2	Copernic building of Paris Est – Marne la Vallée University (Upemlv)	242	256	54
9	7.3	3 university buildings & business and research laboratory zone	205	642	116
10	15.6	North: university building (IUT). South: business zone	411	395	52
11	16.1	2 companies: training and teritary real estate	741	427	73
Site	148.5	Iris Descartes & Iris Nesles Sud	3062	4498	51

Figure 15.19 Comparison of supply & demand (a) in 2008, (b) in 2030.

The survey questionnaire was divided into four sections relating respectively to the car driver, his or her journey, the parking place found and his or her perception of the system.

Car driver characteristics. Respondents were fairly balanced in sex: 54% male and 46% female. For the age split, the 20–30 year decade significantly dominated in zones 8 and 11 (university students), whereas for the zones as a whole, the following three decades were fairly balanced (slightly more 40–50s than 30–40s, of which slightly more than 50–60s). The breakdown by socio-professional category featured numerous students in zones 8 and 11, a majority of general employees in zone 2 (RER station) and 10% to 20% managers.

Journeys to Cité Descartes. The journeys in the survey were mainly motivated by work (43%) and study (39%), way ahead of shopping (8%) and business (7%). Only 3% were home journeys due to the dominant functions on the site and the time and places surveyed. The trip origins were split between the Parisian suburbs, with most coming from Seine et Marne (77) which includes Champs sur Marne (57%), and a significant proportion from Seine St Denis (93) which includes Noisy le Grand (23%), a minority from Val de Marne (94) which includes Bry sur Marne (12%) and marginal contributions from other areas: Essonne (2.7%), Paris-75 (2.3%), Hauts de Seine (1.6%), Val d'Oise (0.8%), and Yvelines (0.6%).

Parking locations and behaviour. Depending on the operating mode, the places chosen were 47% free parking along the road, 32% in the employer's car park, and 7% in a paying place (shopping centre car park). The reasons for these choices were mainly proximity to the final destination (44%), way above price (17%) and time spent looking for a place (9%). This time spent looking, which is valued subjectively, was less than one minute for 30%, from 1 to 5 minutes for 34%, and equal or above 5 minutes for 32%. In parallel, 33% of respondents perceived obtaining a place as difficult. The subsidiary journey time between the parking location and the final destination varied significantly depending on the zone: less than 2 minutes for 95% in zone 8, less than 5 minutes for two-thirds of the cases in other zones; but a non-negligible minority (5% to 15% according to the zone) declared between 5 and 10 minutes, and others more than 15 minutes (20% in zones 2 and 3, i.e. RER users). Lastly, occupancy times in places were very long: commuter usage of over six hours for 80%, compared to 9% from 2 am and 6 am, and 11% for less than 2 hours. This corroborates the breakdown according to trip purpose.

Perceptions of progress. Car drivers that had been using the site for more than 3 years observed a greater parking pressure, with some finding it increasingly difficult to secure a place close to their destination: to succeed, some left earlier in the morning (zones 2 and 11). The reason given was the rising numbers of students with cars. Some also said that space was insufficiently used for parking. Propositions for improvement included creating free places in car parks or along the road, stricter regulations and improved access conditions to places.

d) Interaction between supply and demand in 2010

Parking supply is abundant and well distributed in space: in a 200 m radius around a building, only the ends of residential roads contain less than 100 places. Along passing roads this indicator constantly exceeds 700 and reaches as much as 2500.

However, this capacity is significantly occupied from the morning, by commuters arriving early and up until the end of the day.

Close to Noisy-Champs RER A station, the situation is under pressure: the station's car park is saturated and overflows into neighbouring avenues, including unsuitable places. This saturated pocket is developing in a sprawl. To reserve a place, some users make the journey earlier in the morning.

However, the shopping centre located 200 m from the station contains a large car park that is fairly empty because there is a charge after the first 30 minutes: this time-related pricing creates a barrier to commuter parking. We also know that numerous users of the Noisy-champs RER station come from neighbouring towns because this is the last station in the price zone – they feed into it here to avoid paying for an additional zone.

e) Reconstitution of the 2008 situation and scenarios for 2030

We devised demand and supply scenarios to simulate their interaction and explore supply policies. The original scenario, which we call state 0, corresponds to 2008 in order to reproduce the demand in the reference state based on a general census of the population: i.e. 1,900 inhabitants, 6,300 jobs and 8,000 students, of which 9% are housed on site, for the Descartes Iris. Applying the model presented at the start of this section to the ratios taken from the 2001 Enquête Globale de Transports de Voyageurs (travel survey – EGT), we observe a demand for this neighbourhood of 5,340 places, comprising 60% workers and 17% students, all of them commuters, and 17% residents, 6% feeding into the station and 5% business visitors.

For 2030, in a first project scenario, demand is simulated based on land use hypotheses in the urban planning project (+22% inhabitants and +56% jobs) and maintaining the 2008 ratios. The load demand would therefore come to 7,860 places.

A second scenario was defined based on a strong hypothesis by the urban planning consultant (Atelier Yves Lion), i.e. that the additional housing would be occupied by people working on site. In this case, occupancy requirements increase if households have a car, but drop if they do not. In addition, the level of parking facilities in the new houses is a decision variable for the developer, which interacts with the local parking capacity and inhabitants' possession of a car.

We focus here on the first scenario, which includes fewer uncertainties. The load demanded has been compared to an increased supply of public car parks (from 350 to 2,230 places) but a reduced supply along the road (from 2,700 to 2,350), compared to the situation observed in 2010. Putting to one side residential demand and its private supply, at 2,640 places, public parking load demand comes to 5,220 places for a supply of 4,500 places, or an excess demand of 700 cars – which would form a 3 km line of unauthorized parking along the roads. This shows the importance of locating parking capacities: we propose developing supply on 3 chosen sites, cf. Figure 15.19.

Thus a quantitative expansion of parking supply would not be sufficient to respond to the developing demand, beyond modal transfer. This demand for the most part comprises commuters: a fairly easy target to oust locally but quick to transfer to a neighbourhood close by ...

The best parking policy at neighbourhood scale would be to oblige establishments to equip themselves, based on target ratios of car use for each activity motive. These ratios should be established according to the quality of service at the site by public transport modes, both collective and individual (free service or taxi). Yet some establishments are already located on the site and equipped, and others will locate throughout the course of the project, even though the public transport mode coverage will not have yet achieved its long-term targets. To avoid overinvesting in places in the mid term that will become surplus in the long term, local clubs could be encouraged to pool places between neighbouring establishments, both existing and planned, and to develop public capacity to fill the supply deficit after pooling establishments' private capacities.

CONCLUSION

Two types of lesson can be drawn from this study: factual and methodological. On the factual level, Cité Descartes as a generally extended neighbourhood possesses good accesses to the rest of the metropolitan area, but fewer local amenities than most of its neighbours, due to a smaller, less dense resident population. Professional specialization has not resulted in many amenities. The urban planning project, which will do little to build up the population but greatly increase jobs, while making the Descartes neighbourhood into the core of a more coherent urban centre, ultimately tends to reproduce the pattern of a working core centre with residential suburbs – except that "its suburbs" are in fact more attracted by other polarities across the metropolitan area. Reinforcing the urban aspect would involve putting more emphasis on a campus effect, like the Quartier Latin. A family dimension could also be added by encouraging housing for young families professionally involved in the scientific and technical cluster or the competitiveness cluster. This kind of scenario would require equipping new residential buildings with parking places.

On the methodological level, Geographic Information Systems make complex spatial requests appear simple: of order 0 to connect an object with its physical or postal address, of order 1 to connect two entities to the same address, and of order 2 to connect two entities each with their own address. This spatial modelling is indispensable for detailed treatment of localized entities.

We have developed usage demand models for physical and socio-economic aspects: for the floor area occupied by activities according to the number of jobs, for energy, for parking demand and its occupancy load. These models are based on a simple, common principle: multiplying a specific generator by a typical unit coefficient. An inverse model was used to infer the residential function of a building, after deducting professional functions. In this common form, all of these usage demand models are around an order of 1 in spatial complexity: the order of the urban particle, which is the establishment of an activity in a building. In the next phase of the Chair, we will develop a model of neighbourhood usage of order 2+. This will involve connecting activities established in the neighbourhood as individuals' activities develop, thus as activity purposes respectively at the origin and destination of trips within the neighbourhood. This model of local demand will be completed with a multi-modal model of mobility, broadly inspired by the ParkCap parking model. It will allow us to deal with the neighbourhood's fundamental challenge: the self-reinforcement of opportunities, by superposition based on juxtaposition. This specific neighbourhood effect is a gauge of urban coherence as much as a factor of social cohesion.

ACKNOWLEDGEMENTS

This chapter puts into perspective a series of studies carried out by Houda Boujnah as part of her thesis, Walid Chaker and Natalia Kotelnikova in their respective post-doctoral studies at LVMT, and students' projects at Ecole des Ponts ParisTech on the MASYT, TUSMUR and COSMI courses. We thank all of these participants for the contribution they made to general progress. In addition, most of the analyses are based on the MOS database kindly made available to us by IAU-IDF, and some on the MODUS model and its information bases, made available by the DRIEA. We warmly thank these organizations and their representatives.

REFERENCES

Antoni, J.-P., Fléty, Y., Vuidel, G. & Sède-Marceau (de) M.-H. (2009). Vers des indicateurs locaux de performance énergétique : les étiquettes énergétiques territoriales. Une première approche à partir de l'estimation des mobilités quotidiennes. Neuvièmes Rencontres de Théo Quant, Besançon, 4–6 March 2009.

Boujnah, H. & Chaker, W. (2010). Etude du stationnement à la Cité Descartes Elargie. Enpc working document, 26 pages.

Enpc (2011). Évaluation prospective du stationnement au cluster Descartes à l'horizon 2030. Report of current project TUSMUR 2010–2011, 75 pages.

Enpc (2013). Etude de l'usage du sol, de l'économie territoriale et du transport pour la Cité Descartes dans le cadre du CDT Cluster Descartes. Rapport de projet du cours TUSMUR 2012–2013, 87 pages.

Gallez, C. & Hivert, L. (1995). "Qui pollue où ? Analyse de terrain des consommations d'énergie et des émissions polluantes de la mobilité urbaine", Transports Urbains, No. 89, 15–22.

Gallez, C. & Hivert, L. (1998). "Mode d'emploi, synthèse méthodologique pour les études" budget-énergie-environnement des déplacements", INRETS report, 85 p.

Leurent, F., Boujnah, H. & Chaker, W. (2010). Le stationnement automobile au Cluster Descartes: une analyse offre-demande. Enpc working document, 9 pages.

Orfeuil, J.-P. (1984). "Les budgets énergie-transport : un concept, une pratique, des résultats". Revue RTS n°2, pp. 23–29.

Salamonowicz, P. & Vilaplana-Muller, M. (2012). Transports pour la Cité Descartes. Scientific project report directed by N. Coulombel and F. Leurent. 44 pages.

Chapter 16

Comparison of urban morphologies using life cycle assessment

Eric Vorger, Grégory Herfray & Bruno Peuportier
MINES ParisTech

Marne la Vallée's urbanization is currently fairly patchy and lacks real coherence: it includes blocks of tertiary buildings, social housing neighbourhoods, educational institutions and wooded areas. A long-term project is under study to organize and urbanize this territory. It comprises different phases spread over time. The first, very short-term stage has already delivered design studies. The decision was therefore made to apply the tools developed within the Chair to the project's subsequent stages to feed into long-term thinking.

This chapter shows how a life cycle assessment can help reduce the environmental impacts resulting from this kind of project, taking the example of a portion of land adjacent to the future Boulevard du ru de Nesles. The short-term aim is to situate the project's performance in terms of best European and global practices, i.e. comparison with two morphologies taken from the Vauban eco-neighbourhood in Freiburg (Bade Wurtemberg, Germany).

MODELLING PROJECTS

a) Presentation of the Vauban neighbourhood taken as a reference

Freiburg im Breisgau is a town with a population of 210,000 located in the state of Baden-Württemberg in southeast Germany. The municipality is a pioneer in terms of ecology and sustainable development, including the Vauban eco-neighbourhood, launched in 1996 on the site of a former French military barracks. The 4 hectares of existing buildings were renovated and the remaining 34 hectares were totally restructured.

The neighbourhood is located 4 km from the centre of town and comprises 2,000 houses for total of 5,000 inhabitants, 600 jobs, a school, shops and a student residential development.

The project is the fruit of a public participation process lasting several years and associating a broad range of stakeholders, from the municipality to architect firms, and including citizens and students. This may be the reason behind the undeniable success of the Vauban neighbourhood.

The urban area is particularly coherent because the entire design in terms of architecture and urban planning takes a sustainable development approach.

For transport, the focus is on trams and bicycles and keeping car use to a minimum (multiple-storey car parks at the entrance to the neighbourhood, car rental system, drastic speed limit reduction within the neighbourhood, pedestrian streets, local shops and services, emphasis on short distances, etc.).

A cogeneration plant using gas and wood chips from the neighbouring Black Forest fuels a district heating system. Combined with omnipresent solar panels, it means that the neighbourhood produces 65% of its electricity consumption.

Plants are everywhere, pre-existing trees have been preserved, parks punctuate the neighbourhood and buildings are fitted with roof gardens.

Buildings are varied in terms of morphology and energy performance. However, when it comes to the architecture, they feature recurrent characteristics corresponding to eco-design fundamentals: south-facing, southern façades with significant glazing and protective caps for summer, highly insulated envelopes, efficient windows, natural ventilation or heat recovery exchangers, thermal solar and/or photovoltaics, energy-saving equipment, use of ecological materials, rainwater recuperation systems, etc.

The technical specifications require maximum primary energy consumption of $65\,kWh/m^2$ per year. Some parcels comprise buildings with "passive" labels whose heating consumption must not exceed $15\,kWh/m^2$.

To the east of the neighbourhood lies the "solar city" designed by the architect Rolf Disch. This housing project comprises "positive energy" houses that produce more energy than they consume thanks to a combination of energy efficiency and high solar production. This parcel shall be the object of our study and be used for comparison with the urban morphologies we analyze.

Throughout the study, the housing project inspired by Rolf Disch's "solar city" is referred to as "Positive Energy Neighbourhood" or PEN, and the parcel representing the whole Vauban neighbourhood is referred to as "Low Energy Neighbourhood" or LEN. The third neighbourhood, modelled from a sketch by the architect Christian Binetrui[1], is referred to as Cité Descartes.

Below, we set out the hypotheses chosen at each stage of the modelling.

b) Alcyone, 3D building modelling

The buildings were first entered into the graphic modeller, Alcyone.

Comment: The PEN buildings were modelled from photographs[2] using Google Earth software for sizing.

Different types of building were modelled in each neighbourhood, cf. Table 16.1.

c) Construction data

PEN construction data are known [Heinzel, 2009], and the same values were used for Cité Descartes. For the LEN, construction data were chosen to correspond to the requirements of the label "Passiv Haus"[3]. They are slightly less efficient than those of the other two neighbourhoods, cf. Table 16.2.

---

[1]Architect-urban planner, ACT Consultant
[2]www.plusenergiehaus.de
[3]www.passiv.de – www.igpassivhaus.at – www.passivhouse.com

Figure 16.1 Sketchup sketch (left) and Alcyone modelling (right) of type 1 building (Cité Descartes).

Table 16.1 Number of buildings of each type for the three original neighbourhoods.

PEN neighbourhood	LEN neighbourhood	Cité Descartes
1 type 1 (housing)	2 type A (housing)	(All buildings include housing, offices and shops except type 3-b)
1 type 2 (housing)	2 type B (housing)	3 type 1
3 type 3 (housing)	1 type C (housing)	1 type 2-a
2 type 4 (housing)	School	1 type 2-b
3 type 5 (housing)	Tertiary building	1 type 2-c
Sonnenschiff (housing, offices and shops)	Car park (ground-floor supermarket)	1 type 3-a
Car park (ground-floor supermarket)		1 type 3-b (housing)
		1 type 3-c

Table 16.2 Thermal performance of buildings.

	PEN and Cité Descartes	LEN
Surface characteristics	U in $W/(m^2 \cdot K)$	
Glazing	0.70 (wood triple glazing)	0.87 (wood triple glazing)
Outside walls	0.12 (external insulation)	0.16 (external insulation)
Bottom floors	0.16	0.16
Roofs	0.11	0.11
Thermal bridge coefficient of bottom floor in $W/(m \cdot K)$	0.05	0.10

The intermediate floors and load-bearing partitions, including in wood-framed buildings (PEN type B) are made of 16 cm-thick raw concrete. The inside partition walls are covered with plaster and comprise 4 cm of rock wool for thermal and acoustic insulation.

d) Meteorological data

The objective of the study is to assess the environmental performance of the future Cité Descartes, in comparison with PENs and LENs that represent good practices. The three neighbourhoods are therefore compared in the same meteorological conditions corresponding to the Paris region, where Cité Descartes will be built (Marne la Vallée – Val de Marne).

The weather file used for the Dynamic Thermal Simulation (DTS) of all the buildings was taken from Trappes weather station in the Parisian suburbs.

e) Definition of thermal zones

The DTS is based on a multi-zone modelling of the buildings. Each zone is assumed to have a uniform temperature and exchange heat with other zones through conduction, convection and radiation to the walls of the zone as well as through air movements.

How the zones are defined influences the results of the simulation, and a compromise needs to be found to facilitate input and reduce calculation times without generating too many errors. In addition, each zone is characterized by a single scenario (daily, weekly and annual) for each of the following criteria: occupancy, internal gains, ventilation, heating and air conditioning. Thus, the zones must be defined to encompass rooms with similar behaviour, particularly in terms of occupancy and orientation.

To define the best compromise for dividing buildings into zones, we compared the simulation results for the same building successively divided into 3, 1 and 9 zones.

The building considered is PEN type 3. Among the characteristics likely to influence the "zoning", we observe that the subject is a 3-floor building, south facing, comprising 4 identical homes, each split into three levels. The south wall is largely glazed and the ground floor (kitchen, living room) and 2nd floor (converted attic space) have no partition walls. The 1st floor comprises two rooms on the south side and service rooms.

The 3D representation of the building shown below (Figure 16.2 & 16.3) indicates the close masks and glazed surfaces.

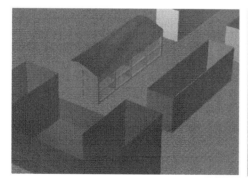

Figure 16.2 Southwest view.

Figure 16.3 Northeast view.

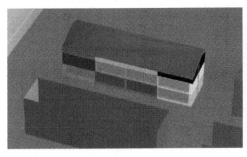

Figure 16.4   9 zones.

Figure 16.5   3 zones (per floor).

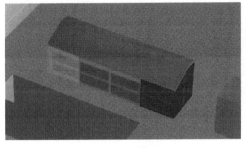

Figure 16.6   3 zones (per apartment).

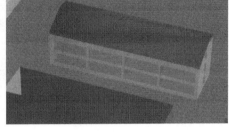

Figure 16.7   1 zone.

The different splits are visible on the following representations (Figures 16.4 to 16.7) in which each thermal zone corresponds to a colour.

The operating scenarios are identical for the four cases studied and identical from one zone to another (see Table 16.3).

Table 16.3   Operating scenarios, type 3 PEN building.

Occupancy rate	0.02 occupants/m² uniformly distributed
Occupancy scenario	Unoccupied (0%) from 10 am to 6 pm Monday to Friday
	Occupied (100%) the rest of the time
Heating setting	19°C, reduced to 16°C (same scenario as "occupancy")
	18°C from 18 h to 19 h to level out re-ignition
Air conditioning setting	None
Controlled ventilation	Extracted airflow = 195 m³/h per apartment
	Heat recovery exchanger, $\varepsilon = 0.80$
	$\Rightarrow 0.10$ vol/h
Infiltrations	0.20 vol/h
Ventilation scenario	Constant, equal to 0.30 vol/h
Internal gains	16 kWh/m²/yr
	Uniform distribution in space
	Distribution over time according to the French thermal regulation for housing

Considering a higher number of zones logically tends to increase heating loads. In fact, solar and internal gains give a better return with few zones because when they are in excess in one part of the building, they are likely to be used to compensate another deficient part at the same time.

This is particularly the case when moving from a single zone to treating zones by level. Solar gains from the large bay windows on the ground floor and 1st floor are used more efficiently by the 2nd floor, which has a large heat-loss surface and no south-facing windows. Thus, at certain times of the year, the 2nd-floor zone needs to be heated in a model split by floor, whereas the single zone does not. This is clearly illustrated in Figure 16.8 below, which shows the temperatures in the three zones corresponding to a floor each and in the single zone.

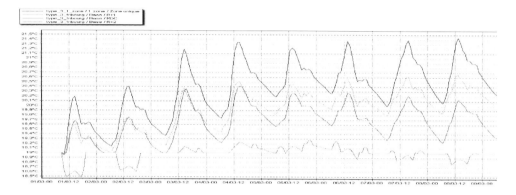

Figure 16.8 Temperature of the different floors and the single zone in mid season.

Comment: the temperature is higher on the ground floor than on the 1st floor, which might seem illogical since the ground floor has a heat-inefficient floor and is more masked by surrounding buildings. The reason is the large overhanging roof (covered with solar panels and not visible on the 3D representation along with the balconies) which constitutes a highly influential solar mask (Figure 16.9) for the bay windows on the 1st floor during the period considered (early March).

We thus observe an 8.8% increase in heating loads with the move from one to three zones (i.e. per floor), and a 3.7% increase when moving from three (per floor) to nine zones. With a direct move from one to nine zones, the difference is 16%.

When we divide the building into 9 zones, heating loads increase, which can be explained by the lower use of gains. On the other hand, the heating power value necessarily decreases in comparison with its value for the model with one zone per floor (Table 16.4).

Table 16.4 Simulation results.

	Unit	1 zone	3 zones (per floor)	3 zones (per apart)	9 zones
Heating load	kWh/yr	9,143	10,099	9,357	10,403
	kWh/m^2/yr	17	19	17	19
Heating power	kW	15,055	15,843	13,575	14,166

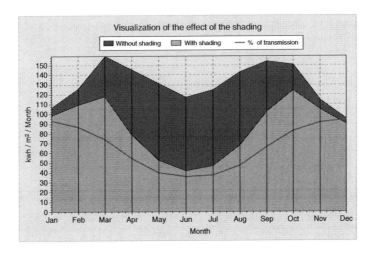

Figure 16.9 Effect of the "roof overhang" solar mask.

The reason is that the apartments located at the east and west ends (called 1 and 4) lose more heat than apartments 2 and 3 at the centre of the building. During the coldest periods, the temperature can therefore drop below 18°C between 6 pm and 7 pm in these zones, which sets on the heating. Thus, the air and inert surfaces in these zones are heated before 7 pm, and the 7 pm power peak is reduced.

The graph below shows the temperature and power demanded for the single zone and for the zones apartment 1, apartments 2–3 and apartment 4. It illustrates the progression in power demand for apartments 1 and 4 (orange and pink curves).

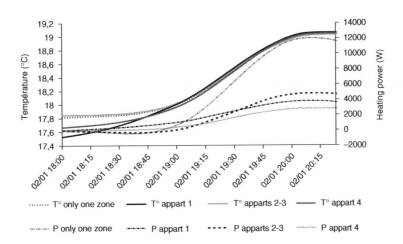

Figure 16.10 Temperature and power demanded for the 3 zones split into apartments and for the single zone, at around 7 pm during the heating season.

The more detailed the split, the closer the results match reality. However, to simplify data input and clarify the analysis and interpretation of results, it is interesting to use a small number of zones. The split should thus be justified and minimized.

In Table 16.5, each model is compared with the 9-zone model, which is the most realistic one, in order to appraise its validity.

Table 16.5 Comparison of different splits with the 9-zone model.

		1 zone	3 zones (per floor)	3 zones (per apart)
Heating loads	kWh/an	−12.1%	−2.9%	−10.1%
	kWh/m²/an	−10.5%	0.0%	−10.5%
Heating power	kW	6.3%	11.8%	−4.2%

The differences measured are judged acceptable only in the case where the building is split into three zones each corresponding to one level, given that the output used for the LCA corresponds to heating loads and not maximum power. In addition, when buildings comprise several wings, these are split beforehand, in particular since most of the time they accommodate different uses.

The "zoning" principle used for all of the buildings is therefore a split by wing and level.

f) Dynamic Thermal Simulation, Pleiades+Comfie tool

Each thermal zone is characterized by five scenarios corresponding to occupancy, internal gains, ventilation, heating and air conditioning. These scenarios are defined hourly for each day of the week then each week of the year. When measured values are not available, the chosen scenarios are taken from Th-CE[4] rules, in other words, they correspond to the values used by the French thermal regulations in force.

The Table 16.6 sums up the values chosen for the occupancy periods.

Table 16.6 Summary of scenarios.

	Housing	Offices	Shops	Schools
Occupancy (occ/m²)	PEN: 0.02 LEN & Cité Descartes: 0.03	0.07	0.07	Classrooms: 0.33 Canteen: 0.64
Internal gains	16 kWh/m²/yr	14 W/m²	14 W/m²	Classrooms: 7 W/m² Canteen: 14 W/m²
Heating	19°C, reduced to 16°C	19°C, reduced to 16°C	19°C, reduced to 16°C	19°C, reduced to 16°C
Air conditioning	None	26°C	26°C	26°C
Ventilation	0.30 vol/h	0.40 vol/h	0.75 vol/h	Classrooms: 2.40 vol/h Canteen: 1.20 vol/h

[4]ThCE rules, defined by the Centre Scientifique et Technique du Bâtiment, are the rules in force for French thermal regulation.

A number of hypotheses and explanations are developed below.

– Comments on occupancy

Shops:
Saturday was added in relation to the ThCE scenarios.

The number of shop occupants is calculated by applying a ratio (0.07 occupants per m^2 according to ThCE rules), corresponding to average occupancy during opening hours, and encompassing customers and workers.

Housing:
For the PEN housing, we know that the occupancy rate is 0.02 occ/m^2 (average of 2.9 habitants and 138 m^2 per house, [11]). This is a relatively low rate, and so for the LEN and Cité Descartes housing we instead opted to use the average French rate (INSEE) of 0.03 occ/m^2.

This difference in occupancy between the PEN on one hand and the LEN and Cité Descartes on the other, impacts on the results of the neighbourhood LCA as we shall see below.

– Comments on internal gains

Internal gains correspond to the heat generated by joule effect from electrical equipment, lighting, computers, hi-fis, domestic appliances, etc.

For housing, they match those defined by the ThCE rules in as much as the time variations are maintained, but the sizes are adjusted based on the values supplied by measurement campaigns carried out on the Solar City in Freiburg.

These values give an average electricity consumption of 21 kWh/m^2/yr. The internal gains considered are set at 75% of this value to take into account the heat evacuated by e.g. clothes washers, dryers and dishwashers, which evacuate hot water out of the system.

Electricity consumption measured in the houses studied was 55% lower than the consumption considered in the ThCE rules. This is due to the higher-than-average energy efficiency of the equipment installed and, no doubt, economical use by occupants who are particularly sensitive to energy savings.

– Comments on ventilation

Controlled ventilation flows are constant in housing and 50% lower during unoccupied periods in shops, offices and schools.

Housing, offices, shops and school canteens are equipped with heat recovery ventilation systems. The principle involves crossing inflows of new air with exhaust air extracted in a heat exchanger so that the new air is pre-heated during heating periods and cooled down during very hot periods. The exchanger is 80% efficient, which is a standard value for this technology. Classrooms are not equipped with this system since their high occupancy rate generates sufficient heat in a very insulated building. In addition, the building remains unoccupied during the summer, when the apparatus works in cooling mode.

Night free cooling

The apartments in the buildings studied are all with openings on opposite facades. This means that natural ventilation is easy to put in place and efficient. We assume that for all housing, occupants open their windows during summer nights. Thus, the building's inert masses cool down and summer comfort is easier to attain. This is facilitated by external insulation of the facades, which makes use of the walls' inertia, and the choice of flooring and heavy weight-bearing partitions with little covering. An average air renewal of 4 vol/h was assumed, using natural free cooling in the housing.

For offices and shops, security measures mean that windows cannot be opened at night. However, heat recovery ventilation systems provide technologies to short circuit the exchanger and programme ventilators so that cool air inflows can be forced in the right conditions. Thus, a free cooling flow of 2 vol/h was assumed for shops and offices at night.

This technique, practised intuitively in houses and ensured by mechanical ventilation in other buildings, is highly efficient: e.g. the annual cooling load of Sonnenschiff, the PEN building that comprises shops, offices and housing, drop from 96 MWh ($14 kWh/m^2$) to 8 MWh ($1 kWh/m^2$). Free cooling at night can almost cancel this building's cooling load.

From an architectural point of view, this underlines the importance of including natural ventilation in buildings to ensure summer comfort, as well as inertia and making the most of it, especially by external insulation.

Infiltrations

Air infiltrations are responsible for a very large share of the heat loss in highly insulated buildings, and this proportion is even greater in buildings equipped with heat recovery ventilation. We considered an infiltration value of 0.20 vol/h in houses and offices and 0.30 in shops to reflect more frequent door opening.

As an example, with an infiltration rate of 0.35 vol/h (corresponding to an average apartment in the existing stock) instead of 0.20 vol/h, the PEN type 1 residential building would see its annual heating consumption rise from 13 to $20 kWh/m^2$, or a 54% increase.

From an architectural point of view, compact buildings are preferable because they include fewer surfaces sensitive to air infiltrations.

From the point of view of neighbourhood morphology, this parameter is highly dependent on the management of wind corridors, through street orientation and building layout [Huang, 2008].

These different scenarios can be used to evaluate the heating and cooling load of buildings in the neighbourhoods studied.

g) Life cycle assessments of buildings, Equer tool

The main hypotheses are as follows.

– Duration of the analysis
    The study considers that all buildings have a life span of 80 years, at the end of which they are destroyed.
– Materials
    • To simplify, all materials are considered as ending up in (inert) landfill at end of life, except for metals, which are recycled,
    • Surplus materials at the worksite: 5%,
    • Transport distance: 100 km,

- Transport distance from the site to the inert landfill at end of life, 20 km,
- Lifespan of doors and windows: 30 years,
- Lifespan of coatings: 10 years,
- No replacement during the ten last percent of the building's lifespan of items whose lifespan is greater than this ten percent.

- Energy
  - French mix for basic electricity production:

Nuclear: 78%   Hydroelectric: 14%   Gas: 4%   Coal: 4%   Fuel oil: 0%

At the end of the calculation, solar production is subtracted from the neighbourhoods' consumption and impacts are added related to the production, replacement and end of life of modules and other items of the installation (cf. § "LCA neighbourhood additions").

- Electricity grid loss: 9%
- Type of heating energy: district heating, with the heat production mix mentioned above, i.e.
  Waste: 0%, Gas: 20%, Fuel oil: 0%, Electricity: 0%, Coal: 0%, Wood: 80%, Geothermal: 0%

This energy mix corresponds to the district heating network in the Vauban eco-neighbourhood, which includes the PEN and the LEN. It is fuelled by a cogeneration plant whose thermal and electricity yields are respectively 61% and 26%. Network losses of 3% are assumed. It is only at the end of the calculation that the cogeneration as such is taken into account (cf. § "LCA neighbourhood additions").

To fulfil the objective of comparing morphologies, and without yet knowing which energy will be chosen to heat Cité Descartes, we considered a heating network fuelled by cogeneration with the same mix and yields: at this stage the aim is to compare urban morphologies all else being equal.

- Type of energy for domestic hot water (DHW): 50% ensured by thermal solar installations, the other 50% by the district heating system.

The LCA of thermal solar thermal collectors was carried out at a later stage (cf. § "LCA neighbourhood additions").

- Air conditioning: an average annual performance coefficient of 3 was considered.
- Additional electricity consumption: 500 Wh per person per day in housing to reflect water heated using electricity (e.g. for washing machines, dishwashers), which is evacuated from the building and therefore not included in the internal gains defined for the thermal simulation.

  - Water

  The ratios of water consumption are taken from studies by suppliers[5] or state organizations[6].

---

[5] Joint local authority for water resource management for the Gironde region (Smegreg): "Principaux ratios de consommation d'eau", 2007.
[6] Cemagref, Strasbourg National School for Water Engineering: "La consommation des ménages en France, état des lieux", 2002.

- Water supply yield: 80%
- Hot water requirements in l/pers/day:
  Housing: 40   Offices: 4   Shops: 0   School (with canteen): 20

As seen above, 50% of DHW needs are supplied from the thermal solar system. This amount is thus accounted for as cold water.

In the current version of Equer, we are only able to input a single whole number per building for hot water consumption, in l/pers/day. For buildings with mixed uses, such as almost all of the buildings in Cité Descartes, we thus input DHW volumes as cold water and added the heating energy afterwards.

- Cold water requirements in l/pers/day:

  For single-use buildings (cold water + solar DHW):
  Housing: 100+20   Offices: 50+2   Shops: 43+0   School (with canteen): 27+10

  For mixed-use buildings (cold water + all DHW):
  Housing: 100+40   Offices: 50+4   Shops: 43+0   School (with canteen): 27+20

- Waste

Neighbourhood morphology does not influence occupants' behaviour in terms of creating and managing household waste.

- Waste is therefore not included in the first stage, but is the subject of a sensitivity study (cf. § "Impact of waste from activities and transport").

  - Occupants' everyday transport

- Occupants' everyday transport is not included in this stage because the link with the transport simulation has not yet been implemented. A sensitivity study has however been carried out with two public transport scenarios (bus and train, cf. § "Impact of waste from activities and transport").

These hypotheses make it possible to carry out a life cycle assessment of the different buildings.

h) Life cycle assessment of neighbourhoods

The main hypotheses and data concerning the different items modelled are set out below.

- Site on which the neighbourhood is located

  - Annual rainfall: 800 l/m^2 (average for Ile de France)
  - Drinking water supply:

Consideration of construction.

Share of metal: 50%   –> Maintenance after: 50 yrs   Lifespan: 100 yrs
Share of polyethylene: 50% –> Maintenance after: 40 yrs   Lifespan: 80 yrs

- • Wastewater network:
- • Consideration of construction.

Loss: 3%   Maintenance after: 40 yrs   Lifespan: 80 yrs

- • District heating system:
- • Consideration of construction.

Maintenance after: 40 yrs   Lifespan: 65 yrs
- Buildings

  - • Use of original terrain:
    One of the indicators assessed at neighbourhood scale concerns land use and transformation. It is thus worth mentioning the initial use. The PEN and LEN neighbourhoods were an extension of the urban fabric (both neighbourhoods were built on the site of former military barracks).
    Cité Descartes is located in a green urban area: Cité Descartes will be built on an area currently made up of grass, trees and copses. Some of the trees will be preserved.

- Public spaces
  The compositions of the different surfaces of public spaces (streets, alleys, car parks) are those predefined by default in the software, e.g. for streets: 10 cm concrete aggregate and 5 cm asphalt concrete. The main hypotheses are given in the Table 16.7.

Table 16.7 Characteristics of outside spaces.

	Road	Street	Alley, car park	Permeable pavement	Parks, Green spaces
Waterproof quality	95%	90%	85%	60%	40%
Snow clearing	0	0	0	–	–
Mowing	–	–	–	–	5 per yr
Lighting	4.8 kWh/m²/yr	4.8 kWh/m²/yr	4.8 kWh/m²/yr	0	–
Water (watering, cleaning)	10 l/m²/yr	9 l/m²/yr	9 l/m²/yr	3 l/m²/yr	60 l/m²/yr
Waste	None	None	None	None	None
Lifespan	100 yrs	100 yrs	100 yrs	100 yrs	–

i) LCA neighbourhood additions

We have seen that a number of elements required for a neighbourhood LCA are not yet included in the current version of Equer. These various elements, described below (see Figures 16.11 to 16.13), were thus added in an Excel file.
  The starting point for this file corresponds to the results of the LCA.

- Addition of diverse elements, "Basic" option

Figure 16.11   Balconies and staircase on type 1 and 2 PEN buildings.

Figure 16.12   Staircases and passageways on a Type B LEN building.

Figure 16.13   Car park (split between the PEN and the LEN).

Firstly, to obtain the so-called "basic" version, a number of elements that have not yet been enumerated are added.

This involves, e.g. the car park at the PEN and LEN of which we were only able to model the ground floor, which accommodates a supermarket. The higher floors, which are supported by pillars and have no side walls, could not be entered into Alcyone-Pleiades.

With help from an expert, we thus had to assess the quantities of concrete, steel and wood based on photos, and include them as impacts of building and demolishing the neighbourhood. Lighting requirements were also added to the use stage. Other additions at this stage were balconies and terraces, outside staircases, passageways, foundations and lifts, along with missing DHW energy consumption. Assessing the quantities of materials and energy involved in the life cycle from these various elements involved research[7].

As an example, for the PEN, the progress of impacts following the consideration of these various elements is shown in the bar chart in Figure 16.14.

Comment: the "Equivalent number of inhabitants" unit enables us to normalize the results. For each indicator, the normalization coefficient corresponds to a French person's average impact over a year. The only indicator that is not normalized due to the lack of a reference value is odour.

We observe an average 10% increase for the indicators as a whole. In reality, these changes are more irregular than indicated by the bar chart depending on the indicator observed.

For example, 83% of the increase in primary energy consumption is based on the use stage, whereas 49% of the increase in human toxicity is based on the construction stage. This last result is mainly due to the manufacturing process of steel, which is particularly present in the added elements.

– Consideration of cogeneration

Cogeneration fuels the district heating network, which supplies 50% of DHW needs and all of the heating needs. The thermal and electricity yields are respectively 61% and 26% and the network's heat loss is 3%.

We assumed that consumption corresponds to the satisfaction of heating loads, which represents routine practice.

Thus, for each neighbourhood whose overall heating loads (heating + DHW) come to Eh, the total energy consumed by the cogeneration plant is $Et = Eh/0.61$, and electricity production is $Eel = Eh/0.61 * 0.26$. In comparison with the district heating consumption calculated by Equer, we must therefore add $Esup = Et * (1 - 0.69)$ converted in line with the heating mix (20% gas and 80% wood). In the hypothesis that the electric energy produced avoids production from the grid, we should also subtract Eel with the corresponding conversion into the network's electricity mix (78% nuclear, 14% hydroelectric, 4% gas and 4% coal) and related impacts.

---

[7]Information was found on constructors' websites (www.Kone.com, www.arcelormittal.com) and the site www.energieplus-lesite.be (developed by the University of Louvain in Belgium) to ascertain the electricity consumption of lifts and car park lighting.

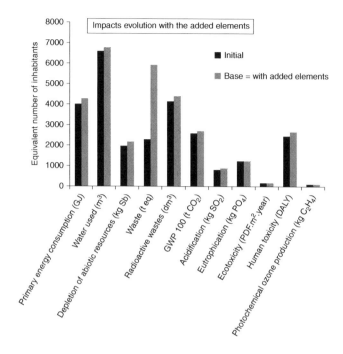

Figure 16.14 Progress of impacts following the addition of diverse elements for the PEN.

Due to a lack of information on the plant itself, we did not consider the construction, renovation and demolition stages. From this point of view, our case is that of a standard district heating system with a boiler room located outside the boundaries of the study.

Adding the impacts of cogeneration onto the "basic" version of Cité Descartes, the indicators progress as in Table 16.8.

We observe that primary energy consumption decreases. Although the power plant's total yield of final energy is only 87% (gas condensing boilers attain 97%), when the electricity produced is converted into primary energy, the primary energy balance is better than that using a standard boiler house.

Given the yields considered, the quantity of gas and wood consumed increases by $0.39/0.61 = 64\%$, and the additional energy produced mainly replaces energy from nuclear sources. This explains the increase of almost all impacts except for the water used (because water is consumed to cool electric power plants) and the nuclear waste produced.

– Consideration of solar thermal power

The hypotheses are as follows:

- 50% of DHW requirements are provided by the thermal solar system. This corresponds to the in Table 16.9 values.

Table 16.8 Effects on the different impacts of cogeneration compared to a standard boiler house for Cité Descartes.

Impact category	Total
Energy consumed (GJ)	−4%
Water used (m^3)	−2%
Depletion of abiotic resources (E-9)	1%
Inert waste produced (t eq)	−1%
Radioactive waste (dm^3)	−18%
Greenhouse effect (t CO^2)	1%
Acidification (kg SO$_2$)	6%
Eutrophication (kg PO$_4$)	2%
Aquatic ecotoxicity (PDF · m^2 · year)	15%
Human toxicity (DALY)	12%
Photochemical ozone (kg C$_2$H$_4$)	2%
Odour (Mm^3)	16%

Table 16.9 DHW requirements provided by thermal solar installations.

Housing	Offices	Schools
20 l/j/pers	2 l/j/pers	5 l/j/pers
330 j/yr	230 j/yr	180 j/yr

- Insolation: 1200 kWh/m²/yr (average for Paris region),
- Overall yield of the installation (collectors, storage, circulation): 35%,
- Coefficient relative to position of collectors: 98% (south facing, slope between 15° and 40°),
- Lifespan of collectors: 25 years
- 2 replacements (no replacement during the last ten per cent of the building's life),

The consideration of thermal solar power leads to an average 3% reduction of all impacts for the 3 neighbourhoods.

The progress of the different stages and some of the indicators is shown on Table 16.10 for the Cité Descartes (the effects are similar for the PEN and LEN).

Table 16.10 Progress of 5 indicators for each stage following consideration of thermal solar power for Cité Descartes.

Basic + thermal solar	Construction	Use	Renovation	Demolition	Total
Energy consumed (GJ)	2%	−6%	10%	0%	−5%
Depletion of resources (kg Sb)	2%	−7%	10%	0%	−4%
Greenhouse effect (t CO$_2$)	2%	−6%	23%	0%	−3%
Acidification (kg SO$_2$)	4%	−8%	19%	0%	−4%
Human toxicity (DALY)	4%	−12%	26%	0%	−7%

The reduction in environmental impacts is due to the use stage, because manufacturing and replacing panels generates higher impacts in the construction, renovation and demolition stages.

The renovation stage is particularly affected because it includes replacing 2 panels (manufacture and landfill).

This progress is however relative to the respective weight of each stage. Thus, for e.g. the greenhouse effect, a 6% reduction in the use stage is sufficient to result in an overall 3% drop despite a rise of 2% and 23% for the construction and renovation stages.

The following bar chart (Figure 16.15) illustrates the weight of each stage in the progress as a whole.

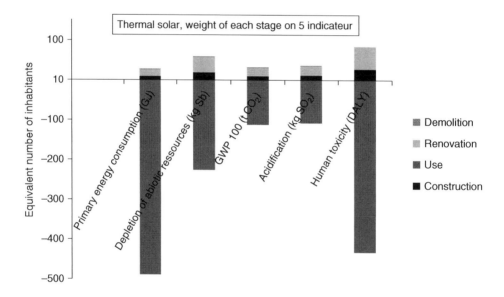

Figure 16.15 Progress linked to consideration of thermal solar power for Cité Descartes. Respective weight of each stage on 5 indicators.

– Consideration of photovoltaic solar power

The main hypotheses and calculation data are as follows.

• The electricity produced by the photovoltaic installations is subtracted from the electricity drawn from the network.
• Solar radiation: 1200 kWh/m²/yr (average for Parisian region),
• Yield from panels: 15% (mono-crystalline),
• Coefficient of performance (loss from shading, temperature, etc.): 97%
• Yield of the invertor: 87%
• Configuration: Integrated into roof for Cité Descartes and PEN, installed on top of roof for LEN,

- Coefficient relative to position of modules: 98% (south facing, slope between 15° and 40°),
- Lifespan of modules: 25 years
- 2 replacements (no replacement during the last ten per cent of the building's life).

For each neighbourhood, we use architect's sketches or photographs to ascertain the surface areas available to install solar modules. The modules' surface areas are obtained by subtracting the surfaces of the thermal solar collectors. For Cité Descartes, for example, the total surface area of the panels is 6,899 m², of which 557 m² is for thermal and thus 6,342 m² for photovoltaics.

Given the large surface areas of the modules installed, the construction and renovation stages are highly impacted. The total weight of solar panels for Cité Descartes comes to 95 tonnes (i.e. 15 kg per m²) and the manufacturing process for one m² requires among other things 1.1 kWh primary energy and 1.5 m³ of water and produces 200 kg eq CO_2.

However, the overall effect of photovoltaic installations is highly beneficial. Despite the spectacular increase in impacts during the construction and renovation stages, the electricity saved during the use stage makes the overall life cycle balance largely positive.

Of the 12 indicators, 4 see their total value rise (inert waste, ecotoxicity, photochemical ozone and odour), but these impacts are low after normalization. The other 8 go down, by as much as −42% for primary energy and −63% for radioactive waste (see Table 16.11 and Figure 16.16).

Table 16.11 Progress of 6 indicators for each stage following consideration of photovoltaic power for Cité Descartes.

Basic + photovoltaic	Construction	Use	Renovation	Demolition	Total
Energy consumed (GJ)	36%	−50%	145%	0%	−42%
Water used (m³)	10%	−7%	111%	0%	−6%
Inert waste produced (t eq)	38%	−20%	83%	0%	5%
Radioactive waste (dm³)	40%	−65%	351%	0%	−63%
Greenhouse effect (t CO_2)	22%	−32%	314%	0%	−6%
Aquatic ecotoxicity (PDF · m² · year)	102%	−9%	516%	0%	29%

It logically emerges that the use stage predominates all of the indicators apart from inert waste. We can also see that the ecotoxicity indicator, which increases sharply, is in reality at a very low level when normalized by the number of equivalent inhabitants. This illustrates that the neighbourhood's impact on ecotoxicity, i.e. biodiversity, is low (sectors like farming have a much greater impact on this indicator).

It is however worth emphasizing that these results are obtained from approximations that are likely to overestimate the advantages of photovoltaic energy. In fact, the calculation method involves counting annual PV production negatively, converting it according to the network's (constant) mix.

In reality, though, this mix evolves over the year, and e.g. the share of fossil energy is greater during the heating season. Improvements are being developed to integrate the electricity mix's temporal variation into the Equer LCA tool.

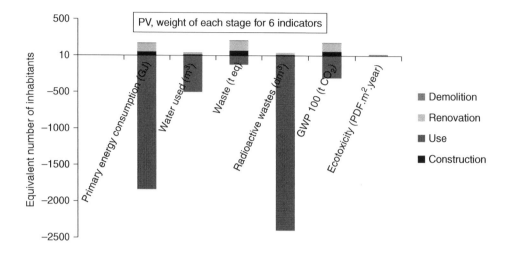

Figure 16.16 Progress linked to consideration of photovoltaic energy for Cité Descartes. Respective weight of each stage for 6 indicators.

In addition, PV production is not constant over the year either, since it is proportional to insolation. It is therefore at a maximum during the summertime, when electric demand is at its lowest, and at a minimum in the winter, when it is highest.

– Overview

The full version corresponds to the "basic" neighbourhood to which cogeneration, solar thermal and solar photovoltaic have been added (see Figure 16.17).

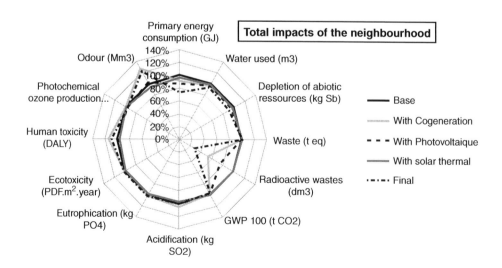

Figure 16.17 Radar chart comparing each version with the basic version.

It is much more efficient in terms of the impacts of primary energy consumed and nuclear waste produced. However, it is less efficient for a number of indicators, like human toxicity, photochemical ozone and aquatic ecotoxity.

However, these indicators evolve around low values as shown by Figure 16.18 below, converted into number of equivalent inhabitants, which illustrates that neighbourhoods have little influence on these issues.

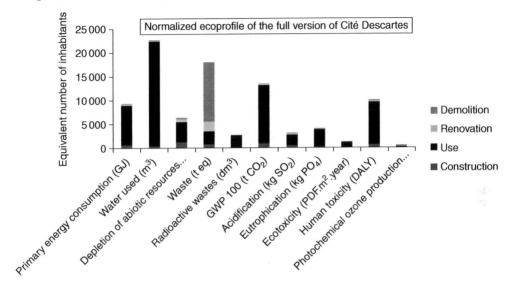

Figure 16.18 Normalized ecoprofile of the full version of Cité Descartes.

COMPARISON OF URBAN MORPHOLOGIES

The Cité Descartes project aims to be exemplary in terms of sustainable development. It is therefore interesting to position its performances in relation to the best practices presented above, and if possible attain even lower environmental impacts.

The methodology (cf. chapter 2) requires defining a functional unit common to the different alternatives or solutions compared. Defining the functional unit is complex because the different neighbourhoods do not comprise the same number of buildings of each type (residential, tertiary). Yet the environmental impacts of offices, housing, schools and shops differ widely during the use stage since the occupancy rate, energy and water needs and required comfort level, etc. are all parameters connected to the activity accommodated.

Neither a functional unit in number of inhabitants that considers all occupants independently from their nature, nor a functional unit simply related to useful m², is suitable in this case.

Thus, to define a pertinent functional unit common to the three neighbourhoods, we have two possibilities:

- The first requires establishing a relationship between types of occupants so as to be able to calibrate them. This therefore involves establishing coefficients of proportionality to express each type of occupant according to a reference type. For example, if we define as a reference the resident of a house corresponding to the requirements of the Thermal Regulations, the idea is to express, for each stage p and each indicator i, an office occupant as $c_{b,p,i} \times$ "typical inhabitant", a student as $c_{e,p,i} \times$ "typical inhabitant", etc.

- The second involves adapting the neighbourhoods studied by adding buildings or slightly adjusting surface areas, so that they present the same number, or at least the same proportion, of occupants of each type. The difficulty is to remain faithful to the urban architect's work, thus maintaining the spirit of the urban space, the density, proportions of public spaces, etc.

The first method has the following disadvantages:

- The results depend on the choice of typical occupants and their activities (established, for example, in relation to the scenarios of ThCE rules). This choice can increase or reduce the weight of one use compared to another.
- The different life cycle stages are not treated in the same way, creating a distortion that modifies the share of each stage in the overall impact.
- The method would need to be validated by several projects.

We therefore chose to adopt the second approach.

a) Neighbourhood "adaptation"

The "solar city" parcel (Positive Energy Neighbourhood, PEN) accommodates less housing and more offices than the Vauban neighbourhood (Low Energy Neighbourhood, LEN). However, we added positive energy houses and only considered a fraction of the "Sonnenschiff" office building. Maintaining the spirit of the urban form, the PEN was thus adapted to present exactly the same number of occupants of each type as the LEN.

For the comparison with Cité Descartes, the method was slightly different because the neighbourhoods were adapted to have the same proportion of each use, rather than the exact number of occupants.

- Description of the original neighbourhoods

As we can see in Table 16.12, the uses are split very differently. The PEN comprises significant tertiary buildings, whereas the LEN is mainly made up of housing. Cité

Table 16.12 Number of each type of occupant in the original neighbourhoods.

Original	Housing	Offices	Shops	School
PEN	168	361	215	0
LEN	364	72	122	120
Cité Descartes	698	710	356	0

Descartes accommodates a high proportion of offices and shops. The LEN is the only neighbourhood with a school.

Comment: We opted to reason in terms of inhabitant numbers rather than surface area to account for the fact that the occupancy rate is lower in the PEN, which means that for the same housing surface, the PEN has a greater advantage in terms of heating and DHW.

For acoustic comfort, the mixed buildings (PEN and Cité Descartes) comprise offices and shops in the area parallel to the road, and housing elsewhere. This sound barrier is necessary on the main axes of these two neighbourhoods, particularly in Cité Descartes, which is located close to a very busy motorway junction.

This underlines the fact that morphology is mainly defined by the urban context into which a neighbourhood fits. In this case, the orientation of some buildings is not optimized in line with strictly energy-related considerations, which are here only one aspect of the decision.

– Adapting the neighbourhoods to base the comparison on the same function unit

To compare the morphology of the three neighbourhoods (see Figure 16.13) based on the same functional unit, they were adapted so as to present the same proportion of occupants for each use.

To obtain similar proportions for each neighbourhood, we:

– Added houses to the PEN,
– Added tertiary buildings to the LEN, slightly adapting the surface areas of the offices and shops on the ground floor,
– Doubled the surface area of school buildings in the PEN and LEN,
– Added buildings and replaced some office surfaces with shops in Cité Descartes.

The final numbers of occupants of each type are shown in Table 16.14.

In each case, the proportion in number of occupants is 39% housing, 32% offices, 19% shops and 10% schools.

To compare the neighbourhoods with identical numbers of occupants, the LCA impacts are thus multiplied by the following coefficients of proportionality:

– 2.02 between the PEN and Cité Descartes,
– 1.99 between the LEN and Cité Descartes.

The surface areas of external areas also need to be adapted to maintain the building density and, as a general rule, the proportion of surfaces of each type (see Table 16.15). The increased ground space of buildings in comparison with the original neighbourhoods was used to define coefficients of proportionality. All external area surfaces and network lengths were then multiplied by these coefficients (see Tables 16.16 and 16.17).

As a reminder, in the neighbourhood LCA, public spaces are included in construction, use (lighting, watering, mowing, snow clearing, cleaning, etc.), upkeep (maintenance works) and demolition.

Table 16.13 Overall view of the 3 original neighbourhoods.

Positive Energy Neighbourhood (Architect Rolf Disch)	Low Energy Neighbourhood (Vauban neighbourhood)	Cité Descartes

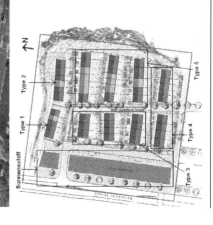

Table 16.14 Number of occupants of each type in adapted neighbourhoods.

Adapted	Housing	Offices	Shops	School
PEN	441	361	215	120
LEN	443	373	217	120
Cité Descartes	890	731	431	240

Table 16.15 Coefficients of adaptation for external surface areas.

	PEN	LEN	Cité Descartes
Building ground space, original n/hood	5,860 m^2	5,480 m^2	10,060 m^2
Building ground space, adapted n/hood	10,080 m^2	7,803 m^2	13,400 m^2
Coefficient	1.72	1.42	1.33

Table 16.16 Surfaces of external spaces in m^2 for adapted neighbourhoods.

	Surface of block	Buildings	Road	Street	Pedestrian alley	Permeable paving	Parks and green spaces
PEN	35,948	10,079	2,236	2,064	6,536	0	14,104
LEN	33,796	7,782	852	3,124	4,118	4,118	13,802
Cité Descartes	95,627	13,380	7,315	5,985	18,620	12,635	38,038

Table 16.17 Network lengths for original and adapted neighbourhoods.

	PEN		LEN		Cité Descartes	
	Original	Adapted	Original	Adapted	Original	Adapted
Network lengths	650 m	1,120 m	850 m	1,210 m	1,850 m	2,460 m

The heating, water supply and sewers systems are assumed to be of the same length. The result is a single "network length" value per neighbourhood.

We thus obtain neighbourhoods that can be compared in a pertinent manner. The functional unit is thus "a neighbourhood of 887 housing occupants, 734 office occupants, 432 shop occupants and 240 school occupants, in Paris, considering a building lifespan of 80 years."

b) Results of dynamic thermal simulations

The results are summed up in Table 16.18, which gives the total heated surface and energy requirements of all buildings for each neighbourhood in kWh/m^2/year.

For an equal number of inhabitants, the PEN has a higher heated surface area because the occupancy rate of its housing is lower than that of the LEN and Cité Descartes.

Table 16.18 Total heated surface and total energy requirements for the 3 neighbourhoods.

	Heated surface	Energy requirements in kWh/m²/year			
		Heating	Cooling	DHW	Internal gains
PEN	61,269 m²	12.1	0.8	5.0	22.7
LEN	48,303 m²	11.4	1.2	6.3	24.7
Cité Descartes	48,349 m²	9.4	0.8	6.1	25.2

Similarly, internal gains are slightly lower because the supplement corresponding to the quantity evacuated in the form of hot water is added into Equer in Wh per person and consequently depends on the occupancy rate.

For the same reason, DHW requirements per m² are lower for the PEN.

Heating loads are higher for the PEN, despite being south facing. The explanation for this is that the PEN houses are less compact, the occupancy rate is lower (and thus the associated gains), and the buildings mask each other due to their proximity. Thus, type 4, which is identical to type 3 but not masked on the south side because located at the end of the row, requires 10 kWh/m²/yr for heating compared to 18 kWh/m²/yr for type 3. Opting for density at the expense of sun here penalizes the buildings' performance.

The most efficient neighbourhood is Cité Descartes, in which buildings tend to be south facing with similar compactness to the LEN

Cooling loads are similar in size for the 3 neighbourhoods. They are "diluted" here because the housing surfaces are taken into account. When related to the surfaces of shops and offices only, they are about 3 kWh/m²/yr for the 3 neighbourhoods.

c) LCA results

We observe that the environmental impacts of the 3 neighbourhoods are similar apart from primary energy consumed and radioactive waste produced.

The radar chart below (Figure 16.19) gives an overall comparative view. We will now study several indicators in more detail.

Comment: Buildings are responsible for most of the impact. Public spaces represent 1% of water consumption for the 3 neighbourhoods. For electricity, the street and car park surfaces in Cité Descartes result in a 9% share for public areas in this neighbourhood, compared to only 2% for the LEN and PEN.

Primary energy indicators

The significance of the construction and renovation stages (see Figure 16.20) is directly related to the quantity of solar panels installed (see Table 16.19).

Concerning the use stage, the differences between the three neighbourhoods are explained by the photovoltaic solar installations. For the PEN, this stage only represents 12% of the total, compared to 87% of the total for the LEN and 74% for Cité Descartes.

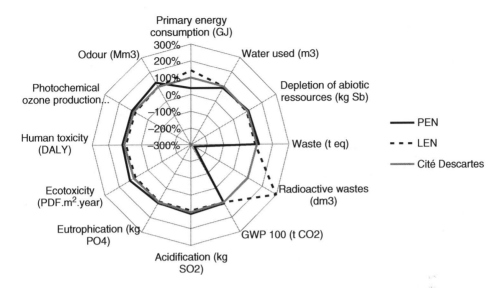

Figure 16.19 Comparison of the 3 neighbourhoods in line with the 12 Equer indicators.

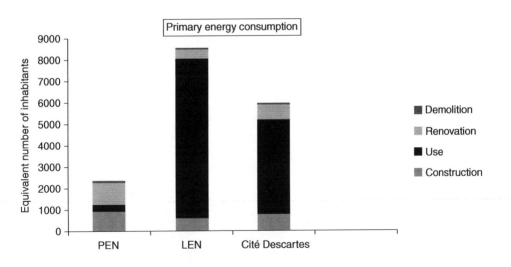

Figure 16.20 Comparison of primary energy consumption for the entire life cycle.

Primary energy consumption is $23\,kWh\ EP/m^2/yr$ for the PEN, $107\,kWh\ EP/m^2/yr$ for the LEN and $75\,kWh\ EP/m^2/yr$ for Cité Descartes.

Cogeneration and thermal solar power help bring down primary energy consumption, but in identical proportions for the 3 neighbourhoods, unlike photovoltaic solar power (see Table 16.20).

Table 16.19 Surface areas of each type of panel for the 3 neighbourhoods.

	Thermal	Photovoltaic	Total
PEN	702 m²	14,853 m²	15,556 m²
LEN	697 m²	4,039 m²	4,736 m²
Descartes	702 m²	8,681 m²	9,383 m²

Table 16.20 Progress in consumption of primary energy compared to a basic situation without cogeneration, solar thermal or photovoltaic power.

	PEN	LEN	Cité Descartes
Cogeneration	−6%	−5%	−5%
Solar thermal	−5%	−5%	−5%
Photovoltaic	−73%	−23%	−46%
Total	−85%	−33%	−56%

Contribution to the greenhouse effect

We can see (see Figure 16.21) the significance of the construction and renovation stages (66% for the PEN, 39% for the LEN and 51% for Cité Descartes).

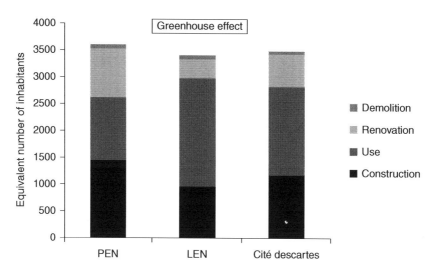

Figure 16.21 Comparison of impacts on the greenhouse effect over the entire lifecycle.

Total greenhouse gas emissions come to 6 kg CO_2 eq/m² for the PEN, 8 kg CO_2 eq/m² for the LEN and 8 kg CO_2 eq/m² for Cité Descartes (the PEN has a greater surface area).

Radioactive waste

Radioactive waste production is negative for the PEN because its photovoltaic production exceeds its consumption of electricity from the grid.

Depletion of abiotic resources

The 3 neighbourhoods have very similar total impacts but very variable respective shares for each stage (see Figure 16.22).

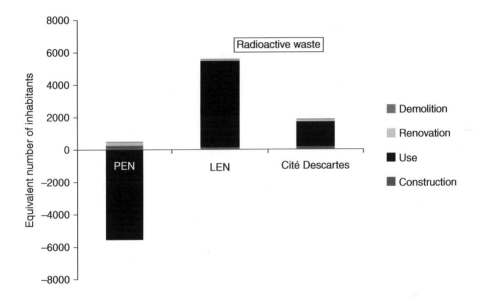

Figure 16.22 Comparison of radioactive waste production over the entire life cycle.

d) Discussion concerning the adaptation method

The method used meant that we could compare the characteristic morphologies of the 3 neighbourhoods but not the actual neighbourhoods. By adding houses to the PEN, for example, we reduced its building density because the average height of buildings was modified. This operation also sharply increased the surface area of photovoltaic modules in proportion to the number of inhabitants.

In addition, obtaining identical proportions for each use was a trial and error process, and preserving the "spirit" of the morphology could be considered as subjective.

e) Impact of waste from activities and transport

As we have seen, an eco-design approach can be used to attain the level of the best European practices. The LCA results are similar for the 3 neighbourhoods studied and

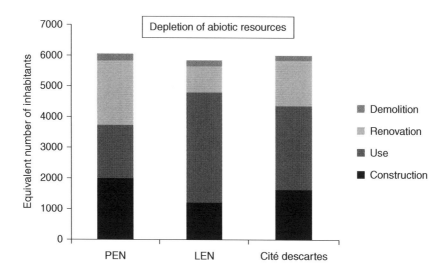

Figure 16.23 Comparison of the depletion of abiotic resources indicator over the entire life cycle.

this leads us to think that efforts in designing eco-neighbourhoods should also relate to transport facilities and the treatment of waste from activities. To give an idea of the part played by these sectors in the overall environmental impact, we considered waste (see Table 16.21) then transport with two alternatives for each.

Table 16.21 Scenarios for waste.

	Recycling		Distances in relation to site (km)			Non-recycled waste	
	Glass	*Cardboard*	*Landfill*	*Incinerator*	*Recycling*	*Incinerated*	*With energy recovery**
Good practices	80%	80%	1	10	10	80%	yes
Bad practices	20%	20%	1	10	10	50%	no

*Recovered energy substitutes natural gas with an 85% yield.

In Figure 16.24, we note a significant change in nearly all of the indicators with, for example, a 27% increase in energy consumed compared to the "no waste" option for the "good practices" option, and 48% for the "bad practices" option. For the greenhouse effect, the increase is 74% for "bad practices" and 85% for "good practices" because of the quantity of waste incinerated.

Considering transport would require a specific study, nevertheless an assessment was carried out to get a better idea of the importance represented by this aspect in the overall balance. Concerning everyday transport, we took the residence to be 1 km away from the shops, 500 m from the public transport network and 5 km from the workplace. The neighbourhood comprises bicycle lanes and for public transport we have two options: "bus" and "train". We consider that 80% of residents make a

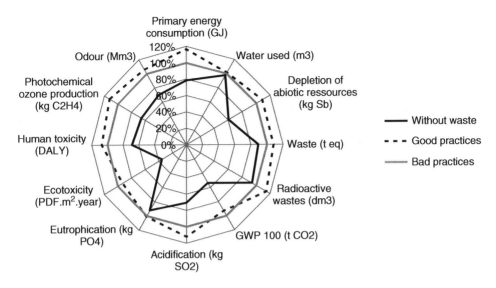

Figure 16.24 Progress of impacts following consideration of waste from activity.

home-work journey 5 times a week in public transport and a home-shop journey once a week in a car. The neighbourhood considered is Cité Descartes.

Transport thus also has a very heavy impact on the neighbourhood's LCA results (see Figure 16.25). The difference between transport by bus and by train highlights

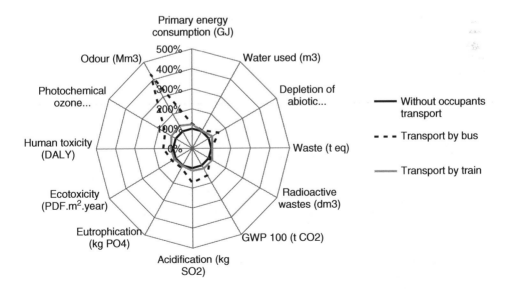

Figure 16.25 Progress of impacts following consideration of transport.

the importance of choices in terms of transport facilities and the neighbourhood's environmental quality.

The train also benefits from the French electricity mix, which mainly impacts nuclear waste.

The following graph (Figure 16.26) considers waste (bad practices) and transport (bus). The result is explicit since waste and transport represent about 60% of total greenhouse gas emissions, 40% of the energy consumed and 60% of depletion of resources. This illustrates the advantage of taking an LCA approach at neighbourhood scale at a time when thermal regulations relate exclusively to buildings' energy performance.

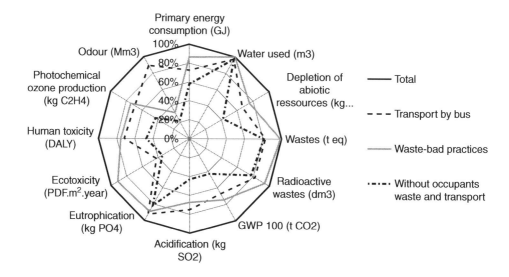

Figure 16.26 Addition of the impact of waste and transport.

CONCLUSIONS AND PERSPECTIVES

This chapter shows that a life cycle assessment can be applied at the scale of a block of buildings or small neighbourhood in order to compare different urban morphologies and different concepts, such as passive or positive energy neighbourhoods. In the case studied here, performances corresponding to best practices, i.e. those of the Vauban neighbourhood in Freiburg, were attained with the same functional units (i.e. same number of inhabitants, same climate, same use scenario). However, this comparison to references required adapting the neighbourhood in order to harmonize the functional units. A comparison of alternatives of the same programme is thus made easier and totally corresponds with the eco-design approach presented in this book.

Different avenues could be explored following this study. Sensitivity studies on uncertain parameters, e.g. the duration of the analysis period considered, could improve the reliability of the results. Implementing a dynamic LCA would refine the consideration of interactions between the neighbourhood and the background system, in particular for energy. In fact, the temporal variation of electricity consumption and production could be studied more precisely by adapting the production mix over time, or even considering a marginal mix. The influence of urban morphology, in particular the organization of parking, on the choice of modes of transport is an open subject: it would be useful to link the LCA to a transport simulation. Lastly, research could be done on improving environmental performance using low-impact materials to make even greater progress in building practices.

REFERENCES

Heinzel, M. and Voss, K. 2009: Goal: zero energy building, Exemplary Experience Based on the Solar Estate Solarsiedlung Freiburg am Schlierberg, Journal of Green Bilding, Vol. 4, n°4, 2009.

Huang et al., 2008: (CERMA Laboratory, UMR CNRS 1563, Architecture School of Nantes, Nantes, France): Towards urban design guidelines from urban morphology description and climate adaptability, PLEA 2008 – 25th Conference on Passive and Low Energy Architecture, Dublin, 22nd to 24th October 2008.

Vorger, E., Application de l'analyse de cycle de vie à la comparaison de morphologies urbaines, rapport Master de Sciences et Technologies de l'UPMC, June 2011.

Chapter 17

Study of Cité Descartes: Application of tools to consider biodiversity for urban developments

Alexandre Henry & Nathalie Frascaria-Lacoste
AgroParisTech, Laboratory of Ecology, Systemics and Evolution,
UMR CNRS/UPS/AgroParisTech, Paris Sud University, ORSAY cedex

PRESENTATION OF CITÉ DESCARTES

Depleting natural resources and climate change have made sustainable development a major challenge for many of those involved in urban planning, in particular construction companies and local authorities (Houdet, 2008; Natureparif, 2011; Natureparif, 2012). At present, biodiversity is not a priority component of urban planning, partly because those involved lack the necessary knowledge, and partly due to a dearth of tools that are easy to understand and use by non-specialists in biodiversity (Henry & Frascaria-Lacoste, submitted). New, simpler, decision-aid tools need to be developed so that professionals can take sustainable development issues into account in their practices.

Tools like this are being developed at the ParisTech-VINCI Chair on "Eco-design of Buildings and Infrastructure" and will be applied to a case study: Cité Descartes.

Cité Descartes (Figures 17.1, 17.2, 17.3 and 17.4) is located in the municipalities of Champs-sur-Marne (77) and Noisy-le-Grand (93), east of Paris, and covers 123 hectares. It forms a research and university cluster of 15,000 students. In an urban extension project, EPAMARNE (Marne-la-Vallée's public planning organization), which is responsible for developing the site for the state and local authorities, commissioned Ateliers Lion to produce a guideline plan, design public areas, and ensure operational follow-up of the constructions with the aim of producing an eco-neighbourhood. Their approach involves developing living areas, places for meeting and exchanging knowledge, and developing commercial hubs as an alternative to the region's commercial centres. This will involve restructuring the site and constructing new buildings (Ateliers Lion, 2010).

BIODI(V)STRICT

a) Presentation of the tool

The Profil-Biodiversité tool only gives a brief overview of a site's environmental characteristics, and we wanted to develop another tool to bring us closer to the state of biodiversity and permit more pertinent responses. As seen above (Henry *et al.*, submitted), the main four measures that can foster a site's ecological operation are: complementation of diverse habitats, connection between biodiversity reservoirs, a less

Figure 17.1 Aerial view of Cité Descartes (77).

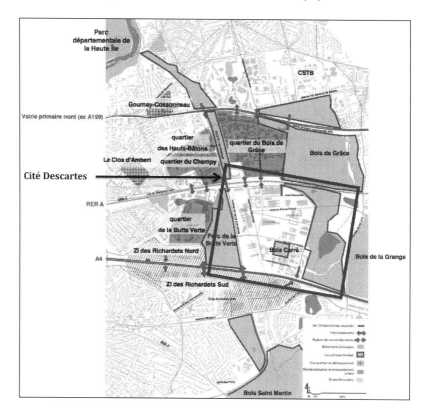

Figure 17.2 Representation of the Descartes Cluster, breaks and continuity (Source: Ateliers Lion, Sol Paysage).

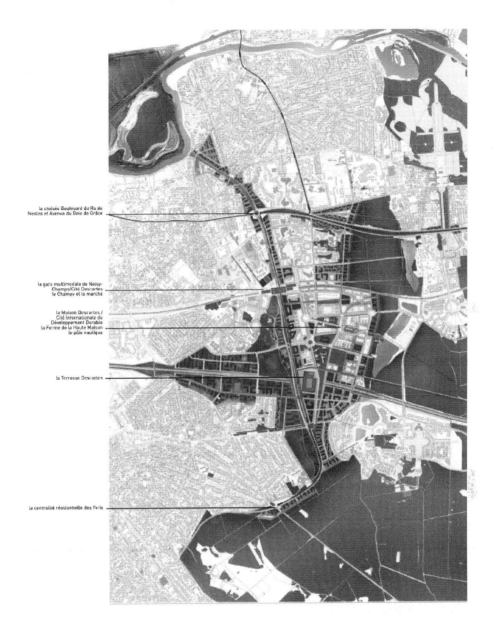

la croisée Boulevard du Ru de
Nesles et Avenue du Bois de Grâce

la gare multimodale de Noisy-
Champs/Cité Descartes
le Champy et le marché

la Maison Descartes /
Cité Internationale du
Développement Durable
la Ferme de la Haute Maison
le pôle nautique

la Terrasse Descartes

la centralité résidentielle des Yvris

Figure 17.3 Programme plan of Cité Descartes. Existing buildings are shown in grey. Planned buildings are shown in red, blue and purple (Source: Ateliers Lion).

anthropocentric approach to nature, and adaptive management of green spaces. We wanted this new tool to be able to consider these elements.

We took inspiration from a method developed by Hermy and Cornelis (2000), who proposed a tool to generally monitor biodiversity in urban and peri-urban parks,

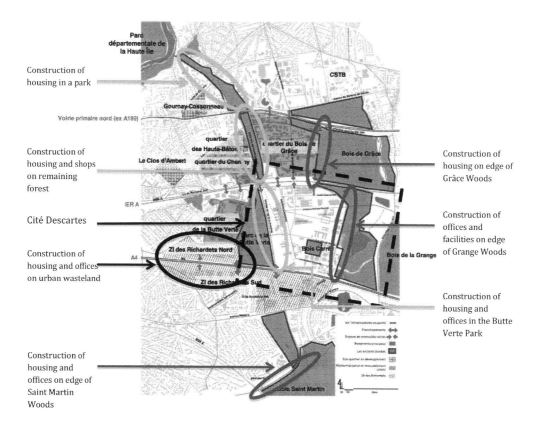

Figure 17.4 Representation of Descartes Cluster and zones impacted by the construction project (based on Ateliers Lion & Sol Paysage). Green circles: green areas to be built on; red circles: forest borders to be built on; black circle: urban wasteland to be developed.

taking into account the complexity of habitats and species. This tool is based on: (1) the diversity of habitats and (2) the diversity of species. Its vocation is also to provide a fast and inexpensive aid to urban planning that can give an insight into the site's putative ecological functioning.

– Presentation of habitat diversity

A positive correlation exists between habitat diversity and specific diversity (Tews *et al.*, 2004), which shows the advantage of using indicators on habitats in the first part of the tool. However, defining habitats often requires long and precise local investigations. We intentionally simplified these habitats by assimilating them to a series of "elements" measurable on site. To do this, we modified the list established by Hermy and Cornelis (2000), with 59 elements classed into 3 categories (Annex 1):

- *Surfacic elements*: these are elements with a surface area above $100\,\text{m}^2$ and a length/width ratio over 10. These figures were arbitrarily chosen so as to class into this category elements with a large surface area. These could be e.g. units of forest vegetation, buildings, car parks, orchards or meadows. Each element is represented by its surface area expressed in square metres.

- *Linear elements*: these are elements with a length/width ratio under 10. This figure was chosen arbitrarily so as to class elements with a long length and a short width in this category. Examples are rows of trees, roads, roadsides, rivers and riverbanks. Each element is represented by its length in metres.
- *Isolated elements*: these elements have a surface area of less than 100 m². This figure was chosen arbitrarily so that elements with a small surface area would be classed in this category. Examples are isolated trees and bushes, small ponds and fountains. Each element is characterized by its quantity.

Another distinction is made between "green elements" and "grey elements". Examples of "grey elements" are roads, paths, car parks and buildings. "Green elements" are what are commonly called green spaces, like parks, forests and gardens, and also include aquatic zones like rivers and ponds (Annex 1).

- **Presentation of species diversity**

The second part of this indicator is based on a measure of specific diversity. A comprehensive inventory of fauna and flora is not obligatory for this methodology, which must remain simple, fast and inexpensive for the planner. Focusing on groups of species indicating the state of the ecosystem struck us as an effective approach in this context. The only constraint is that data on these specific groups must be easy to obtain, and species must be easy to recognize. Hermy and Cornelis (2000) chose four groups of species: plants, butterflies, amphibians and nesting birds. Scientists' long-term interest in these species has led to significant knowledge of their biology and distribution, and local associations and participative sciences have completed this information. Several programmes have been set up in recent years, in particular Vigie-Nature (http://vigienature.mnhn.fr), with its common bird monitoring scheme (STOC), butterfly monitoring scheme (STERF), photographic monitoring of pollinating insects (SPIPOLL), monitoring of amphibian populations (POPAMPHIBIEN), and Vigie-flore flora surveillance project. These programmes rely on voluntary specialists and non-specialists to build up databases and study the evolution of the different species or groups of species targeted. In this method, to consider specific diversity we will therefore focus on four groups indicating ecosystem quality:

- *Vascular plants*: these are the basis of a site's architecture. An initial appreciation of the environment's quality will consider the presence of invasive species or protected species, as well as other species sensitive to air or land pollution.
- *Butterflies*: these are important pollinators and central to trophic networks. Their presence depends on plants to feed their larvae and nectar-rich plants for imagos. They are interesting indicators because they have fairly short generation times and lifecycles, so that they rapidly react to environmental changes (Heikkinen, 2010). Some species are studied to assess the state of forests (such as the Silver-washed Fritillary, *Argynnis paphia*), wetlands (like the Large Copper, *Lycaena dispar*) and other milieus. They are indicators of ecosystem quality and climate change (Bergerot *et al.*, 2012).

- *Amphibians*: they are sensitive to perturbations and modifications of the environment. Some species (like the Green Frog, *Rana esculenta)* reproduce and develop in the same sites, but others (like the Common Frog, *Rana temporaria)* live partially on the ground and must therefore migrate to reach aquatic breeding sites. They indicate water quality, pollution, eutrophication of lentic environments, climate variations and landscape connectivity (Noos *et al.*, 1992).
- *Nesting birds*: they are often at the top of trophic networks, and their presence depends on several factors, such as the site's tranquillity, the structure and age of trees, the management mode and heterogeneity of habitats. They indicate the overall evolution of species and environments and the diversity of habitats.

b) Methodology

– **Habitat diversity**

We aim to appraise the diversity of habitats and their distribution in Cité Descartes. Starting with a satellite photograph of the site taken from Google Earth, the surfacic, linear and isolated habitats (full list in annex) are mapped using a GIS (Geographic Information System). For this study, we used Quantum GIS (http://www.qgis.org), a free, multi-platform software programme. We chose this method to reduce the cost of using the tool.

We calculate the surface area (square metres) of each surfacic element, the length (metres) of each linear element, and the number of each isolated element.

The diversity of habitat units is then calculated using the Shannon-Wiener diversity index (H) (Shannon, 1948):

$$H = -\sum_{i=1}^{s} \frac{n_i}{N} \ln \frac{n_i}{N}$$

i: habitat unit i
s: number of habitat units
n_i: surface, length or quantity of habitat unit i
N: total surface, length or quantity on the site

This index can be used to express diversity by considering the number of habitat units and the abundance of elements within each of the habitat units. Thus, a site dominated by a single habitat unit will have a lower coefficient than a site where all habitat units dominate jointly. The index value varies from 0 (a single habitat unit or a habitat unit that greatly dominates all the others) to ln(s) (where all habitat units have the same abundance). This index is interesting because it can be used to quantify the heterogeneity of habitats.

Eight diversity indices (H) are calculated: (1) total surfacic elements, (2) total linear elements, (3) total isolated elements, (4) total elements, (5) "green" surfacic elements, (6) "green" linear elements, (7) "green" isolated elements, and (8) "green" elements.

A site's diversity index cannot be clearly interpreted if it is not compared to other sites or if the threshold values have not been previously defined. Since we have not yet measured this index on other sites, we chose to attach the value of H to a saturation index, or the Piélou equitability index, for each of the categories studied (Piélou, 1966). This index will help us translate the distribution of habitat abundances on the site (Peet, 1974). It is useful for comparing the dominance of habitats between sites. We will only apply it to our site in an initial stage. Its value should obviously also be compared to other sites.

This saturation index will be presented as a percentage and will initially allow us to determine whether the site is as diversified as it could be, i.e. composed of a maximum possible number of different habitat elements. This involves calculating a maximum diversity index (H_{max}) for each of the 6 categories. It is calculated from the list of potential habitats (Annex 1):

$$H_{max} = - \ln \frac{1}{s_{max}} = \ln s_{max}$$

S_{max}: total number of different habitat units

We can then calculate the saturation index (percentage):

$$S = \frac{H}{H_{max}} \times 100$$

For the "green" elements, we calculate a first saturation index (S1) using the maximum diversity (H_{max}) of green elements in order to appraise the intrinsic saturation of these elements. Then a second saturation index (S2), using the maximum diversity of total elements to appraise the saturation of green elements in relation to all of the site's elements.

This index varies from 0 (dominance of an entity) to 100 (equal distribution of entities).

The total saturation index (S_t):

$$S_t = \frac{S_{su} n_{su} + S_{li} n_{li} + S_{po} n_{po}}{n_t}$$

S_{su}: index of saturation of surfacic elements
N_{su}: number of surfacic elements
S_{li}: index of saturation of linear elements
n_{li}: number of linear elements
S_{po}: index of saturation of isolated elements
n_{po}: number of isolated elements
n_t: total number of habitat units

We must consider the diversity index (H) and the saturation index (S) to appreciate an environment's state, respectively indicating the heterogeneity of habitats and their equitability on the site.

– **Diversity of specific groups**

For fauna and flora inventories, several databases are available to the public, from e.g. national botanic protection boards, the national inventory of natural heritage (http://inpn.mnhn.fr) and data collected by local naturalist associations.

- *Vascular plants*

  We used the database available on the website of the Conservatoire Botanique National du Bassin Parisien (http://cbnbp.mnhn.fr). Most of Cité Descartes is located in the municipal area of Champs-sur-Marne, and so we decided to use data from this town. Based on this list of species, we consider the presence of protected species, invasive species and species that indicate the environment's quality.

- *Butterflies, Amphibians and Nesting Birds*

  For groups of animal species, we chose to use data taken from a report produced by Ecosphère in 2010 and commissioned by the Syndicat d'Agglomération Nouvelle (SAN – new town association) of Val Maubuée, which wanted a study of the natural heritage on its territory. The study was carried out over the territory comprising the six towns that make up the SAN of Val Maubuée: Champs-sur-Marne, Croissy-Beaubourg, Emerainville, Lognes, Noisiel and Torcy. Cité Descartes is located on this territory. We used the data from this inventory. The fauna explorations ran from July 2008 to March 2009. The data used are not only for Cité Descartes, but also for Val Maubuée.

We calculated 3 saturation indexes (butterflies, amphibians and nesting birds). This involves the number of species actually present on the site in relation to the number of species potentially present. For species potentially present throughout the site, we chose to consider species present in the *département* (administrative area, county).

$$S = \frac{\textit{Number of species observed on the site}}{\textit{Number of species observed in the } \text{département}} \times 100$$

We used the number of species present in the *département* of Seine-et-Marne: for amphibians and birds, these data are taken from the INPN website; for butterflies from the website http://www.lepinet.fr.

APPLICATION TO CITÉ DESCARTES

a) Habitat diversity at Cité Descartes

After geo-referencing a satellite photo of Cité Descartes (Figure 17.5) taken from Google Earth, we placed the surfacic elements (Figure 17.6), the linear elements (Figure 17.7) and the isolated elements (Figure 17.8).

The surfacic, linear and isolated elements present at Cité Descartes were listed (Table 17.1).

Using these data, we were able to calculate the following indicators (Table 17.2):

Out of the 59 possible habitat elements (Annex 1), we listed 29. Taking into account the proportions of each type of element, we calculated a diversity index of 2.01. The linear and surfacic elements have saturation indices of 68.1% and 61.5%, and the saturation index of the isolated elements is 42.3%.

Figure 17.5 Satellite photo of Cité Descartes (source: Google Earth).

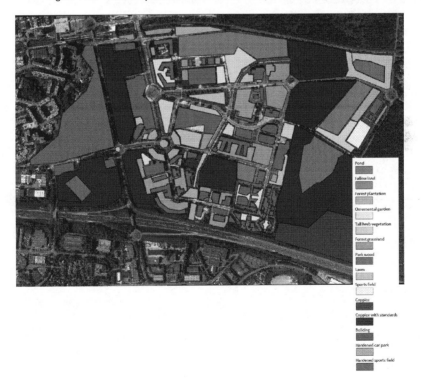

Figure 17.6 Surfacic elements at Cité Descartes.

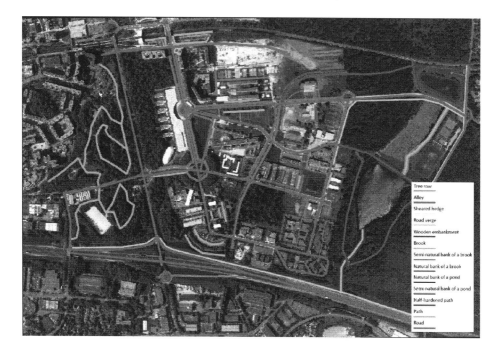

Legend:
- Tree row
- Alley
- Sheared hedge
- Road verge
- Wooden embankment
- Brook
- Semi-natural bank of a brook
- Natural bank of a brook
- Natural bank of a pond
- Semi-natural bank of a pond
- Half-hardened path
- Path
- Road

Figure 17.7 Linear elements at Cité Descartes.

Legend:
- Pool
- Single tree

Figure 17.8 Isolated elements at Cité Descartes.

Table 17.1 Habitat elements at Cité Descartes.

Surfacic elements	Surface (m²)
Pond	46342
Fallow land	189293
Tree plantation	30303
Ornamental gardens	19284
Tall herbaceous plants	39147
Meadow with trees	26400
Parkland	94392
Lawns	51596
Grass sports field	8661
Coppice	61234
Coppice with standards	358121
Building	150710
Car park	61035
Waterproof sports ground	13114

Linear elements	Length (m)
Tree row	1545
Double tree row	3325
Clipped hedge	489
Non-asphalted roadside	3449
Planted road bank	1123
Stream	643
Semi-natural riverbank	708
Natural riverbank	594
Semi-natural water body bank	625
Natural water body bank	645
Semi-permeable path	2329
Path	2504
Roads	11508

Isolated elements	Quantity
Small ponds	8
Isolated trees	6

Table 17.2 Indicators of habitat diversity at Cité Descartes.

	Number of categories	Habitat diversity (H)	Saturation index 1 (S1)	Saturation index 2 (S2)
Surfacic elements (max = 34)	14	2.17	61.5%	
Linear elements (max = 20)	13	2.04	68.1%	
Isolated elements (max = 5)	2	0.68	42.3%	
Total (max = 59)	29	2.01	63.1%	
Green surfacic elements (max = 30)	11	1.89	55.6%	53.6%
Green linear elements (max = 14)	10	2.02	76.5%	67.4%
Green isolated elements (max = 3)	2	0.68	61.9%	42.3%
Total (max = 47)	23	1.84	65.2%	51.7%

Concerning green habitats, i.e. natural and semi-natural, we listed 23 out of a possible 45. The diversity index is 1.84. Compared to the maximum diversity of green elements, the saturation indices (S1) of isolated and linear green elements are 61.9% and 76.5%, whereas the green surfacic elements have a saturation index of 56%. In comparison with the total maximum diversity, the saturation indices (S2) of the green surfacic, linear and isolated elements are 53.6%, 67.4% and 42.3%.

b) Diversity of species at Cité Descartes

- **Plants**

216 species were observed on the territory of Champs-sur-Marne. They feature 5 protected species:

- Narrow-leaved bittercress (*Cardamine impatiens* L.)
- Green hellebore (*Helleborus viridis* L.)
- Hard Shield Fern (*Polystichum aculeatum* (L.) Roth)
- Broad-leaved Helleborine (*Epipactis helleborine* (L.) Crantz)
- Common Twayblade (*Listera ovata* (L.) R.Br.)

Six invasive species were also observed:

- Panicled Aster (*Aster lanceolatus* (Willd.) G.L.Nesom)
- Butterfly Bush (*Buddleja davidii* Franch.)
- Small Balsam (*Impatiens parviflora* DC.)
- Black Locust (*Robinia pseudoacacia* L.)
- Narrow-leaved Ragwort (*Senecio inaequidens* DC.)
- Canada Goldenrod (*Solidago canadensis* L.)

Based on the list used (http://cbnbp.mnhn.fr), we note the presence of species that indicate hydromorphic soils, such as five species of rush (*Juncus bufonius, J. conglomeratus, J. effusus, J. inflexus* and *J. tenuis*), one species of reed (*Calamgrostis epigejos*) and two species of cattail (*Typha angustifolia* and *T. latifolia*), as well as the presence of Yorkshire Fog (*Holcus lanatus*) which indicates hay meadows.

- **Butterflies, Amphibians and Nesting birds**

In Val Maubuée, 41 species of butterfly were identified (Annex 2) out of 117 species present in Val-de-Marne (77), which is a saturation index of 35% (Table 17.3). Among these, 7 are ZNIEFF (protected area) determinants, and 2 are protected at regional level: the Poplar Admiral (*Limenitis populi* L.) and the Large Tortoiseshell (*Nymphalis polychloros* L.)

Twelve species of amphibian were listed on the territory (Annex 3) out of a possible 18 species in the *département*, i.e. a saturation index of 72% (Table 17.3). Among these, 11 are protected, of which one, the Crested Newt (*Triton cristatus* Laurenti), is listed in Annex II of the "Habitats" directive and two are ZNIEFF determinants: the European Tree Frog (*Hyla arborea* L.) and the Natterjack Toad (*Epidalea calamita* Laurenti).

Table 17.3 Indicators of diversity of fauna species at Cité Descartes.

Species group	Number of species in département	Number of species observed locally	Saturation index
Butterflies	117	41	35.0%
Amphibians	18	13	72.2%
Nesting birds	248	96	38.7%

Of the 248 species of nesting birds identified in Val-de-Marne, 96 were observed on the territory studied (Annex 4), i.e. a saturation index of 39% (Table 17.3). 18 of these species are ZNIEFF determinants in Ile-de-France and/or listed in Annex I of the "Birds" directive.

DISCUSSION – CONCLUSION

a) Overview of Cité Descartes

● **Current state**

Considering the total number of types of element and their respective proportions (surface area, length or number), the site's overall diversity index (2.01) is higher than the diversity index of green spaces (1.84). This biodiversity index is particularly useful for comparing several sites. Here, because we have only analyzed one site, we cannot draw any significant conclusions. The overall saturation (63.1%) shows that the elements of the site ("green" or "grey") are fairly well represented but perfectible. If we look at the diversity of "green" elements, we listed 23 types of element out of a possible 47. Thus, the saturation index intrinsic to green habitats (S1) is 65.2%. We might deduce that green spaces are fairly satisfactorily represented. However, when comparing with the site's total elements, we find a saturation of 51.7%, which is a fairly average value. More specifically, the "green" linear elements have a fairly satisfying saturation rate (67.4%), whereas the "green" surfacic elements have an average rate of 53.6% and the "green" isolated elements have a lower rate (42.3%).

These results might simply lead us to conclude that the grey habitat element, i.e. impermeable built surfaces that do not participate in the functioning of the ecosystem, represent a significant portion of the diversity of the site's habitats.

At the same time, if we choose to set a 50% average threshold for the animal species saturation index, then the amphibian saturation index (72.2%) shows that the quality of the aquatic sites is fairly good (Table 17.3). However, the rather low saturation indices of butterflies (35%) and nesting birds (38.7%) (Table 17.3) point to a potential dysfunction in the ecosystem for three main reasons. (1) Looking at the habitat diversity results, we could say that the rather poor saturations of birds and butterflies could be due to both the low diversity of green spaces making up the site, and the very average representativeness of each of the areas. (2) The poor results for a number of butterflies and birds could probably also be due to the site's excessive fragmentation due to transport facilities and buildings. (3) The site's environmental management is undoubtedly intensive, meaning that green spaces are mown too frequently so that plant species cannot terminate their life cycle (flowering and seed production), which has an impact on pollinators like butterflies.

Regarding plant diversity, we did not have access to a more exhaustive inventory to match the protocol defined by Hermy and Cornelis (2000). We were therefore unable to calculate the plant diversity index. We chose instead to focus on indicator species, in particular connected to their specific features (Godefroid & Koedam, 2003). Some plant species can be characteristic of a type of soil or farming practice (hydromorphic soils or hayfields). We also looked at invasive species and protected species.

Invasive species have different impacts depending on the environment they develop in. They can occupy a different ecological niche from local species and thus coexist with little interaction, but nevertheless they can have a considerable impact. Alternatively, they can occupy the same ecological niche and be dominated or dominant (MacDougall, 2009). Invasive species indicate a potentially perturbed ecosystem. They move into natural environments deteriorated by human activities. They often proliferate because the environment has been altered (destruction of riverbeds, water pollution, etc.). The presence of several invasive species on our study site reflects this alteration. The evolution of these populations will need to be monitored to judge whether they present a danger to the ecosystem and whether specific management measures should be taken, such as elimination by different means.

On the contrary, the presence of protected species will constrain planners to maintain the site or part of the site in its original state to obey regulations. However, this presence shows that the ecosystem has significant heritage value, but that measures should be taken to preserve these populations in line with regulations (Article L411-1 of the French Environment Code). Since the recent Article 47 of the Grenelle 2 Act (29 June 2010), attempts to encroach on protected areas are punishable.

- **Potential state**

The site's current state is fairly poor in terms of diversity of habitat and species. In addition, in their proposed planning project for Cité Descartes, Ateliers Lion aim to construct new buildings in locations that are important for biodiversity (e.g. the Butte Verte and Grâce Woods). Thus, the site's state could deteriorate further. In the face of this observation, planning proposals could emerge aimed at improving the neighbourhood's biodiversity. Given the site's biotic and abiotic characteristics, we suggest making modifications to the urban development plan and its management to move towards a putative ideal state.

The habitat diversity index considers both the number of element types and their proportion. Thus, to optimize this index, potential options include creating new areas on the site, like an orchard, arboretum, community garden and green roofs.

Differentiated management measures can also be applied to make a neighbourhood or town into an environment that encourages biodiversity, and practices can aim to respect and preserve natural environments. In addition to ensuring the site's security and attractiveness, differentiated management can implement alternative, non-polluting techniques that do not endanger health. These techniques include mechanical and thermal weed control; using large grassy surfaces for grazing cattle, horses and sheep; adapting maintenance periods to the fauna and flora present, and leaving refuge

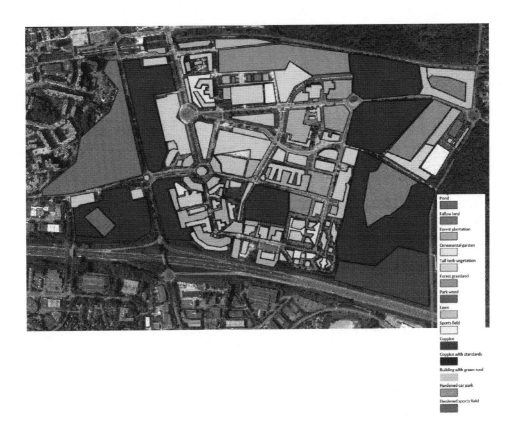

The legend reads:
Pond
Fallow land
Forest plantation
Ornamental garden
Tall herb vegetation
Forest grassland
Park wood
Lawn
Sports field
Coppice
Coppice with standards
Building with green roof
Hardened car park
Hardened sports field

Figure 17.9 Cité Descartes with green roofs on all of the buildings.

zones. These different techniques are aimed at rational environmental management of the site, restoring, preserving and managing biodiversity, and improving the quality of life and usage of the site by diversifying landscape qualities. Urban planners will find useful advice in the *Guide de Gestion Différenciée à l'Usage des Collectivités* produced in 2009 by Natureparif and the Loing Valley and Fontainebleau Massif Naturalist Association (ANVL).

To make up for a lack of connectivity within a neighbourhood, one solution is to install green roofs and walls on existing buildings. The presence of diversified green roofs added to existing green areas could reinforce local green and blue corridors (Henry & Frascaria-Lacoste, 2012). Installing intensive green roofs (Figure 17.9) planted with local species could improve the site's ecological functioning, in particular by reducing its fragmentation. We have calcuated the potential gain of this kind of action (Table 17.4). The diversity index of green elements would increase from 1.84 to 1.91, and the saturation index (S2) from 51.7% to 55.4%. This progress shows that these indices can be appreciably increased by modifying a single type of element. This implies that the potential for improving the site could also be significant.

Table 17.4 Diversity indicators of habitats in Cité Descartes with green roofs installed on all buildings.

	Number of categories	Habitat diversity (H)	Saturation index 1	Saturation index 2
Surfacic elements (max = 34)	14	2.17	61.5%	
Linear elements (max = 20)	13	2.04	68.1%	
Isolated elements (max = 5)	2	0.68	42.3%	
Total (max = 59)	29	2.01	63.1%	
Green surfacic elements (max = 30)	12	2.01	59.7%	57.6%
Green linear elements (max = 14)	10	2.02	76.5%	67.4%
Green isolated elements (max = 3)	2	0.68	61.9%	42.3%
Total (max = 47)	24	1.91	66.9%	55.4%

b) Discussion on the tool

The general idea of developing a new tool was to supply building professionals with the incentive of a simple, fast, inexpensive method to make an initial characterization of biodiversity and its functioning on a site, and to prompt the frequently overlooked issue of nature in the town. We have suggested using bioindicator species whose biology and diversity are well known and for which data are easily available. Other bioindicators obviously exist, but we chose to limit their number in this tool to ensure that it remains easy to use because its primary aim is to act as a decision aid tool for planners, which rules out exhaustive data at this early stage. The list of surfacic, linear and isolated elements could be expanded following future applications in other geographic regions where plant formations might be different.

By integrating these elements, our tool can be used to consider four main measures that should foster the improved ecological functioning of a site (Henry *et al.*, submitted): (1) the complementation of habitats, evaluated by the indices of the site's habitat diversity, which depend on the number of habitat types and their proportions; (2) the connection between biodiversity reservoirs can be evaluated from the indices of the saturation of bioindicator animal species; (3) a less anthropocentric approach to nature, represented by the diversity and heterogeneity of habitats chosen with a concern for their actual representativeness rather than just human wellbeing (greenwashing); and (4) the adaptive management of green areas could involve a broad choice of potential habitats sometimes defined according to their management mode (see Annex 1 for these modes) and by the presence of endangered or invasive species that would restrict managers' practices. The tool can be used not only to appraise biodiversity, but as a pedagogical tool so that urban planners can understand ecological functioning when developing a territory.

One of the tool's limitations relates to the classification of habitat elements. We could take the distinction between "grey" and "green" elements further. This is because not all of the elements that we have listed as "green" have the same potential and role in the ecosystem's functioning. A frequently mown lawn should not be considered in the same way as a vegetable garden or a mixed forest. The results on the diversity and saturation of habitats should therefore be interpreted with caution. The tool's validity should involve comparing numerous sites to calibrate it.

Classifying the ecological state based on habitat diversity, and choosing the acceptable saturation threshold for animal species are both crucial to using the tool efficiently. To compare a site's diversity with that of a much larger geographic zone, we could have chosen other scales than the *département*, such as local authority associations or regions. *Départements* struck us as the most suitable choice because they cover reasonably sized territories compared to the surface of a neighbourhood or town. A distortion can nevertheless appear in *départements* comprising very diverse ecosystems, like the Alpes Maritimes (06), which has a highly contrasting topography ranging from the coastline to high altitudes in the Alps (over 3000 metres). In this type of *département*, we expect to find very high specific diversity, so that fixing the average level of saturation indices at 50% is no longer suitable. Thus, we need to calibrate the saturation calculations, either by considering broad specific diversities, or by changing the scale, or by modifying the saturation acceptability threshold.

Our methodology does not make sufficient use of plant diversity. It would be interesting to use the properties of some indicator species, such as their intolerance of types of pollution, to improve the appraisal of the site's quality.

The tool gives rapid results thanks to the simplicity of the indicators used. Site mappings can be produced from photographs available on the Internet using free GIS software. Internet databases can be used for the fauna and flora inventories. However, these databases are still incomplete for some geographic zones due to the recent set-up of participative sciences and the heterogeneous numbers of observers depending on the region. The more observers there are, and so observations, the more exhaustive the information will be. In the absence of available data, an ecologist will need to make an inventory. Given the specific groups studied, the inventory could be produced fairly quickly.

The tool that we have developed could be used at the start of a project as a decision aid tool for urban development, to produce a site characterization in order to determine whether it is sufficiently diverse in terms of habitat and species; and if not, development leads could be given as guidelines for the project. The tool could also be used to monitor biodiversity during the worksite phases and throughout the neighbourhood's operation in order to follow up on the development's efficiency and the evolution of the presence of indicator species. The neighbourhood's residents could participate in monitoring specific groups in a participative science approach. Management modes could be modified if indicators showed that the level of biodiversity was still too low.

A more precise consideration of the site's level of fragmentation could be made using GIS. Lastly, a cross-reference chart of results between habitats and species should be produced to make it easier to use and draw conclusions.

CONCLUSION

To consider biodiversity in urban developments, managers like architects and urban planners lack not just awareness of urban biodiversity, but also appropriate tools. We started this chapter by presenting the Profil-Biodiversité tool developed by Frank

Derrien. Its advantage is that it comprises fairly varied indicators based on the main threats to biodiversity. However, the rating grid for indicators is fairly arbitrary and the tool makes a very indirect and approximate report of the ecosystem's functioning, with measurements remaining partly subjective. We therefore developed a new tool, BioDi(v)Strict, based on measuring the diversity of habitats in connection with four groups of indicator species. This allows us to obtain an initial idea of the ecosystem's state, mainly considering the heterogeneity of habitats and their representativeness on the site in terms of fauna/flora indicators. The tool was devised to be fast and easy to use by non-specialists, and inexpensive. This first approach is a means to produce a rapid overview of a site's biodiversity and then monitor it. Spatial planners who have not yet included biodiversity in their specifications will find it an important diagnosis tool.

When we applied these two tools to Cité Descartes, we were able to show that the site's current ecological state is fairly mediocre. Despite the fact that Ateliers Lion's project is referred to as a future "eco-neighbourhood", the proposed perspectives do not move towards improved consideration of biodiversity and bring the risk that the site's quality may deteriorate further. This new example confirms our conclusions on eco-neighbourhoods, produced in the first part of our thesis, which show that consideration of biodiversity in eco-neighbourhoods focuses on the aesthetic aspect and rarely translates into functionality and sustainability.

REFERENCES

Ateliers Lion. 2010. Cité Descartes. Cœur du Cluster Descartes. EPAMARNE – Avril 2010 – Concours de maîtrise d'œuvre urbaine pour l'aménagement de la Cité Descartes sur les communes de Champs-sur-Marne et Noisy-le-Grand.

Bergerot B, Merckx T, Van Dyck H, Baguette M. 2012. Habitat fragmentation impacts mobility in a common and widespread woodland butterfly: do sexes respond differently? BioMed Central Ecology. 12: 5.

Ecosphère. Diagnostic écologique du territoire du SAN Val Maubuée – Tome1: Etudes. Avril 2010.

Godefroid S, Koedam N. 2003. Identifying indicator plant species of habitat quality and invasibility as a guide for peri-urban forest management. Biodiversity and Conservation. 12: 1699–1713.

Heikkinen RK, Luoto M, Leikola N, Pöyry J, Settele J, Kudrna O, Marmion M, Fronzek S, Thuiller W. 2010. Assessing the vulnerability of European butterflies to climate change using multiple criteria. Biodiversity and Conservation. 19: 695–723.

Henry A, Frascaria-Lacoste N. 2012. The green roof dilemma – Discussion of Francis and Lorimer (2011). Journal of Environmental Management. 104: 91–92.

Henry A, Roger-Estrade J, Frascaria-Lacoste N. The eco-district concept: effective for promoting urban biodiversity? soumis à Cities.

Henry A, Frascaria-Lacoste N. Biodiversity in decision-making for urban planning: Need for new improved tools. soumis à Journal of Urban Planning and Development.

Hermy M, Cornelis J, 2000. Towards a monitoring method and a number of multifaceted and hierarchical biodiversity indicators for urban and suburban parks. Landscape and Urban Planning. 49: 149–162.

Houdet J. 2008. Intégrer la Biodiversité dans les stratégies des entreprises : Le Bilan Biodiversité des organisations. FRB. Paris: Orée. 393p.

MacDougall AS, Gilbert B, Levine JM. 2009. Plant invasions and the niche. Journal of Ecology. 97: 609–615.

Natureparif. 2011. Entreprises, relevez le défi de la biodiversité. Un guide collectif à l'usage des acteurs du monde économique. Victoires éditions. 142p.

Natureparif. 2012. Bâtir en favorisant la biodiversité. Un guide collectif à l'usage des professionnels publics et privés de la filière du bâtiment. Victoires éditions. 205p.

Noos FR, Cline SP, Csuti B, Scott JM. 1992. Monitoring and assessing biodiversity. In: Lykke E (Ed), Achieving Environmental Goals, the Concept of Pratice of Environmental Performance Review. Belhaven Press, London, pp. 67–85.

Peet RK. 1974. The measurement of species diversity. Annual Review of Ecology and Systematics. 5: 285–307.

Piélou EC. 1966. Measurement of diversity in different types of biological collections. Journal of Theoretical Biology. 13: 131–144.

Shannon CE. 1948. A mathematical theory of communication. The Bell System Technical Journal. 27: 379–423 and 623–656.

Tews J, Brose U, Grimm V, Tielbörger K, Wichmann MC, Schwager M, Jeltsch F. (2004), Animal species diversity driven by habitat heterogeneity/diversity: the importance of keystone structures. Journal of Biogeography. 31:79–92. doi:10.1046/j.0305-0270.2003.00994.x

ANNEXES

Annex 1: List of habitats. (in *italics*: "green element"; in regular font: "grey element")

1. Surfacic elements

1.1. Forest stand: unit composed of more or less natural forest vegetation

1.1.1. Deciduous forest: forest in which most of the trees are deciduous

1.1.1.1. Coppice: forest in which trees are regularly cut down to stumps for new growth (1)

1.1.1.2. Coppice with standards: coppiced forest including standard trees (2)

1.1.1.3. Parkland: forest of tall herbaceous vegetation, scattered large trees and scrub (3)

1.1.1.4. Regular high deciduous forest: forest stand of regular high deciduous trees (4)

1.1.2. Coniferous forest: forest stand of conifers (5)

1.1.3. Mixed forest: forest comprising deciduous and coniferous trees (6)

1.2. Plantation: unit comprising planted trees

1.2.1. Orchard: enclosed unit planted with fruit trees (7)

1.2.2. Tree-lined meadow: meadow planted with forest trees (8)

1.2.3. Tree gallery: linear plantation of trees with no undergrowth (9)

1.2.4. Arboretum: plantation of different species of tree for exhibition or study (10)

1.2.5. Plantation forest: plantation of forest stands (<3 m) (11)

1.3. Labyrinth: unit composed of dense hedges in the form of a labyrinth (12)

1.4. Shrub plantation: unit composed of shrubs (13)

1.5. Meadow: unit composed of herbaceous species

1.5.1. Grassland: meadow usually cut (14)

1.5.2. Sports field: field usually mown used as a sports ground (15)

1.5.3. Hayfield: meadow used for hay (16)

1.5.4. Pastureland: meadow for grazing animals (17)
1.5.5. Grazing hayfield: meadow used for grazing after hay harvest (18)
1.6. Tall herbaceous vegetation: unit composed of wild grasses including reeds (19)
1.7. Heather moorland: unit composed of heather (20)
1.8. Agricultural zone: unit composed of arable crops (21)
1.9. Fallow: temporary unit composed of fallow land (22)
1.10. Garden: closed unit composed of fruit, vegetables or ornamental plants
1.10.1. Vegetable garden: garden composed of fruit and vegetables (23)
1.10.2. Herb garden: garden composed of medicinal plants (24)
1.10.3. Rose garden: garden composed of roses (25)
1.10.4. Ornamental garden: garden composed of other ornamental plants (26)
1.11. Ornamental plantation: open unit composed of ornamental plants (27)
1.12. Ornamental pond: unit composed of water
1.12.1. Moat: aquatic element surrounding a historic building (28)
1.12.2. Pond: unit of water without buildings (29)
1.13. Building: unit composed of buildings, including the limited space between buildings (30)
1.14. Car park: unit composed of parking places for vehicles
1.14.1. Semi-permeable: car park with a surface that is not totally waterproof (31)
1.14.2. Permeable: car park without a waterproof surface (32)
1.15. Semi-permeable sports ground (tennis courts, hard stadiums, etc.) (33)
1.16. Green roof (34)

2.   Linear elements

2.1. Alley: double row of trees, including verges (35)
2.2. Row of trees: row of trees (36)
2.3. Hedge: linear woody vegetation
2.3.1. Trimmed hedge: regularly trimmed hedge (37)
2.3.2. Untrimmed hedge: hedge that is not regularly trimmed (38)
2.3.3. Planted bank: hedge on an artificial bank (39)
2.4. Roadside: non-bituminous band along the edge of a road (40)
2.5. Bank: band of earth along the edge of a water body or watercourse
2.5.1. Water body bank: bank of a moat or pond
2.5.1.1. Natural: bank without manmade consolidation (41)
2.5.1.2. Semi-natural: bank with manmade consolidation where vegetation is still possible (42)
2.5.2. Watercourse bank: bank along a trench, stream or river?
2.5.2.1. Natural: bank without manmade consolidation (43)
2.5.2.2. Semi-natural: bank with manmade consolidation where vegetation is still possible (44)
2.6. Watercourse: linear element used to evacuate water
2.6.1. Swale: ditch with a max. width of 1 m that can contain water (45)
2.6.2. Stream: watercourse with a max. width of 3 m that always contains water (46)
2.6.3. River: watercourse over 3 m wide (47)
2.7. Road infrastructure: band used for and prepared for pedestrians and car traffic

2.7.1. Road: road infrastructure over 2 m wide

2.7.1.1. Semi permeable: road with a surface that is not totally waterproof (48)

2.7.1.2. Permeable: road without a waterproof surface (49)

2.7.2. Sunken road: "sunken" road infrastructure and side slopes (50)

2.7.3. Lane: road infrastructure under 2 m wide

2.7.3.1. Semi permeable: lane with a surface that is not totally waterproof (51)

2.7.3.2. Permeable: lane without a waterproof surface (52)

2.8. Wall: linear masonry used as a fence (53)

2.9. *Green wall (54)*

3. Isolated elements

3.1. Isolated tree or bush: tree or bush that is not surrounded by other trees or bushes (55)

3.2. Small pond: stagnant body of shallow water smaller than 100 m^2 (56)

3.3. Icehouse: building where ice used to be stored (57)

3.4. Tumulus: mound of earth and stones (58)

3.5. Infrastructure elements: human constructions (well, fountain, kiosk, chapel, monument, statue, bridge, aviary, etc.) (59)

Annex 2: List of butterflies taken from the Ecological Diagnosis of the SAN Val Maubuée territory – Vol. I of ECOSPHERE, April 2010.

	Evaluation of the regional species scarcity	Species frequencies on the site
Protected Species (PN)	– espèces Protégées Nationales (Arr. du 22.07.93) – espèces inscrites á la Directive "Habitats" (Annexe 2 ou 4), – espèces inscrites á la Convention de Berne (Annexe II),	**0 PN**
Rare species (R)	– espèces a priori non revues en Ile-de-France après 1970 = NRR (Non Revues Récemment), – espèces inscrites sur la Liste Rouge de la Faune menacée en France: ED = En Danger; VUL = Vulnérables; R = Rares, – espèces déterminantes de ZNIEFF en Ile-de-France (espèces très localisées, avec des effectifs faibles á très faibles).	**8 R**
Uncommon species (PC)	– espèces déterminantes de ZNIEFF en Ile-de-France (espèces á répartition limitée, absentes de certains départements franciliens, peu communes dans d'autres). – espèces liées à des types de milieux reliques ou peu fréquents en Ile-de-France: tourbières, coteaux calcaires …,	**13 PC**
Common species (C)	– espèces ne bénéficiant d'aucun statut de protection particulier du fait de leur large distribution. – espèces ubiquistes (capables de peupler un grand nombre de types de milieu de diverse qualité). – espèces à populations abondantes sur l'ensemble de la région IDF.	20 C
	Total	**41 species**

Families	French name	Scientific name	Regional rarity	Species following the French ZNIEFF	National or Regional Protection	Remarks
HESPERIIDAE	Grisette	Carcharodus alceae	R	X		Contactée à l'unité à Emerainville en 2009
HESPERIIDAE	Hespérie de la Mauve	Pyrgus malvae	PC			Contacté à l'unité à Croiss-Beaubourg (Lamirault) en 2009
HESPERIIDAE	Sylvaine*	Ochlodes venatus	C			
LYCAENIDAE	Argus bleu	Polyommatus icarus	C			
LYCAENIDAE	Argus frêle*	Cupido minimus	R	X		
LYCAENIDAE	Azuré bleu céleste*	Polyommatus bellargus	PC	X		
LYCAENIDAE	Azuré de l'Ajonc*	Plebejus argus	PC	X		
LYCAENIDAE	Azuré des Anthyllides*	Cyaniris semiargus	R (NRR)	X		
LYCAENIDAE	Collier-de-corail	Aricia agestis	C			
LYCAENIDAE	Cuivré commun	Lycaena phlaeas	C			
LYCAENIDAE	Thécla de la Ronce	Callophrys rubi	PC			
LYCAENIDAE	Thécla du Bouleau*	Thecla betulae	R	X		http://pagesperso-orange.fr/renard-nature-environement/BiotopeEtgBBg.htm
LYCAENIDAE	Thécla du Chêne*	Neozephyrus quercus	PC			
LYCAENIDAE	Thécla du Coudrier	Satyrium pruni	PC	X		Contacté à Croissy-Beaubourg (Lamirault) et Emerainville (Célie) en 2009
LYCAENIDAE	Thécla du Prunellier*	Strymonidia spini	R			
NYMPHALIDAE	Amaryllis	Pyronia tithonus	C			
NYMPHALIDAE	Carte géographique	Araschnia levana	C			
NYMPHALIDAE	Demi-deuil	Melanargia galathea	PC	X		
NYMPHALIDAE	Fadet commun	Coenonympha pamphilus	C			
NYMPHALIDAE	Grand Mars changeant*	Apatura iris	R	X		
NYMPHALIDAE	Grand Sylvain*	Limenitis populi	PR (R)	X	PR	
NYMPHALIDAE	Grande Tortue*	Nymphalis polychloros	PR (R)	X	PR	
NYMPHALIDAE	Mégère, Satyre	Lasiommata megera	C			
NYMPHALIDAE	Myrtil	Moniola jurtina	C			
NYMPHALIDAE	Némusien, Ariane*	Lasiommata maera	PC			

Famille	Nom français	Nom scientifique	Statut	
NYMPHALIDAE	Paon du jour	Inachis io	C	
NYMPHALIDAE	Petit Mars changeant	Apatura illa	PC	×
NYMPHALIDAE	Petit Sylvain	Ladoga camilla	PC	
NYMPHALIDAE	Petite Tortue*	Aglais urticae	PC	
NYMPHALIDAE	Robert-le-Diable	Polygonia c-album	C	
NYMPHALIDAE	Tircis	Pararge aegeria	C	
NYMPHALIDAE	Tristan	Aphantopus hyperantus	C	
NYMPHALIDAE	Vanesse des Chardons	Cynthia cardui	C	
NYMPHALIDAE	Vulcain	Vanessa atalanta	C	
PAPILIONIDAE	Machaon*	Papilio machaon	PC	
PIERIDAE	Aurore	Anthocharis cardamines	C	
PIERIDAE	Citron	Gonepteryx rhamni	C	
PIERIDAE	Piéride de la Rave	Pieris rapae	C	
PIERIDAE	Piéride du Chou	Pieris brassicae	C	
PIERIDAE	Piéride du Navet	Pieris napi	C	
PIERIDAE	Souci	Colias crocea	PC	

* données bibliographiques

Annex 3: List of amphibians taken from the Ecological Diagnosis of the SAN Val Maubuée territory – Vol. 1 of ECOSPHERE, April 2010.

Evaluation of the regional scarcity After: l'Atlas de répartition des Amphibiens et Reptiles de France – S.H.F., 1989			Species frequencies on the site	
			Reptiles	Amphibia
Very rare species	**TR**	*I à I5% des 34 cartes IGN au I/50 000*	*0*	*0*
Rare species	**R**	*I5 à 30%* ”	*0*	*I*
Rather rare species	**AR**	*30 à 45%* ”	*I*	*3*
Rather common species	**AC**	*45 à 55%* ”	*0*	*4*
Common species	C	55 à 70% ”	0	2
Very common species	TC	70 à 100% ”	I	2
Introduced species	INT	–	0	I
		Total =	**2**	**I3**

French name	Scientific name	Regional rarity	Specific species of the ZNIEFF nomenclature	National protection	National red list	Habitat directive	Remarks
Alyte accoucheur*	*Alytes obstetricans*	**AC**		**PN ind + hab**	**Préoccupation mineure**	**Ann 4**	http://pagesperso-orange.fr/renard-nature-environment/BiotopeEtgBbg.htm.
Crapaud calamite**	*Bufo calamita*	**AR**	**X**	**PN ind + hab**	**Préoccupation mineure**	**Ann 4**	Habitat disparu à Croissy-Beaubourg
Crapaud commun	*Bufo bufo*	C		PN ind	Préoccupation mineure		
Grenouille agile	*Rana dalmatina*	TC		PN ind + hab	Préoccupation mineure	Ann 4	
Grenouille rieuse	*Rana ridibunda*	INT		PN ind	Préoccupation mineure		
Grenouille rousse	*Rana temporaria*	**AC**			**Préoccupation mineure**		
Grenouille verte	*Rana kl. Esculenta*	TC			Préoccupation mineure		
Rainette verte*	*Hyla arborea*	**AR**	**X (sites non forestiers)**	**PN ind + hab**	**Préoccupation mineure**	**Ann 4**	Donnée issue du transport d'individus à la migration prénuptiale à Croissy-Beaubourg (RENARD)
Salamandre tachetée**	*Salamandra salamandra*	**AR**		**PN ind**	**Préoccupation mineure**		Donnée issue de Croissy-Beaubourg
Triton alpestre	*Triturus alpestris*	R	**X**	PN ind	Préoccupation mineure		Densité importante sur la commune d'Emerainville (forêt de Célie)
Triton crété	*Triturus cristatus*	**AC**		**PN ind + hab**	**Préoccupation mineure**	**Ann 2 et 4**	Densités importantes relevées à Torcy (golf) et présence notable à Emerainville, Lognes et Croissy-Beaubourg
Triton palmé	*Triturus helveticus*	C		PN ind	Préoccupation mineure		
Triton ponctué	*Triturus vulgaris*	**AC**		**PN ind**	**Préoccupation mineure**		

*données bibliographiques
**espèce non revue récemment, présence et localisation á actualiser

Annex 4: List of nesting birds taken from the Ecological Diagnosis of the SAN Val Maubuée territory – Vol. 1 of ECOSPHERE, April 2010.

Breeding species evaluation in Ile de France (based on the evaluation of the number of breeding pairs)		Rarity of breeding species
Rarity degree	Ile de France classes	Val Maubuée territory
OCC (occasional breeding species)	espèces nicheuses occasionnelles	0
TR (very rare species)	1 à 20 couples nicheurs en Ile-de-France	2
R (rare species)	21 à 100 couples en IDF	4
AR (rather rare species)	101 à 500 couples en IDF	14
AC (rather common species)	501 à 2000 couples en IDF	11
C (common species)	2 001 à 20 000 couples en IDF	32
TC (very common species)	plus de 20 000 couples en IDF	27
INT (introduced species)	espèces nicheuses introduites	5
	Total =	96 species

French name	Scientific name	Regional rarity	Specific species of the ZNIEFF nomenclature	National protection	National red list UICN 2008	Bird directive	Remarks
Accenteur mouchet	Prunella modularis	TC		X	Préoccupation mineure		
Alouette des champs	Alauda arvensis	TC			Préoccupation mineure		
Bécasse des bois*	**Scolopax rusticola**	**R**	**X**		**Préoccupation mineure**		Donnée bibliographique au sein de la forêt de Célie, Emerainville
Bergeronnette des ruisseaux	**Motacilla cinerea**	AR	**X (5 couples)**	**X**	**Préoccupation mineure**		Plusieurs couples nicheurs notés au niveau d'infrastructures hydrauliques d'étangs (exutoires): Champs-sur-Marne, Croissy-Beaubourg
Bergeronnette grise	Motacilla alba	C		X	Préoccupation mineure		
Bergeronnette printanière	Motacilla flava	C		X	Préoccupation mineure		
Bernache du Canada	Branta canadensis	INT		X	Non applicable (introduite)		
Blongios nain*	**Ixobrychus minutus**	**TR**	**X**	**X**	**Quasi menacé**	**Annexe I**	Nicheur régulier depuis 1990 sur plusieurs étangs de Croissy-Beaubourg (Delapré, com. pers.)
Bondrée apivore*	**Pernis apivorus**	**AR**	**X (10 couples)**	**X**	**Préoccupation mineure**	**Annexe I**	Donnée bibliographique des ensembles forestiers de Croissy-Beaubourg (Ferrières, Armainvilliers)
Bouscarle de Cetti*	**Cettia cetti**	**R**	**X**	**X**	**Préoccupation mineure**		Donnée bibliographique de nidification certaine mais irrégulière à Croissy-Beaubourg (Delapré, com. pers.)
Bouvreuil pivoine	Pyrrhula pyrrhula	C		X	Vulnérable		
Bruant des roseaux	Emberiza schoeniclus	C		X	Préoccupation mineure		nicheur au sein des ceintures hélophytiques de certains étangs, noté également au sein d'une prairie de la pice de Lamirault, Croissy Beaubourg
Bruant jaune	Emberiza citrinella	C		X	Quasi menacé		

French name	Scientific name	Regional rarity	Specific species of the ZNIEFF nomenclature	National protection	National red list UICN 2008	Bird directive	Remarks
Buse variable	**Buteo buteo**	**AR**		**X**	**Préoccupation mineure**		Nicheur régulier à Emerainville, Noisiel et Croissy Beaubourg
Caille des blés	**Coturnix coturnix**	**AR**			**Préoccupation mineure**		Nicheuse en 2009 à Croissy-Beaubourg (Pièce de Lamirault)
Canard mandarin		INT					
Canard colvert	Anas platyrhynchos	C			Préoccupation mineure		
Chardonneret élégant	Carduelis carduelis	C		X	Préoccupation mineure		
Choucas des tours	Corvus monedula	C		X	Préoccupation mineure		
Chouette hulotte	Strix aluco	C		X	Préoccupation mineure		
Carbeau freux	Corvus frugilegus	C			Préoccupation mineure		
Corneille noire	Corvus corone	C			Préoccupation mineure		
Coucou gris	Cuculus canorus	C		X	Préoccupation mineure		
Cygne tuberculé	Cygnus olor	INT		X	Non applicable (introduite)		
Effraie des clochers*	**Tyto alba**	**AR**		**X**	**Préoccupation mineure**		Donnée bibliographique d'un couple nichant à Croissy-Beaubourg (Renard, Lamirault)
Epervier d'Europe	**Accipiter nisus**	**AR**		**X**	**Préoccupation mineure**		Nicheur probable en forêt de Célie, Emerainville
Etourneau sansonnet	Sturnus vulgaris	TC			Préoccupation mineure		
Faisan de Colchide	Phasianus colchicus	INT			Préoccupation mineure		
Faucon crécerelle	Falco tinnunculus	C		X	Préoccupation mineure		
Faucon hobereau	**Falco subbuteo**	**R**	**X**	**X**	**Préoccupation mineure**		Nicheur probable à Croissy-Beaubourg en 2009

Nom français	Nom latin	Abondance		Statut		Remarques
Fauvette à tête noire	*Sylvia atricapilla*	TC		Préoccupation mineure	X	
Fauvette babillarde	*Sylvia curruca*	**AR**		**Préoccupation mineure**	**X**	Nicheur en 2009 à Torcy et Croissy-Beaubourg
Fauvette des jardins	*Sylvia borin*	TC		Préoccupation mineure	X	
Fauvette grisette	*Sylvia communis*	TC		Quasi mennacé	X	
Foulque macroule	*Fulica atra*	**AC**		**Préocccupation mineure**		Plusieurs couples nicheurs sur les étangs du SAN
Fuligule milouin*	*Aythya ferina*	**TR**	**X**	**Préocccupation mineure**		Nicheur irrégulier à Croissy-Beaubourg (Delapré, com. pers.)
Gallinule poule d'eau	*Gallinula chloropus*	C		Préocccupation mineure		
Geai des chênes	*Garrulus glandarius*	C		Préocccupation mineure		
Gobemouche gris	*Muscicapa striata*	C		Vulnérable	X	
Grèbe castagneux*	*Tachybaptus ruficollis*	**AR**		**Préoccupation mineure**	**X**	Nicheur régulier à Croissy-Beaubourg (Delapré, com, pers.)
Grèbe huppé	*Podiceps cristatus*	**AC**		**Préocccupation mineure**	**X**	Plusieurs couples nicheurs sur les étangs du SAN
Grimpereau des jardins	*Certhia brachydacyla*	TC		Préocccupation mineure	X	
Grive draine	*Turdus viscivorus*	C		Préocccupation mineure		
Grive musicienne	*Turdus philomelos*	TC		Préocccupation mineure		
Grosbec casse noyaux	*Coccothraustes coccothraustes*	**AC**		**Préoccupation mineure**	**X**	Nicheur probable en forêt de Célie, Emerainville ainsi qu'à Ferrières
Héron cendré*	*Ardea cinerea*	**AR**		**Préocccupation mineure**	**X**	Nicheur irrégulier à Croissy-Beaubourg (Delapré, com. pers.) ; Donnée bibliographique à Croissy-Beaubourg (Renard)
Hibou moyen-duc*	*Asio otus*	**AR**		**Préocccupation mineure**	**X**	

French name	Scientific name	Regional rarity	Specific species of the ZNIEFF nomenclature	National protection	National red list UICN 2008	Bird directive	Remarks
Hirondelle de fenêtre	Delichon urbica	TC			Préoccupation mineure		
Hirondelle rustique	Hirundo rustica	TC		X	Préoccupation mineure		
Hypolaïs polyglotte	Hippolais polyglotta	C		X	Préoccupation mineure		
Linotte mélodieuse	Carduelis cannabina	C		X	Vulnérable		
Locustelle tachetée	**Locustella naevia**	**AC**		**X**	**Préoccupation mineure**		Nicheur en 2009 à Croissy-Beaubourg (Pièce de Lamirault)
Loriot d'Europe[*]	**Oriolus oriolus**	**AC**		**X**	**Préoccupation mineure**		Nicheur irrégulier à Croissy-Beaubourg et Emerainville
Martinet noir	Apus apus	TC		X	Préoccupation mineure		
Martin-pêcheur d'Europe[*]	**Alcedo atthis**	**AR**	**X (5 couples)**	**X**	**Préoccupation mineure**	**Annexe I**	Donnée bibliographique d'un couple nichant à Croissy-Beaubourg (Delapré., com. pers.) et Torcy (Barth., com. pers.)
Merle noir	Turdus merula	TC			Préoccupation mineure		
Mésange à longue queue	Aegithalos caudatus	TC		X	Préoccupation mineure		
Mésange bleue	Parus caeruleus	TC		X	Préoccupation mineure		
Mésange boréale[*]	Parus montanus	C		X	Préoccupation mineure		
Mésange charbonnière	Parus major	TC		X	Préoccupation mineure		
Mésange huppée[*]	Parus cristatus	C		X	Préoccupation mineure		
Mésange nonnette	Parus palustris	TC		X	Préoccupation mineure		
Moineau domestique	Passer domesticus	TC		X	Préoccupation mineure		

Nom français	Nom latin		X	Statut	Notes
Perdrix grise	*Perdix perdix*	TC		Préoccupation mineure	
Perdrix rouge	*Alectoris rufa*	INT		Préoccupation mineure	
Phragmite des joncs*	***Acrocephalus schoenobaenus***	**R**	**X**	**Préoccupation mineure**	
Pic épeiche	*Dendrocopos major*	C	X	Préoccupation mineure	
Pic épeichette	*Dendrocopos minor*	C	X	Préoccupation mineure	
Pic mar*	***Dendrocopos medius***	**AC**	**X (30 couples)**	**Préoccupation mineure Annexe I**	Nicheur récent à Croissy-Beaubourg (parc de Croissy et Beaubourg) depuis 2006 (Delapré, com. pers.)
Pic noir	***Dryocopus martius***	**AR**	**X (10 couples)**	**Préoccupation mineure Annexe I**	Plusieurs couples nicheurs à Emrainville (forêt de Célie), Champs-sur-Marne (bois de la Grange), Croissy-Beaubourg (parc de Beaubourg)
Pic vert	*Picus viridis*	C	X	Préoccupation mineure	
Pie bavarde	*Pica pica*	TC		Préoccupation mineure	
Pigeon biset "sauvage"	*Columba livia*			En danger	
Pigeon colombin	***Columba oenas***	**AC**		**Préoccupation mineure**	Nicheur à Croissy-Beaubour en 2009 (Lamirault)
Pigeon ramier	*Columba palumbus*	TC		Préoccupation mineure	
Pinson des arbres	*Fringilla coelebs*	TC	X	Préoccupation mineure	
Pipit des arbres	*Anthus trivialis*	C	X	Préoccupation mineure	
Pipit farlouse	***Anthus pratensis***	**AC**	**X**	**Vulnérable**	Plusieurs couples nicheurs en 2009 à Emerainville (aérodrome) et Croissy-Beaubourg (Lamirault)

French name	Scientific name	Regional rarity	Specific species of the ZNIEFF nomenclature	National protection	National red list UICN 2008	Bird directive	Remarks
Pouillot fitis	Phylloscopus trochilus	C		X	Quasi menacé mineure		
Pouillot véloce	Phylloscopus collybita	TC		X	Préoccupation mineure		
Râle d'eau*	**Rallus aquaticus**	**AR**	**X (2 couples)**		**Données insuffisantes**		Nicheur régulier à Croissy-Beaubourg (Delapré., com. pers.)
Roitelet huppé	Regulus regulus	C		X	Préoccupation mineure		
Roitelet triple-bandeau	**Regulus ignicapillus**	**AC**		**X**	**Préoccupation mineure**		Nicheur en 2009 à Croissy-Beaubourg, Champs-sur Marne et Emerainville
Rossignol philomèle	Luscinia megarhynchos	C		X	Préoccupation mineure		
Rougegorge familier	Erithacus rubecula	TC		X	Préoccupation mineure		
Rougequeue à front blanc*	**Phoenicurus phoenicurus**	**AC**	**X (25 couples)**	**X**	**Préoccupation mineure**		Donnée bibliographique de 2001 de couples nicheurs à Emerainville (CORIF)
Rougequeue noir	Phoenicurus ochruros	TC		X	Préoccupation mineure		
Rousserolle effarvate	Acrocephalus scirpaceus	C		X	Préoccupation mineure		
Rousserolle verderolle	**Acrocephalus palustris**	**AR**	**X (15 couples)**	**X**	**Préoccupation mineure**		plusieurs couples nicheurs en 2009 à Croissy-Beaubourg (Lamirault) et Torcy (le Couvent)
Serin cini	Serinus serinus	C		X	Préoccupation mineure		

Nom français	Nom latin			Statut	Remarque
Sittelle torchepot	*Sitta europaea*	TC	X	Préoccupation mineure	
Tarier pâtre	***Saxicola rubicola***	**AC**	**X**	**Préoccupation mineure**	Nicheur en 2009 à Croissy-Beaubourg (Lamirault)
Tourterelle des bois	*Streptopelia turtur*	C		Préoccupation mineure	
Tourterelle turque	*Streptopelia decaocto*	C		Préoccupation mineure	
Troglodyte mignon	*Troglodytes troglodytes*	TC	X	Préoccupation mineure	
Verdier d'Europe	*Chloris chloris*	TC	X	Préoccupation mineure	

* données bibliographiques

Conclusions and perspectives

Bruno Peuportier[1], Fabien Leurent[2] & Jean Roger-Estrade[3]
[1]*MINES ParisTech,*
[2]*Ecole des Ponts ParisTech*
[3]*AgroParisTech*

Five years of activity and knowledge production at the ParisTech Chair, in association with VINCI "eco-design of buildings and infrastructures", have allowed us to develop operational tools, some of which have been tested extensively on the Cité Descartes urban development project in Marne la Vallée.

The objectives pursued from the start of this Chair mostly focused on supporting research operations, and can be summed up in four points:

- Establish measures for making eco-design progress by consolidating suitable methods, and in particular the Life Cycle Assessment (LCA),
- Improve existing modelling tools by extending their domain of application to neighbourhood scale and applying an integrated approach to buildings (new and existing), transport and green areas in neighbourhood design,
- Take better account of environmental aspects when designing and managing eco-neighbourhoods and transport infrastructures, in particular issues relating to the preservation of biodiversity,
- Contribute to creating a scientific community focusing on eco-design, by pooling the efforts of the three ParisTech universities and their associated research laboratories, and encouraging genuine debate at national level.

These objectives establish a major ambition, since the object concerned – a large-scale set of buildings including infrastructures – is an eminently complex system. Models and methods already operational for individual buildings, i.e. at an elementary scale, need to be applied to a set of elements, which brings the complexity of abundance. What is more, the change of scale from a building to a large built-up area radically increases both the usages (and so the functions rendered) and the impacts. The configuration of elements inside a neighbourhood influences both physical and ecological performance as well as occupants' exposure to local impacts. Attributing different functions to the buildings will determine the activities inside the neighbourhood, and thus the usage of each building, as well as individual chains of activities between buildings and with the outside, with incidences on the forms of mobility, journey requirements and means of transport used. The detailed layout of transport stations within the neighbourhood – whether these are public transport stations or parking lots for cars or motorbikes, private or shared – similar to the neighbourhood's situation in relation to major transport infrastructures at built-up area scale, determines as much if not more the use of transport means and thus the environmental performance of the mobility system associated with the neighbourhood – a performance that is as important as the environmental performance "intrinsic" to the buildings as a whole.

Similarly, the specific composition of places and their layout in the neighbourhood interact with biodiversity. This will depend, for example, on the place given to green and blue corridors, interacting with the grey corridors of roads and technical networks; it will depend on the revegetation of ground areas, and possibly buildings (on the roof or walls of the envelope) or even on the inside, perhaps for urban agriculture ... Lastly, beyond the complex abundant elements and aspects, and the complex meanings in the phenomena and their interaction, a large built-up area has an overall dimension, which we could call a significant mass, that is several orders of magnitude greater than that of a building: it is therefore less marginal on an overall level, and it is useful to devise some "virtuous" measures with which to equip it, in close interaction with their equivalents outside the system. This might typically concern questions on the energy sources and distribution networks and their dynamic conditions.

The Chair gave its three research teams the opportunity to work together to set out these challenges and develop a general understanding of a qualitative and systemic nature and demonstrate diverse methods in a joint case study. The teams' diverse fields of competence, i.e. building physics and LCA evaluation for Mines; ecology and ecosystem design for Agro; physico-economics of transport and land use for Ponts, involved complementary features, mutual assistance rather than competition. The diversity resulting from their respective specialities was reinforced by their diverse approaches to evaluation, ranging from the LCA method clearly set out and implemented by Mines, to Agro's studies of local impacts and biodiversity, and the cost-benefit analysis and socio-economic evaluation done by Ponts. This diversity was respected because their respective foundations are well established and justified and not in contradiction with each other. The Chair helped federate and allowed them to make progress together through the driving force of confrontation and emulation; it also encouraged them to work on a joint case study. This mutual input has proved efficient and is worth continuing.

In addition to this mutual input, the Chair gave rise to a series of developments, and conceptual and methodological progress on evaluation in the broad sense, including indicators, evaluation procedures, simulation models and decision aid measures.

In terms of buildings, the main progress relates to LCA methodology and specific treatment at neighbourhood scale:

- The life cycle assessment tool for buildings was thus extended to study positive-energy neighbourhoods, integrating models on electricity production and renewable energy. The LCA's contribution is crucial to decision-making, but it needs to be used with other methodological tools. In particular, synergy with impact studies could extend the scope of information provided by an LCA and enrich the definition of the functional unit, in particular in relation to eco-systemic services.
- At neighbourhood scale, it is indispensable to tackle the existing situation. Considering buildings' initial state means being able to reconstitute the characteristics of the existing situation (i.e. nature and state of materials). To do so, it is useful to reduce/identify models to make up for the lack of data. A first attempt led to a better understanding of the problem and should lead to practical recommendations.
- Dealing with the existing situation brings up the issue of taking efficient technical measures in line with budget constraints. An approach to managing a set of

property was defined thanks to the LCA simulation capacities. This approach can be developed to produce a methodological guide.

- A dynamic LCA can consider time changes in the energy mix. This theoretical effort was indispensable to approach the energy balance from a realistic point of view, in contrast with the current averaged-out approach to positive-energy buildings, since environmental impacts do not offset each other annually. This question was crucial to translate the successive transformations of neighbourhoods following the introduction of renewable energy.

- An LCA's usefulness is not so much about providing absolute information as its capacity to compare several building configurations. At neighbourhood scale, this is possible for materials, but from a mobility point of view it is indispensable to be able to "trace" journeys within the block's perimeter. It emerged that modelling tools currently correspond to an inter-neighbourhood problematic, but the way they tackle the practices of residents within the actual neighbourhood is not yet satisfactory. An effort should thus be made in this direction.

Still at neighbourhood scale, but in terms of eco-systemic and governance services, the Chair focused on biodiversity and on stakeholder interaction:

- Considering biodiversity in towns in a structured way is still a fairly new area, and most projects that call themselves "eco-neighbourhoods" do not propose truly efficient action in terms of biodiversity. We think that these projects should go beyond simply trying to preserve biodiversity and that they should allow nature to adapt and become resilient, thus creating a sustainable ecological system that can function in the heart of the town and render services to humans. To attain this objective, a tool was developed to consider how ecosystems function within neighbourhoods. This tool is based on habitat diversity and management, along with the presence of indicator species chosen to show how the urban ecosystem is working.

- Another model was developed to represent interactions between natural dynamics and social dynamics and simulate the impact on biodiversity of decisions made by the different stakeholders. This model's vocation is to be used for all types of new building or neighbourhood retrofit projects so as to bring together the main decision-makers to identify urban forms and developments that are not just favourable for biodiversity but that find consensus among stakeholders.

- Lastly, areas in towns are increasingly being developed for farming production. This trend is growing fast for very different reasons depending on the place: major cities in emerging countries, towns confronted with deindustrialization, or urban centres that need to develop service activities. We suggest developing research on urban agriculture that would explore its various forms, the services it can render, the risks in terms of food security, and the issues of managing urban territories. This project would focus in particular on the risks connected to urban traffic (product quality, soil quality) and more broadly to the use of infrastructures (airports, car parks, etc.). These risks would be considered with regard to the potential assets (in terms of social links, local food production).

Regarding land use and mobility, the main progress was as follows:

- A return to basics through systemic cost-benefit analysis and socio-economic evaluation: comparison with the LCA, for the inter-sectorial propagation of impacts, put into perspective the creation and circulation of value flows in urban planning and transport projects.
- An integrated modelling of vehicle parking and traffic for part or all of a built-up area that is sensitive to supply and demand for places at local level, and to capacity constraints, and identifies journeys to find places.
- A refined modelling of pollution emitted by road traffic, applicable to the whole network and to its dynamic functioning in the course of a day.
- A doctrine of the use of public transport to ensure urban mobility, based on a detailed physico-economic model of the supply and demand for journeys on a network, and integrating diverse capacity constraints in the traffic of passengers and vehicles. With a series of design applications, including for high level of service bus lines on urban expressways.
- For land use, the integrated understanding of a territory as a set of interacting systems: system of residential locations, system of activities and jobs, transport system. This understanding takes the shape of theoretical models of the urban economy specifically developed to identify certain systemic effects. A simulation model applicable to all built-up areas is being developed in design and computer phases; it will make it possible to evaluate urban planning operations.

A case study considered ways of combining different analysis tools with the aim of introducing a regulation capacity into building choices contributing to sustainable territorial development. To go beyond multidisciplinarity, the simple juxtaposition of skills requires an acculturation of research teams, or better still their inter-culturation or even trans-culturation. This joint case study made this process easier by illustrating the concrete involvement of the various disciplines. The project focusing on Cité Descartes (in Marne la Vallée east of Paris) concerns buildings, transport, and biodiversity, so that different teams involved in the Chair were able to make a contribution within a broader framework responding to societal issues. This study has shown that a certain number of tools are already operational, and helped highlight their limitations and potential for improvement.

The studies open out several perspectives, in particular:

- Uncertainties in models need to be understood better in order to move towards guaranteed performance, which increasingly features among decision-makers' expectations,
- This requires more detailed modelling of behavioural aspects, in particular in the building and transport domains,
- We suggest exploring diverse forms of urban agriculture and the services they can render, along with risks in terms of food security and urban territory management issues,
- Research should be continued on urban biodiversity through pursuing development of the multi-agent simulation tool and diffusing it widely, especially at VINCI,

- Eco-design requires a detailed study of spatial planning, which strongly influences the choice of transport modes and their respective usage: easy use of pavements and road crossing for pedestrians; bicycle lanes; easy, available parking for two- and four-wheeled vehicles, etc.,
- LCA models applied to a road could be improved by taking better account of traffic, climate conditions and road surface maintenance, which has a significant influence on energy consumption and thus polluting emissions,
- Connecting phenomena at different scales (buildings, neighbourhoods, territories and networks) could usefully improve systems management (notion of smart cities) and integrate a neighbourhood's contribution into the general functioning of a built-up area,
- In the perspective of Climate-Energy Territorial Plans, the aim is to identify eco-design requirements to plan towns and local territorial policy,
- In order to study the application of compensation measures, we propose tackling the issues of the ecological equivalence and rehabilitation of functional connections in a territory,
- Improving our understanding of the consequences of decisions from one sector to the next (notion of consequential LCA) would improve the decision process and move towards greater societal responsibility.

EPILOGUE: THE CORPORATE POINT OF VIEW

The Chair on Eco-design for Buildings and Infrastructures is a base-building project for VINCI, fitting in with joint research activities on eco-design and sustainable towns. The Chair was created in 2008, involving a 5-year partnership between the Group and three ParisTech universities – MINES ParisTech, Ecole des Ponts ParisTech and AgroParis-Tech. The three partner universities develop new reference systems and tools for urban eco-design (i.e. buildings, neighbourhoods, transport and biodiversity). Based on measurements and simulation, these become decision-aid devices for urban stakeholders. The Chair's scientific progress is recognized and the subject of numerous publications (more information at www.chaire-eco-conception.org). The most noteworthy feature in this book: development of a tool to assess urban biodiversity; adaptation of environmental databases in the building sector; work on guaranteeing buildings' energy performance; ecodesigned parking in urban environments; and dynamic eco-design that responds to Smart Grid issues, etc.

I/Issues

Eco-design for buildings and infrastructures is an issue that cannot be resolved by entrepreneurs, who need to open up their areas of competency, in particular to the world of science.

o   Measuring to move forward

One of VINCI's strategic objectives is to accompany its customers in seeking more efficient use of energy and encourage them to adopt eco-responsible behaviour. The tools and reference systems developed by scientists can be used to design projects that respect the Group's rationale, which is to "know what you are talking about" and present valid results. VINCI is the first construction and services group in France to have had its social and environmental data verified as part of its sustainable development policy. Eco-design complements this approach, with a focus on buildings and making progress with measurements.

In terms of environmental performance, we need to propose a solution for reducing the impact on the entire life cycle.

o   Mobilizing several disciplines to change scale

For the first time in this type of partnership, the ParisTech-Vinci Eco-design Chair has brought together a company and three universities.

The eco-design and life cycle assessment approach had already been applied to infrastructures and used on numerous projects on a building and infrastructure scale. The Chair's work allowed us to extend its field of application to the whole neighbourhood, connecting buildings to the urban system in which they fit. This approach, developed for both new buildings and retrofits, is a major challenge and constitutes an inter-sectorial response.

At neighbourhood scale, it is crucial to tackle the question of buildings, transport and biodiversity to provide a systemic response.

o   Modelling

In the existing building stock, the share of CO_2 emissions due to construction is estimated at 10% of emissions over the life cycle. Low environmental impact solutions should therefore aim at the 90% that result from the usage phase, and propose efficient technical solutions to help final users in their behaviour. The choice of materials is an additional lever to improve a project's environmental performance. For new buildings, respecting 2012 thermal regulations and consuming little energy, this share can easily reach 30%, 40% or even 50%.

In the motorway concession trade, multimodal solutions like car sharing are increasingly encouraged. Financial resources to develop new infrastructures are becoming scarce and it is vital to have access to tools and methods for optimizing the use of existing facilities, in particular urban expressways.

For urban biodiversity, now a vital topic, the results obtained from green roofs (see chapter 17) are fairly representative of the need for good eco-design tools and reference systems. It is clearly essential to assess environmental impact in relation to the service initially expected (in general, encouraging biodiversity).

The varied technical problems arising in these domains require detailed modelling: energy performance depending on the placing of materials in the building; considering the level of traffic congestion to model pollution emissions from vehicles; modelling supply and demand to eco-design car parks; deterministic modelling of the impact of infrastructures on biodiversity, etc.

2/Support action

The Chair brings into question the way companies work. In addition to its various discoveries, it has helped improve dialogue between scientists and companies on innovative subjects. During these five years, teams from ParisTech and VINCI have made bridges between the worlds of research and business. Combined with VINCI's research, development and innovation policy, the ParisTech universities' tools and reference systems have been used to develop the Group's green products, like "Oxygen" for buildings, its "Green Motorway Package" and "PRISM" for specialist trades (at Solétanche Freyssinet).

o   Advisers

With a research action focus, the Group's experts and operational staff participate in putting together research subjects and guiding researchers to come up with directly applicable tools. Thanks to mirror groups, they can follow the progress of research in their fields of competence and benefit from the very first results.

Examples are: the optimization of energy renovation at building stock scale proposed by VINCI Facilities; life cycle assessment at neighbourhood scale by VINCI Construction; and car park eco-design by VINCI Park.

o   Evening events

Since 2008, 21 conferences have been organized at VINCI or at the universities to disseminate ParisTech's progress to the Group and its stakeholders (partners, customers, science sphere, etc.). The evenings also have an educational aspect in presenting opportunities and/or questions on the subjects tackled by the Chair.

For example, on 26 March 2013, VINCI Energies and Renault introduced the evening with a talk on the electric vehicle before the presentation by ParisTech's Ecole des Ponts; GTM Bâtiment (subsidiary of VINCI Construction) and the Office Public de l'Habitat of Seine et Marne gave an introduction on eco-design and retrofits for buildings at the event on 12 June 2012 before the presentation on research done by MINES ParisTech, etc.

o   Symposium

To respond to the challenge of fostering an eco-design approach at VINCI companies and their partners, a teaching event was organized with a focus on the field: the Eco-design Chair Symposium. The first Symposium took place on 2 and 3 October at Vaux de Cernay Abbey in the Paris suburbs. The spirit of the event was open innovation and collaboration, based on the principle that the response to such a complex issue can only be built together, involving all urban project stakeholders.

The Symposium was a great success. Almost 90 people took part in the two-day event of theoretical and practical training on eco-design for buildings, neighbourhoods and mobility. Half of the participants came from VINCI, with representatives from all of its domains (Construction, Energy, Concessions, Motorways, Eurovia, Property), and half were outside partners: companies, engineering specialists, networks of partner companies, local authorities and customers.

This approach was a unique step for VINCI and its partner ParisTech. It was the first time that the Group had organized an event on a research subject open to all of its stakeholders. Other ParisTech research chairs organize events, but these have never been used as a tool for differentiation and dialogue with stakeholders from the

Eco-design master class, 2 October 2012 at the Chair's first Symposium.

businesses financing the studies. The Chair's second symposium, which took place from 7 to 8 November 2013, set the pattern for a long-term gathering.

3/First results

This innovative form of partnership means that we can get the most from pooling our efforts and expertise. It delivers more pertinent results than a company/scientist silo mentality.

○   Tools accessible to all

The principle of scientific patronage results in the creation of tools and reference systems that can be used throughout a town's value chain. Based on scientific modelling, they supply valid results and can be used buy all those involved in designing buildings.

By introducing the same platform for dialogue between the different stakeholders, eco-design constitutes a key stage in constructing a living environment that respects the environment.

○   Tools for different planning scales

In 2012, 229 projects were the object of an eco-design study at VINCI. Thanks to the tools developed by the chair, the Group is in a position to suggest projects in which environmental performance is assessed and guaranteed throughout the life cycle. These tools allow VINCI companies to dialogue with their order givers.

As an illustration, the novaEQUER tool (see chapter 2) was used to assess the environmental performance of a block at Lyon Confluence. In this study, the application of the LCA started with a detailed modelling of the block's geometry. The project's architectural data (figure hereunder, left) were transmitted to the graphical input software Alcyone (figure hereunder, right). The land use and characteristics of the building's envelope were defined. The block was modelled in its entirety: building, land use (green spaces, streets, etc.).

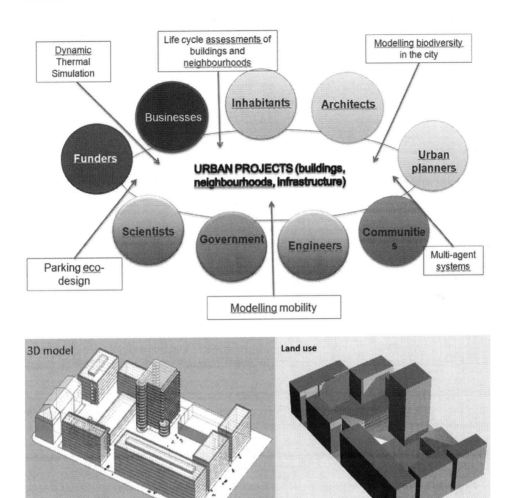

o   The researchers today

Lastly, in addition to its involvement in the ParisTech curriculum (engineering, spe-cialized Master and doctoral students) in the course of these five years, the chair has helped develop a network of experts specializing in eco-design. Several researchers have remained in the science field and become lecturers, others have joined local authorities (CSTB, town associations, etc.), and others still have entered the business sphere at VINCI or with its partners (e.g. SETEC). These key men and women have played a part in technical and conceptual progress. They can now contribute to putting their work into practice in projects: at the earliest stage possible.

During these five years, the work done at the Chair has illustrated the benefits of a partnership between universities and business. What is the point of creating purely academic knowledge if it does not contribute to improving practices? And all the more so, since upstream research clearly benefits from tackling issues faced by building companies and vice versa. Which shows that opportunities are there to be seized. And reproduced.

If you would like to know more, numerous publications and conference proceedings are available for download on the website: www.chaire-eco-conception.org.

Index